INDUSTRIAL HYDRAULICS
third edition

John J. Pippenger

Tyler G. Hicks

Gregg Division
McGraw-Hill Book Company

New York	Mexico
St. Louis	Montreal
Dallas	New Delhi
San Francisco	Panama
Auckland	Paris
Bogotá	São Paulo
Düsseldorf	Singapore
Johannesburg	Sydney
London	Tokyo
Madrid	Toronto

Library of Congress Cataloging in Publication Data

Pippenger, John J
  Industrial hydraulics.

  Includes index.
  1. Hydraulic machinery. 2. Hydraulic engineering.
I. Hicks, Tyler Gregory, [date]   joint author.
II. Title.
TJ840.P52 1979 621.2'6 78-15796
ISBN 0-07-050140-8

Industrial Hydraulics, Third Edition

Copyright © 1979, 1970, 1962 by McGraw-Hill, Inc. All rights reserved. Printed in the United States of America. No part of this publication may be reproduced, stored in a retrieval system, or transmitted, in any form or by any means, electronic, mechanical, photocopying, recording, or otherwise, without the prior written permission of the publisher.

234567890 DODO 7865432109

The editors for this book were Gerald Stoner and Mary Levai, the designer was Tracy Glasner, the art supervisor was George T. Resch, and the production supervisor was Regina Malone. It was set in Palatino by University Graphics, Inc.
Printed and bound by R. R. Donnelley & Sons Company.

Contents

Preface	v
Introduction To Fluid Power	1
1 Fluid-Power Systems	11
2 Hydraulic Symbols	31
3 Fluid-Power Pumps And Motors	65
4 Hydraulic Cylinders And Rams	118
5 Fluid-Power Plumbing	143
6 Pressure Accumulators	165
7 Fluid Reservoirs	186
8 Filtration of Hydraulic Fluids	209
9 Fluid-Temperature Control	232
10 Pressure-Control Valves	247
11 Flow-Control Valves	286
12 Directional-Control Valves	309
13 Electrical Devices for Hydraulic Circuits	341
14 Servo Systems	358
15 Industrial Hydraulic Circuits	377
APPENDIX A: PREFERRED METRIC UNITS FOR FLUID POWER	418
APPENDIX A–1: SELECTED "SI" UNITS FOR GENERAL PURPOSE FLUID-POWER USAGE	432
Index	435

Preface

A growing affluence appears to be spreading to every corner of the globe. Greater demand for manufactured products is evident. Ever-improved transportation for goods and people is becoming essential.

These social changes, which directly affect the world marketplace, have added some new dimensions to the requirements for fluid-power transmission.

Sophisticated controls are now associated with agricultural machinery. High-precision machinery needed to bolster production also demands better-trained personnel within all areas of industry.

Fluid-power systems, particularly hydraulics, are now basic to virtually every major machinery segment. Thus, technical training programs can no longer be considered adequate in major industrial areas if they do not include significant exposure to fluid-power systems and their associated devices.

We are well aware that industrial hydraulics technology is firmly entrenched in our global economy. The usage knows no boundary lines. Because of the importance of industrial hydraulics, there is an ever-growing trend toward international standardization. Technical knowledge about the application and use of liquids as power-transmission agents has been in the hands of a relatively few individuals. Better standardization and even greater use of industrial hydraulics can be expected. We know that there must be more and better educational and training facilities for technicians and operators. Education for the technician, maintenance worker, junior engineer, and machinery operator still lags behind the rapid acceptance of this versatile method of transmitting power.

We have updated this book in the belief that it will assist in filling this gap in the educational field, ranging from the youngest entered apprentice up to and possibly including the component designer. Practical standards cannot be universally satisfactory without basic training programs to acquaint users with the known art of industrial hydraulics.

A second major revision of the symbols and drafting practices to make both of these systems international in character has required some new artwork and modification of other materials in the book to reflect the International Standards and, where appropriate, show those communications devices in general use which may not yet have received international acceptance. The revisions to this book, wherever possible, have been made in cooperation with members of the International Standards Organization (ISO), the international group, to ensure worldwide acceptance with minimum language problems (ISO-1219-1976 E/F).

The original plan of the book has been retained to minimize changes in existing academic programs which have been built around this text.

Several chapters have been specifically and significantly expanded to cover new systems and associated mechanisms.

As with the first two editions, after acquiring a knowledge of elements that make up a hydraulic power-transmission system, the student is introduced to many of the existing tried-and-proven circuits used in modern, hydraulically actuated machinery. In addition, a pattern of functional machine movements is established, correlated with the components needed to provide the movements, so that a number of basic actions can be incorporated into a complete machine circuit.

After completing the study of the textual material in each chapter, students should perform the laboratory experiments when facilities are available. Modifications of the experiments outlined are usually possible when facilities are limited.

Valuable contributions relative to terminology used within the industry were provided by Walter Kudlaty, former chairperson of the National Fluid Power Association Terminology Committee. James L. Fisher, Jr., chairperson, ISO/TC131-SC1, gave valuable assistance in correlating symbol data.

> John J. Pippenger
>
> Tyler G. Hicks

Introduction to Fluid Power

Fluid-power systems are power-transmitting assemblies employing pressurized liquid or gas to transmit energy from an energy-generating source to an energy-use area. Fluid power can be divided into two basic disciplines:

Hydraulics   Employing a pressurized liquid

Pneumatics   Employing a compressed gas

This book is primarily concerned with *hydraulics*. References to pneumatic devices are limited to areas where pressurized gas is used to supplement the hydraulic system, or where combined systems provide significant advantages in power transmission.

Pneumatic systems depend on the compression of a gas. Devices used in pneumatic systems depend on several basic physical laws which are not important in hydraulic systems. For example, pneumatic devices normally operate at pressures less than 200 psi. The natural springiness of the compressed gas provides excellent power storage, but it may limit the needed stiffness of control for other devices. (The term psi, which stands for pounds per square inch, is the basic pressure measurement used in fluid-power systems.)

All hydraulic systems depend on Pascal's law, named after Blaise Pascal, who discovered the law. This law states that pressurized fluid within a closed container—such as a cylinder or pipe—exerts equal force on all of the surfaces of the container.

In actual hydraulic systems, Pascal's law defines the basis of the results which are obtained from the system. Thus, a pump moves the liquid in the system. The intake of the pump is connected to a liquid source, usually called the *tank* or *reservoir*. Atmospheric pressure, pressing on the liquid in the reservoir, forces the liquid into the pump. When the pump operates, it forces liquid from the tank into the discharge pipe at a suitable pressure.

The flow of the pressurized liquid discharged by the pump is controlled by *valves*. Three control functions are used in most hydraulic systems: (1) control of the liquid pressure, (2) control of the liquid flow rate, and (3) control of the direction of flow of the liquid.

The liquid discharged by the pump in a fluid-power system is directed by valves to a *hydraulic motor*. A hydraulic motor develops rotary force and motion, using the pressurized liquid as its energy source. Many hydraulic motors are similar to pumps, except that the motor operates in a reverse manner from a pump.

Where linear instead of rotary motion is desired, a cylindrical tube fitted with a movable piston, called a *hydraulic cylinder,* is often used. When the piston is moved by the pressurized fluid, the piston rod imparts a force or moves an object through a desired distance.

Restricting the movement of the piston in a hydraulic cylinder, as when the piston carries a load, creates a specific pressure relationship within the cylinder. The surface area of the piston face is said to contain a specific number of square inches. The pressure of the pressurized liquid, multiplied by the piston area, produces an output force, measured in pounds, at the end of the piston rod.

The speed of movement of the piston rod depends on how fast the pressurized fluid enters the cylinder. Flow into the cylinder can be directed to either end, producing either a pushing or pulling force at the piston-rod end. A seal around the rod where it passes through the cylinder end prevents leakage of the liquid.

Directional control of the piston depends on which end of the cylinder the liquid enters. As pressurized liquid enters one end of the cylinder, liquid must be drained from the other end. The drained liquid is led back to the reservoir. In a pneumatic system using air, the air in the exhausting end of the cylinder is vented to the atmosphere.

Directional-control valves, also called two-way, three-way, four-way, etc., are named in accordance with their basic function. Pressure-control and simple restrictor valves are usually two-way valves. They provide ON or OFF service. A three-way valve may perform several functions, all associated with the three ports in the valve. For example, the power or pressurized liquid from a pump on a tractor may be sent to the hydraulic system serving the tractor's front-end loader. Or the three-way valve may send the pressurized liquid to a hydraulic motor driving a feed conveyor while the front-end loader is not being used.

Three-way valves may also be used to direct pressurized fluid to a single-acting (i.e., force in only one direction) hydraulic cylinder. As the three-way valve is actuated (operated) it can stop the pressurized flow to the cylinder. Further, the same valve can divert liquid from the cylinder to the reservoir, so the cylinder can retract by gravity or return springs and assume its original position.

A four-way valve has four *ports* or openings. The *pressure port* directs fluid flow to an area where pressurized liquid is desired. One of the other ports can simultaneously drain liquid from a pressurized area. The drain liquid can be directed to the reservoir.

In a fluid-power system, such as those shown in Fig. I-1, the movement of pressurized fluid resembles the action of electric current in an electrical system. In such a system, electrical energy is continually moving when work is being done. The same is true of a fluid-power system.

In a direct-current electrical system, the speed of a device is varied by changing the flow of current to the device. Alternating-current systems use phase shifting to attain similar speed control. Hydraulic systems can

Fig. I-1 Modern equipment using fluid-power systems.

Introduction to Fluid Power **3**

Fig. I-2a A versatile hydraulic power transmission system combines accurate machine guidance with massive energy transfer to provide power for construction machinery.

obtain infinite speed variations by several methods of control. Pneumatic systems, because of the springiness of the gas, have relatively crude speed control.

In fluid-power systems, the pump can be designed so the discharge flow of the pressurized liquid can be varied by manipulating the pump mechanism, giving varying speeds of the hydraulic motor or other output device. A restriction in the pump outlet can limit the fluid flow in a manner similar to that used in a direct-current electrical system. Where desired, hydraulic systems can be designed to use valves which divert flow to different flow channels to give the desired speed control.

The compressibility of hydraulic fluids is slight. But, if desired, this slight compressibility can be used to absorb small, erratic machine movements which cause pressure variations in a fluid-power system. Pneumatic systems have this capability inherently because of the springiness of the air or gas used in the system. Hydraulic systems may use one or more *accumulators*—devices which contain pressurized gas which can accommodate rapid pressure changes in the fluid portion of the system.

Gravity, springs, and compressed gases provide potential energy in many hydraulic systems. Thus, gravity assist is often used to return a part to its original position without the use of power from the hydraulic system. This type of assistance also simplifies the parts in the hydraulic system. One example of gravity assist is found in presses used for stamp-

Fig. I-2b The depth and level of concrete being spread is controlled by a hydraulic system on this paving machine.

ing and forming metal parts. In these presses there is often a long drop of the die segment from the elevated position until resistance is met. Then a large hydraulic force is applied to form the part. Hydraulic circuits for presses are explained in later chapters of this book.

Forces are easily controlled in hydraulic systems. The usual way of controlling force in a hydraulic system is by use of a device in the pump or pipe to reduce fluid flow when the desired pressure is obtained. Only enough extra fluid is discharged to overcome any leakage in the system. Pneumatic systems use a *pressure-regulating valve* to maintain a certain pressure level in the force-producing device.

Many hydraulic systems use pumps which deliver a constant flow of pressurized fluid when the operating speed of the pump is constant. For example, a pump may deliver 1 gallon per minute for each 100 revolutions per minute. Thus, a 1200-rpm electric motor would produce a 12-gpm flow when driving this pump. But if the pump were connected to a 2400-rpm diesel engine, the pump would discharge 24 gpm. With a constant flow from a pump, it may be necessary to divert some of the flow to the reservoir when a specific pressure is reached, if a certain force is needed. A *relief valve* may be used to divert flow to the reservoir. Many relief valves operate automatically when the desired pressure level is reached.

Direction control in an electrical system is usually obtained by reversing an electric motor. In a fluid-power hydraulic system or a pneumatic system, directional-control valves switch the column of fluid to obtain a change in direction. Hydraulic systems may be designed so that the pump can be reversed, if desired.

If an electric power source is disconnected, all potential energy in the system is removed. In a fluid-power system, however, the stored energy may be available even after the pump driving force is disconnected. Since this could be a hazard, much information on the safe maintenance of fluid-power systems is given in later chapters.

Electrical systems have fuses and grounded equipment to prevent overloads, which might lead to excessive heating or fires. Hydraulic and pneumatic systems do not have fuses comparable to those in electrical systems. Line breakage usually results in loss of the hydraulic fluid. If a flammable fluid is used, there is a possibility of ignition of the fluid by hot furnaces or other high-temperature surfaces when a leak occurs. For this reason, fire-resistant hydraulic fluids are often used today.

Many fluid-power systems may be used in a typical large, modern machine employed in factory production. Thus, an injection molding machine uses electric-motor-driven hydraulic pumps for power-transmission purposes. Pneumatic systems eject parts made by the molding machine, close safety gates, and vibrate the feed hopper of the molding machine. Cooling water circulates through the die to remove heat from the molded parts. Electrical solenoids control both the hydraulic and pneumatic directional-control valves used in the fluid-power system. An electronic device, using an electric eye, prevents operation of the machine if the operator has any part of his body in an area where he or she might be injured. Designers of industrial machinery have skillfully integrated electronic and electrical controls into hydraulic and pneumatic power-transmission systems.

The paving machine of Fig. I-2 a and b (see pp. 4–5) follows a surveyor's line system. This machine faithfully spreads concrete in a pattern dictated by the surveyor's line. The mechanical signal received at the wire is turned into motions providing correct direction, level, and position of the material. Two hydraulic variable-flow pumps are teamed with other constant-flow pumps to control all of the power for the output hydraulic motors and cylinders. Valving is employed where needed in the control

Fig. I-3 Hydraulic cylinders provide linear force and precise position.

system. Note the hydraulic motors which provide the interface to the traction drive area.

Figure I-3 shows the hydraulic cylinder, which correctly positions the alternate curb-form device in much the same manner as the leveling mechanism when a roadway is being paved.

Similar machines provide basic grading prior to pouring the concrete.

Auxiliary hydraulic power is available from many machines, so that hydraulically driven tools can be used without need for an air compressor or electric power source. The front-end loader of Fig. I-4 is equipped with a hydraulic hand-held concrete breaker.

Availability of new electrohydraulic mechanisms, which will be studied in this text, have greatly expanded the capabilities of fluid-power-transmission systems. Signals can be accepted from tape controls, surveyor's lines [as in Figs. I-2 and I-3], remote electric sources, beams of light, lasers, fluidic mechanisms, and other complex memory or originating sources.

Many of the items discussed in this introduction are considered in greater detail later in this book. You, as a student, will find the world of fluid power an interesting and challenging one. Once you master the subject, you will be able to apply the principles you have learned when you operate, maintain, design, or assemble an industrial hydraulic system.

Introduction to Fluid Power **7**

Fig. I-4 Mobile construction machinery provides auxilliary power for hydraulically powered tools.

SUMMARY

Hydraulic, pneumatic, and electrical power-transmission systems have much in common. They transmit energy from an energy-generating source to an energy-use area.

Electrical power excels for constant-speed rotary power. Pneumatic systems provide resilient power with fast reversal for either linear or rotary output. The forces developed in a pneumatic system are limited because of the relatively low gas or air pressures used. Hydraulic fluid-power systems excel for infinitely variable speed, precise control of an applied force, and instant reversibility in linear and rotary applications. All three systems considered here are superior to mechanical power-transmission systems because of the relative freedom of position of the power-transfer devices.

REVIEW QUESTIONS

I-1 The basic control functions required in a hydraulic system are pressure, direction, and flow. Why?
I-2 A hydraulic system consists of a pump, a motor, and control valves.

Why might a pneumatic device be included in this hydraulic system?

I-3 What is the hydraulic equivalent of an electric generator?

I-4 What is the purpose of a relief valve?

I-5 Pressure is a function of resistance. Name two ways of controlling the maximum pressure in a hydraulic system.

I-6 Why is the control of flow within a hydraulic system more precise than in an equivalent pneumatic system?

I-7 What is meant by psi?

I-8 If rotary force and motion can be provided in a hydraulic system by a rotary motor similar to the pump, how can linear force and motion be obtained?

I-9 What is the advantage of gravity in a hydraulic system?

I-10 Can the delivery (discharge) of a hydraulic pump be changed? How?

I-11 How does a relief valve work?

I-12 How may energy be stored in a fluid-power system?

I-13 What are the two basic disciplines encompassed in a fluid-power system?

I-14 What is the purpose of using fire-resistant hydraulic fluids?

I-15 Why are pneumatic systems considered resilient?

LABORATORY EXPERIMENTS

I-1 Place two electric fans so that they face each other. Note how the blades of the second fan rotate when the first fan is energized. This experiment illustrates the principles of hydrodynamics used in torque-converter devices.

I-2 Insert a hydraulic jack under a load. Note the solid movement of the piston as the lever is actuated. This is a hydrostatic device. Devices of this type are discussed in later chapters.

I-3 Watch a garage mechanic check out a hydraulic brake system. Note the bleeding of air from the system to prevent springy action. Locate the position in the car of the hydraulic pump and cylinders.

I-4 Watch a bulldozer at work. Note the use of gravity to return the cylinders to their original position. Listen to the high-pitched sound when the cylinder extends to its maximum or encounters a heavy load.

I-5 Note the smooth action of the basket trucks used by utility workers when working on poles. Observe the number of hydraulic units in the basket apparatus.

1
Fluid-Power Systems

Fluid-power systems are widely used in industry to perform hundreds of important tasks. Applications of fluid power include the use of hydraulic jacks for lifting heavy objects, hydraulic rams for moving or compressing materials, hydraulic presses for drawing or forming materials, hydraulic elevators for transporting people or freight vertically, and hydraulic operation and control for machine tools. In this book you will learn how the various components of industrial fluid-power systems are linked to provide a desired output. To begin our study we will examine the various components and fluids used in typical industrial hydraulic systems. Once you have acquired this knowledge, you will be ready to make a detailed study of each of the more complex devices used in various systems. Lastly, you will study a number of practical circuits and learn how fluid-power systems are applied in industry today.

1-1 FLUID MECHANICS

Every hydraulic system uses at least one *fluid;* some systems use more. The primary purpose of the hydraulic fluid is to transmit *energy* from one location to another. Fluids commonly used in industrial hydraulic systems are usually considered to be *incompressible;* their volume does not change markedly when a pressure or force is applied to the fluid.

When you work with the different parts of a fluid-power system, you will meet various units of measurement that are common to all systems. These units of measurement are those applying to pressure, flow rate, area, temperature, viscosity, etc. Since fluid-power systems are used worldwide, the units of measurement can be expressed in two ways: (1)

the United States Customary System (USCS), (2) the System International (SI), also loosely termed the "metric" system. Both systems are presented in this book so you will be able to understand drawings, calculations, and explanations using either system.

To help you understand how both systems are used, Table A-1 in the appendix presents the National Fluid Power Association "Preferred Metric Units for Fluid Power." A total of 38 units of measurement are listed in Table A-1. Look over Table A-1 now so you will be better able to understand both systems of units, the conversion factors given, and the calculations that are worked out in this chapter.

In these calculations, the USCS unit is presented first in the problem statement, followed by the SI unit in parentheses (). Where a numerical value is provided, the USCS value is given first, followed by the SI value and unit, again in parentheses. To save space, the problems are worked only in the USCS units, but the answer is expressed in both USCS and SI units, following the method given above. When you have time, you should solve the problems using both sets of units, because this will enormously improve your comprehension. (When working the problems in SI, you may find a slight variation between your answer and the given answer. This is caused by the rounding off of values.)

Of the 38 units of measurement listed in Table A-1, you will most often encounter 17 units. These units are listed for both systems of measurement in Table A-2 in the appendix.

Study Tables A-1 and A-2 carefully so you understand the relationship

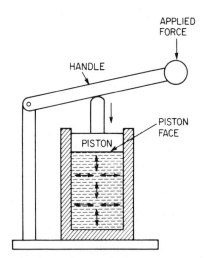

Fig. I-1 A contained fluid exerts equal pressure on each surface of the container.

12 *Industrial Hydraulics*

between the two systems of measurement and the conversion factors used to convert from one system to another. If you start your study of fluid power/industrial hydraulics with a sound understanding of both systems of units, you will be able to work anywhere in the world easily and quickly.

All fluid-power systems follow natural and predictable laws. Pascal, a French scientist, first developed the law underlying the operation of all modern hydraulic devices. Pascal's law states that *pressure applied anywhere to a body of confined fluid is transmitted undiminished to every portion of the surface of the containing vessel.* This law makes the operation of fluid-power systems highly predictable.

When a hydraulic fluid is poured into a container [Fig. 1-1], it assumes the shape of the container. When a *force* is applied to the fluid by a piston or other similar device, the fluid transmits this force equally to all surfaces of the container. The arrows in Fig. 1-1 show how the fluid transmits the applied force.

The piston face in Fig. 1-1 contacts the fluid when force is applied to the handle. To determine the pressure P in pounds per square inch (or in kilopascals, kPa), developed in the container when a known force of F pounds is applied to a piston having an area of A square inches or square cm (cm^2), substitute in

$$P = \frac{F}{A} \qquad [1\text{-}1]$$

Example 1-1 What pressure is developed in the container in Fig. 1-1 if the force applied to the piston is 100 lb (444.8 N) and the piston has an area of 1 in² (6.45 cm²)? Of 4 in² (25.8 cm²)?

Solution Substitute in Eq. [1-1]: $P = F/A = 100/1 = 100$ psi (689.5 kPa). With an area of 4 in², $P = 100/4 = 25$ psi (172.4 kPa). Note that the smaller piston developed a higher pressure in the container when the same force was applied to each piston.

In the above example, *force* was used to develop *pressure*. The reverse of this operation—using *pressure* to develop *force*—can also be performed by a hydraulic system. Assume that an operator moves the handle in Fig. 1-2 forward until a pressure of 1000 psi (6895 kPa) is developed in the cylinder. Pascal's law states that the pressure is transmitted *undiminished* to every part of the container. Therefore a pressure of 1000 psi (6895 kPa) is exerted on the three pistons A, B, and C in Fig. 1-2.

Example 1-2 If piston A [Fig. 1-2] has an area of 0.5 in² (3.23 cm²), what is the pressure developed in the cylinder if the force on this piston is 500 lb (2224 N)?

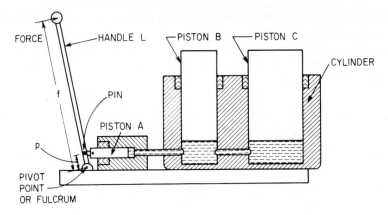

Fig. 1-2 Effective area determines the force developed by a hydraulic piston.

Solution Use Eq. [1-1]: $P = F/A = 500/0.5 = 1000$ psi (6.895 kPa).

Example 1-3 Assume that piston C is held stationary and piston B [Fig. 1-2] is allowed to rise. What force would piston B exert if its area is 10 in² (64.5 cm²), and the fluid pressure is 1000 psi (6895 kPa)? Neglect friction.

Solution Use Eq. [1-1], solving first for F, or $F = PA = (1000)(10) = 10\,000$ lb (44.5 kN).

Example 1-4 Assume that piston B is held stationary and piston C is allowed to rise. What force would C exert if its area is 20 in² (129 cm²), and the fluid pressure is 2000 psi (13 790 kPa)? Neglect friction.

Solution Use Eq. [1-1]: $F = PA = (2000)(20) = 40\,000$ lb (177.9 kN).

Note that by using a long handle L [Fig. 1-2], a small force can be used to develop a high pressure in the cylinder. Thus, if $f = 10$ in (254 mm), and $p = 1$ in (2.54 mm), the arm has a *mechanical advantage* of $10/1 = 10$. An operator exerting a force of 50 lb (222.4 N) on the end of the handle would produce a force of $(50)(10) = 500$ lb (2.224 kN) on piston A. This force can be used to develop a high pressure in the cylinder, as was shown in Example 1-2.

When two pistons have unequal diameters [Fig. 1-2], the piston having the larger area will usually move at a lower fluid pressure.

Several variables are involved in the relative cylinder movements. Friction or resistance to movement is one. Friction or resistance to movement is usually of less significance for larger pistons or rams because of the relatively larger force resulting from the greater effective area of the piston. An assumption of equal frictional resistance for pistons or rams of different areas can be predicated on similar packing, surface finish, and lubrication of the ram. The relationship between the piston's effective surface area and its circumference (where the packing resists the piston movement) changes with the diameter of the piston. The ratio of circumference to area is such that less packing drag is encountered, relatively speaking, as the piston or ram becomes larger in diameter. This provides the basis for the statement that the larger piston usually moves first. We can assume that *fluid under pressure will take the path of least resistance* and start movement of the large piston first.

PISTON MOVEMENT

Example 1-3 showed that the larger the piston area, the greater the force it can exert when a given fluid pressure is acting on it. But what about the motion of two pistons in a cylinder?

Piston Y in Fig. 1-3 uses a small force to develop a high pressure in the cylinder. Piston Z is used to do work by producing a movement against an external load. The ratio of the piston movements, in inches, is inversely proportional to the piston areas, in square inches, or

$$\frac{M_y}{M_z} = \frac{A_z}{A_y} \qquad [1\text{-}2]$$

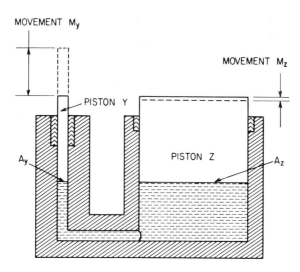

Fig. 1-3 Relative movement of different-size pistons.

Fluid-Power Systems

Example 1-5 How far will piston Z in Fig. 1-3 move if $A_y = 1$ in² (6.45 cm²), $A_z = 50$ in² (322.6 cm²), and $M_y = 12.5$ in (31.75 cm)?

Solution Substitute in Eq. [1-2]: $M_y/M_z = A_z/A_y$, or $12.5/M_z = {}^{50}/_1$; $M_z = 12.5/50 = 0.25$ in (0.635 cm). Thus the small piston must move much farther than the large one. This, in general, is true of all industrial hydraulic systems.

HYDRAULIC PRESS

Figure 1-4 shows a simple hydraulic press used to compress forgings and other materials between two platens. This press uses various simple hydraulic valves. Industrial versions of similar presses can contain relatively complex control assemblies for specialized operating conditions. These controls are discussed in later chapters in this book.

In the press in Fig. 1-4 the ram is raised by actuating the handle of the piston pump to raise the pressure of fluid in the ram cavity. The ram

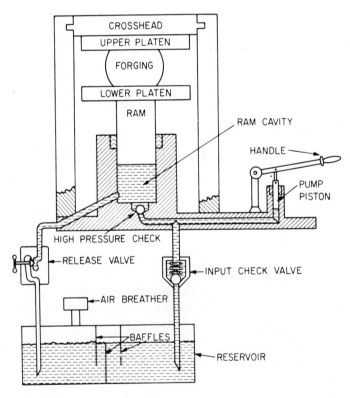

Fig. 1-4 Simple hydraulic press.

16 *Industrial Hydraulics*

moves upward to compress the forging when the fluid pressure is higher than that needed to support the ram weight.

The force developed by the ram is given by

$$\frac{F_r}{F_p} = \frac{A_r}{A_p} = \frac{D_r^2}{D_p^2} \qquad [1\text{-}3]$$

where F_r = ram force, lb (N)
F_p = piston force lb (N)
A_r = ram area, in² (cm²)
A_p = piston area, in² (cm²)
D_r = ram diameter, in (cm)
D_p = piston diameter, in (cm)

Example 1-6 What ram force is developed in a hydraulic press if F_p = 100 lb (444.8 N), A_r = 200 in² (1290 cm²), A_p = 5 in² (32.3 cm²)?

Solution Use Eq. [1–3]: $F_r/F_p = A_r/A_p$: F_r = (100)(200)/5 = 4000 lb (17.79 kN).

Example 1-7 What ram force is developed in a hydraulic press if F_p = 100 lb (444.8 N), D_r = 12 in (30.5 cm), D_p = 4 in (10.2 cm)?

Solution Use Eq. [1-3]: F_r = (100)(12)²/(4)² = (100)(144)/16 = 900 lb (4003 N).

1-2 FLUID COMPRESSIBILITY

All hydraulic liquids can be compressed slightly if sufficient pressure is applied to the liquid. In many practical situations, however, the hydraulic liquid is considered to be incompressible. Oil, which is often used as a hydraulic fluid, will be compressed 1.2 percent of its entire volume when a pressure of 3000 psi is applied to the oil. In high-pressure-system design, compressibility of liquid hydraulic fluids may be particularly important. Where compressibility of the fluid will affect system operation, this fact will be pointed out in later chapters of this book.

Gaseous fluids, which are becoming more popular in certain hydraulic equipment and auxiliary devices, are highly compressible. The compressibility of all gases follows certain basic laws.

Boyle's law states that *the pressure which a given quantity of gas at constant temperature exerts against the walls of the containing vessel is inversely proportional to the volume occupied*, or

$$\frac{P_1}{P_2} = \frac{V_2}{V_1} \qquad [1\text{-}4]$$

where P_1 = initial pressure, psi (kPa)
P_2 = final pressure, psi (kPa)
V_1 = initial volume, in³ or ft³ (cm³ or m³)
V_2 = final volume, in³ or ft³ (cm³ or m³)

Example 1-8 A cylinder having a volume of 100 in³ (1639 cm³), contains a gas at a pressure of 300 psi (2069 kPa). What will the gas pressure be if the cylinder volume is reduced to 50 in³ (819.5 cm³)?

Solution Use Eq. [1-4]: $P_1/P_2 = V_2/V_1$, or $P_2 = P_1V_1/V_2 = (300)(100)/50 = 600$ psi (4138 kPa). Thus, as the volume of the container is reduced, the pressure of the contained gas rises.

Increasing the pressure on a gas increases the gas density. Thus, doubling, tripling, or quadrupling the pressure will double, triple, or quadruple the density. In equation form,

$$\frac{P_1}{P_2} = \frac{D_1}{D_2} \qquad [1\text{-}5]$$

where P_1 = initial pressure, psi (kPa)
P_2 = final pressure, psi (kPa)
D_1 = gas density at initial pressure, pcf (kg/L)
D_2 = gas density at final pressure, pcf (kg/L)

1-3 BERNOULLI'S THEOREM

This is another basic principle affecting fluid-power systems. If a fluid [Fig. 1-5] flows at a constant rate through a pipe, its pressure will be constant unless the pipe diameter changes. If the pipe diameter is reduced, as at *y*, the velocity of the fluid must increase to maintain the

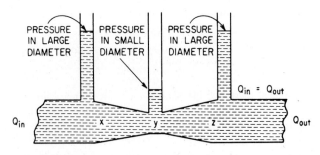

Fig. 1-5 Pressure is less at the minimum diameter of a pinched tube, where the fluid velocity is greater than at either the entry or outlet of the tube.

constant flow rate. When the fluid velocity increases, the fluid pressure decreases, as shown by the vertical pipes in Fig. 1-5. The height to which the fluid rises in each vertical pipe is proportional to the fluid pressure in the main pipe. Where the pipe diameter is increased again to the original diameter, the fluid pressure rises to the original level, as shown at z [Fig. 1-5], if friction in the pipe is neglected.

1-4 TEMPERATURE EFFECTS

There are two laws related to the effect of temperature on a confined gas. These laws are named after their discoverers—Charles and Gay-Lussac. Charles's law states that *if the volume of a gas is held constant,*

$$\frac{P_1}{P_2} = \frac{T_1}{T_2} \qquad [1\text{-}6]$$

where P_1 = initial gas pressure, psi (kPa)
P_2 = final gas pressure, psi (kPa)
T_1 = initial absolute temperature of gas, °R (K)
T_2 = final absolute temperature of gas, °R (K)

Example 1-9 A gas at 50 psi (344.8 kP) has a temperature of 100°F (37.8°C). What will the pressure of this gas be if its temperature is raised to 380°F (193°C) while the volume is held constant?

Solution Use Eq. [1-6]: $P_1/P_2 = T_1/T_2$; $P_2 = P_1 T_2/T_1$. $T_1 = 100 + 460 = 560$°R (310.7 K); $T_2 = 380 + 460 = 840$°R (466 K). Substitute in Eq. [1-6], $P_2 = (50)(840)/560 = 75$ psia (517 kPa).

Gay-Lussac's law states that *if the pressure of a gas is held constant,*

$$\frac{V_1}{V_2} = \frac{T_1}{T_2} \qquad [1\text{-}7]$$

where V_1 = initial volume of gas, in³ or ft³ (cm³ or dm³)
V_2 = final volume of gas, in³ or ft³ (cm³ or dm³)
T_1 = initial absolute temperature of gas, °R(K)
T_2 = final absolute temperature of gas, °R (K)

Example 1-10 A gas has a volume of 10 ft³ (283.2 dm³) at 100°F (37.7°C). What will its volume be if the temperature is raised to 380°F (193.3°C) while the pressure is held constant?

Solution Use Eq. [1-7]: $V_1/V_2 = T_1/T_2$; $V_2 = V_1 T_2/T_1$; $T_1 = 100 + 460 = 560$°R (37.7°C); $T_2 = 380 + 460 = 840$°R (466 K). Substitute in Eq. [1-7], $V_2 = (10)(840)/560 = 15$ ft³ (424.8 dm³).

1-5 HYDRAULIC HORSEPOWER

To pump Q gal (L) of liquid against a pressure rise of p psi (kPa), work must be done in the amount of W ft-lb (J), where

$$W = \frac{231Qp}{12} \qquad [1\text{-}8]$$

If this work must be done in 1 min, the horsepower *input* H required is

$$H = \frac{231Qp}{12(33\,000)}$$

or

$$H = 0.0005833Qp \qquad [1\text{-}9]$$

Example 1-11 What horsepower motor will be required to drive a hydraulic pump handling 100 gpm (378.5 L/min) of 0.85 specific gravity ($=s$) fluid against a pressure of 3000 psi (20 685 kPa)?

Solution Use Eq. [1-9]: $H = 0.0005833Qp = 0.0005833(100)(3000) = 174.99$, say 175 hp. In SI units, power in kW = $QSP/60\,000 = 378.5\,(20\,685)(0.85) = 110.9$ kW.

1-6 FLUID PRESSURE

A column of water [Fig. 1-6] will produce a pressure on its base of

$$P_a = 14.7\ (101.4\text{ kPa}) + 0.433h\ (2.98\text{ kPa}) \qquad [1\text{-}10]$$

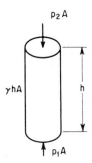

Fig. 1-6 Forces on a fluid column.

where P_a = absolute pressure, psia (kPa)
h = height of water column, ft (mm)

If the atmospheric pressure is zero, or if the pressure is measured by a gage in which the pressure-sensitive element of the gage is subjected to an atmospheric pressure equal and opposed to that acting on the fluid column,

$$P_g = 0.433h \text{ (2.98 kPa)} \qquad [1\text{-}11]$$

where P_g = gage pressure, psig (kPa). In both these equations h is often referred to as the *head* of fluid.

For water, weighing 62.4 pcf (1.028 kg/L), if p is in pounds per square inch,

$$h = \frac{p \text{ psi}}{62.4 \text{ pcf}/144 \text{ in}^2/\text{ft}^2} = 2.32p \qquad [1\text{-}12]$$

For any other fluid with specific gravity s, if p is in pounds per square inch,

$$h = \frac{2.32p}{s} \qquad [1\text{-}13]$$

1-7 MANOMETERS

The basic principle of Pascal's law is used in devices to indicate the fluid pressure existing in a container filled with liquid under pressure. In the piezometer tube of Fig. 1-7, the pressure at point A is

$$p_a = 14.7 \text{ (101.4 kPa)} + 0.433hs \text{ (2.98 kPa)} \qquad [1\text{-}14]$$

A U-shaped tube connected to a vessel under pressure can be used to indicate gas or liquid pressure. The tube is called a *manometer*. In Fig. 1-8,

$$p_a = 14.7 \text{ (101.4 kPa)} + 0.433(2.98 \text{ kPa})h_2s_2 - 0.433(2.98 \text{ kPa})h_1s_1 \qquad [1\text{-}15]$$

Manometers are generally used to measure low pressures. The liquid used in manometers is often mercury.

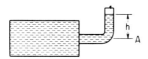

Fig. 1-7 Piezometer.

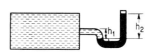

Fig. 1-8 Manometer.

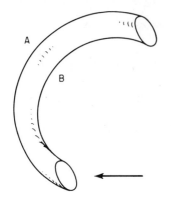
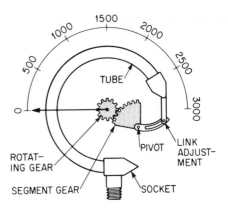

Fig. 1-9 Bourdon-tube section.

Fig. 1-10 Bourdon-tube gauge mechanism.

1-8 BOURDON PRESSURE GAUGE

The Bourdon gauge works on the principle that pressure in a curved tube will tend to straighten it. Thus in Fig. 1-9, pressure acts equally on every square inch of area in the tube, but since the surface area A on the outside of the curve is greater than the surface area B on the shorter radius, the force acting on A will be greater than the force acting on B. When pressure is applied, the tube will straighten out until the difference in force is balanced by the elastic resistance of the material composing the tube. If a pointer is attached to the tube and a scale laid out, the main elements of a Bourdon gauge are obtained.

Figure 1-10 shows the working parts of a Bourdon gauge. The tube is bent into a circular arc and is oval in cross section so that it will tend to straighten more easily when under pressure. The tube works by differential areas, since the area on which the pressure acts outwardly is greater than the area on which the pressure acts inwardly. The open end of the tube passes through a socket which is threaded so that the gauge can be screwed into an opening in the hydraulic system. The closed end of the tube is linked to a pivoted segment gear in mesh with a small rotating gear to which a pointer is attached. Beneath the pointer is a scale reading in pounds per square inch. The gauge is calibrated against known pressures to ensure accurate readings. The working parts are enclosed within a protective case of metal, plastic, or similar material with the dial visible through a clear glass or plastic face. Under pressure the tube tends to straighten and the segment moves about its pivot, rotating the meshing gear and pointer. The pointer assembly is usually pressed on the shaft in such a manner that it is removable for resetting when the gauge is calibrated against a master unit.

1-9 HYDRAULIC FLUIDS

A number of hydraulic fluids are used in industrial hydraulic systems. These fluids include oil, water, compressed air, phosphate esters, aqueous-base ethylene glycol compounds, oil in water, and water in oil.

Probably the least expensive hydraulic fluid is water. Water is sometimes *treated* with chemicals before being used in a fluid-power system. This treatment removes undesirable contaminants. Water which is not treated is passed through a strainer to remove any large solids.

Hydraulic mixtures, emulsions, and combinations may use water-soluble oil, ethylene glycol, and other substances with water. The purpose of these substances is to reduce the corrosive and erosive action of the water on metal machine parts. Additives may also be needed to give enough lubrication for pumps and certain controls.

Water has ideal fire resistance. When water is combined with certain oils, the combination will be fire-resistant until the water evaporates. Then the oil may burn. Other combinations of water and other fluids can eliminate ignition problems but may have mechanical or economic limitations. These fluids are sometimes called *fire-resistant hydraulic fluids.*

Compressed air or gas is used in many fluid-power systems. The cost varies considerably with the type of installation and compressor. In some installations the cost may be comparable to the cost of raw water. Cleanliness is not too much of a problem with compressed air, and lubrication is usually easy to arrange. Central air-compressing stations may be expensive if an extensive piping system is needed. Also, compressed-air systems may be noisy.

Mineral-base oil is perhaps the most widely used hydraulic fluid. It has excellent lubricating properties, does not cause rusting, dissipates heat readily, and can be cleaned easily by mechanical filtration and gravity separation.

When choosing a hydraulic fluid, the following factors must be considered: speed of operation, surrounding atmospheric conditions, heat, economic conditions, availability of replacement fluid, required pressure level, temperature range, contamination possibilities, cost of transmission lines, limitations of equipment in which the fluid is used, lubricity, safety to operators, and expected service life.

1-10 FLUID VISCOSITY

Some hydraulic fluids are thinner than others and flow more easily. The thinner fluids are called *less viscous.* Viscosity is the *resistance a liquid offers to flow.* As the temperature of a liquid decreases, its viscosity (resistance to flow) increases. Cold liquids always flow less easily than hot liquids.

Fluid-Power Systems

A satisfactory fluid for a given hydraulic system must have enough body to give a good seal at pumps, valves, and pistons, but must not be so thick that it offers excessive resistance to flow. Too thick a fluid will add to the load and produce power loss, higher operating temperatures, and excessive wear of parts. On the other hand, too thin a fluid cannot lubricate properly and will also lead to unnecessarily rapid wear of moving parts or of parts which have heavy loads.

1-11 VISCOSITY MEASUREMENT

The instrument most often used by American engineers and technicians to measure the viscosity of liquids is the Saybolt Universal viscometer or viscosimeter [Fig. 1-11]. This instrument measures the number of seconds it takes for a fixed quantity of liquid (60 cm^3) to flow through a small orifice of standard length and diameter at a specified temperature. The viscosity is stated as so many Seconds Saybolt Universal (units SSU) at such and such a temperature. For example, a certain oil might have a viscosity of 80 SSU at 130°F.

The Saybolt Universal viscometer consists of a reservoir for the oil surrounded by a bath heated by heating coils to bring the oil to the temperature at which the viscosity is to be measured. The bottom of the reservoir issues into the standard viscometer orifice. Passage through the orifice is blocked by a cork. The reservoir is filled to a marked level, and a container marked at the 60-cm^3 level is placed under the opening. When the oil to be tested is at the desired test temperature, the cork is removed, and the number of seconds it takes for the liquid to reach the 60-cm^3 level gives the SSU reading.

For quick on-the-job tests it is possible to use comparison devices. Figure 1-12 shows a device consisting of one tube with a known viscosity fluid and a second tube in a parallel plane which is filled with the fluid to

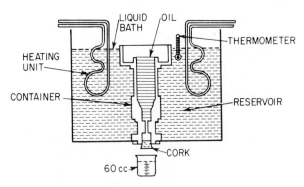

Fig. 1-11 Viscosity testing device using a fixed orifice.

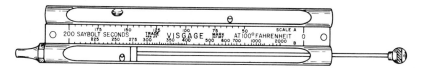

Fig. 1-12 Testing device for comparing the viscosity of fluids.

be tested. By comparing the movement of balls within the two tubes, it is possible to obtain a relatively accurate reading of the viscosity of the fluid being tested.

The commercial classification of oils established by the Society of Automotive Engineers (SAE) relates viscosity range to an SAE number. Thus, SAE 20 oil has a viscosity of 120 to 185 SSU at 130°F; SAE 30 oil has a viscosity of 185 to 255 SSU at the same temperature.

1-12 VISCOSITY INDEX

This value shows how temperature affects the viscosity of an oil. As stated earlier, the viscosity of an oil decreases as the temperature increases, but the variation is greater for some oils than for others. Pennsylvania crude oils (referred to as *paraffinic*) vary comparatively little in viscosity with changes in temperature, while Gulf Coast crudes (*naphthenic*) vary considerably.

To obtain a numerical indication of the degree to which viscosity changes with changes in temperature, these two oils are taken as the basis for a scale. The change in viscosity of Pennsylvania crudes between 100 and 210°F is rated as 100. The change in viscosity of Gulf Coast crudes over the same temperature range is rated as 0. Other oils are then assigned a viscosity index in terms of the degree to which their viscosity changes over this range, as compared with these standard oils.

The lower the viscosity index (V.I.), the greater the variation in viscosity with changes in temperature. Viscosity index figures may range above 100 or below 0 if the viscosity of the oils being measured varies less or more than the standard oils. Thus, viscosity of an oil with a viscosity index of −10 would vary with changes in temperature to a degree even greater than that of the Gulf Coast crudes. An oil with a viscosity index of 120 would show less change in viscosity with changes in temperature than would Pennsylvania crudes.

1-13 POUR POINT

The temperature at which an oil will congeal is referred to as the *pour point*. In testing for pour point, the congealing or solidification is measured with a standard container. Pour-point values vary from oil to oil, according to the nature of the crude oil, the refining methods, and the viscosity of the oil. Chemical additives can be used to lower the pour point. The addition of a *pour-point depressant* will probably affect the viscosity insignificantly through the entire temperature range. The pour point is lowered, but the viscosity continues to increase with decrease in temperature at the same rate it did before the pour point was reached. Any fluid used for hydraulic purposes should have a pour point well below its minimum operating temperature. Special hydraulic fluids have been developed for use on devices that operate through a wide temperature range, such as automobile transmissions and aircraft devices, and for brake fluids and many military applications.

1-14 FLASH POINT

This is the temperature at which a liquid gives off vapor in sufficient quantity to ignite momentarily or *flash* when a flame is applied. A high flash point is desirable because it indicates good resistance to combustion and a low degree of evaporation at normal or working temperatures.

Commercial mineral-base hydraulic oils normally have a high flash point. The temperatures around machine tools are normally close to the surrounding or ambient temperature of the room, but certain types of machines, such as die-cast machines and foundry equipment, may need special fluids because of their proximity to hot metals. In welding machines there must be adequate distances between tank breathers and fill points so that hot metal splatter will not be admitted to the fluid reservoir. Fluid-power lines must be directed around hot manifolds and exhaust lines on tractors and other such equipment.

1-15 FLUID PROPERTIES

In addition to being almost incompressible, a satisfactory hydraulic liquid must have the following properties:

1. *Chemical stability* to prevent formation of sludges, gums, carbon, or other deposits which clog openings, cause valves and pistons to stick or leak, and result in poor lubrication to moving parts.

2. *Freedom from acidity* so that the fluid is noncorrosive to the metals in the system.

3. *Lubricating properties* sufficient to maintain the established metal clearances and prevent galling and scoring. Film strength must be sufficient to prevent the fluid from being wiped or squeezed from between the surfaces when spread out in a thin layer. An oil will not lubricate if the

film strength breaks down, since the motion of part against part wipes the metal clean of oil.

4 *Satisfactory viscosity and viscosity index* so that the oil will perform adequately within the normal temperature ranges of the machinery.

5 *Pour point* well below the minimum temperature expected in normal operation.

6 *Flash point* as high as possible, for safety and to prevent possible evaporation.

7 *Minimum toxicity* since some hydraulic fluids, particularly in the fire-resistant class, may be toxic. Suitable protection and care in handling must be provided.

In addition to the above properties, petroleum oils should also have *high oxidation resistance, good water-separating ability, good antirust properties, and good resistance to foaming.*

Lastly, it is important that hydraulic fluids be kept clean at all times. Fluid rarely wears out. Premature failure of fluids and system is more likely to result from careless handling or inadequate protection. Therefore the fluid should be kept in good condition and protected from contamination.

DESIGN CONSIDERATIONS FOR FLUID-POWER SYSTEMS

Pump suction lines are significantly larger in flow area than discharge lines because of limitations associated with atmospheric pressure needed to move fluid into the pumping mechanism.

Hydraulic systems must be operated within specified temperature limits for maximum efficiency. Cold oil is stiff. With equal power input, cold oil flows at a slower rate than does warm oil.

Hydraulic fluids have specific characteristics that can affect the associated components. Seals, hose, reservoir coatings, packings, and elastomeric diaphragms must be made of a material that will not deteriorate in association with the working fluid.

Three major hydraulic fluids will be encountered: (1) petroleum-based oils, (2) phosphate-ester-type synthetic fluids, and (3) raw water or water-based fluids.

The fluids generally cannot be mixed or interchanged. Different seals, hose elastomers, and coatings are needed for each. Careful cleaning is generally essential when changing from one fluid to another.

The need for different fluids in hydraulic systems is usually the result of environmental change, such as a requirement for fire-resistance, adverse temperature considerations, or economic objectives. When machinery is returned to a different environment, an inverse change may be required to save operational costs or because of a less complicated maintenance schedule. Strict adherence to the manufacturer's recommendations is essential. Some elastomers that may have a higher first cost are compatible with more than one type of operating fluid. Long-term economies can result from the use of the more versatile seals and protective coatings. Stainless-steel reservoirs and associated piping may be used in erosive atmospheres.

The following formula is used frequently in fluid-power design activities. The bar is widely used in fluid-power pressure measurement. Some industries prefer the use of the kilopascal as a pressure measurement.

ENGLISH

Hydraulic Power

Horsepower =

$$\frac{\text{flow (gpm)} \times 231 \times p(\text{psi})}{12 \times 33{,}000}$$

METRIC (SI)

$$\frac{\text{flow (L/min)} \times \text{pressure (bar)}}{448}$$

USEFUL METRIC FORMULA

$$kW = \frac{\text{flow (L/min)} \times \text{pressure (bar)}}{600}$$

Watts (W) = flow (m³/s) × pressure (N/m²)

$$kW = \frac{\text{flow (L/s)} \times \text{pressure (bar)}}{10}$$

ROTARY MOTION

Power (Watts) = Torque (Nm×2πn) (rads/sec) or $\dfrac{Nm \times rpm}{9.55}$

Energy (Joules) = $\dfrac{GD^2 \, (Kg \, m^2)^2 \times (rads/s)^2}{8}$ or $\dfrac{(Kg \, m)^2 \times (rpm)^2}{730}$

LINEAR MOTION

Power (Watts) = Force (N) × velocity (m/s)

Energy (Joules) = ½ mass (Kg) × (velocity (m/s))²

PREFERRED UNITS (METRIC) FOR FLUID POWER—Document NFPA/T2.10.1-1978 can be obtained by writing to the National Fluid Power Association, 3333 N. Mayfair Road, Milwaukee, WI 53222. This document is reproduced in part in Appendix A in this book.

SUMMARY

The key to successful application of all fluid-power devices rests in highly predictable natural laws.

> *Pascal's law states:* Pressure applied anywhere to a body of confined fluid is transmitted undiminished to every portion of the surface of the containing vessel.

Boyle's law states: The pressure which a given quantity of gas at constant temperature exerts against the walls of the containing vessel is inversely proportional to the volume occupied.

Bernoulli's principle states: Whenever the velocity of a fluid such as air or water is high, the pressure is low; when the velocity is low, the pressure is high.

Charles's law states: For all gases at constant volume, the pressure is proportional to the absolute temperature.

Gay-Lussac's law states: For all gases at constant pressure, the volume is proportional to the absolute temperature.

The type of hydraulic fluid used is determined by circuit needs, economic factors, and safety considerations, and by the fluid having properties best suited to the particular application.

Fluids must be adequately contained for maximum efficiency as well as for economic reasons.

REVIEW QUESTIONS

1-1 If piston A in Fig. 1-2 has a surface area of 2 in^2, piston B 10 in^2, and piston C 15 in^2, determine the load that B and C can raise if 75 lb of force is exerted on piston A.

1-2 If the area of piston A is reduced to 1 in^2 and the same force is exerted, what load can be raised by B and C?

1-3 How will reduction in the area of piston A affect the speed of movement if the rate of force application remains constant?

1-4 List 10 factors that should be considered when choosing a hydraulic-fluid medium.

1-5 What does the SAE number of an oil refer to?

1-6 What is the flash point of a hydraulic fluid?

1-7 Describe the construction and operation of a Bourdon-tube type of pressure gauge.

1-8 Define *viscosity index*. What is a viscosity index used for?

1-9 Discuss the different types of hydraulic fluids used today. Classify each according to its lubricating qualities, relative cost, fire resistance, and rust-inhibiting properties.

1-10 Describe the construction and use of a viscosimeter.

1-11 What effect does temperature have on a volume of gas?

1-12 Why are additives sometimes used with hydraulic fluids?

1-13 Is hydraulic oil completely incompressible? Explain.

1-14 What is the usual cause of hydraulic-fluid deterioration?

1-15 What is the purpose of the air breather of Fig. 1-4?

LABORATORY EXPERIMENTS

1-1 Make a simple hydraulic jack with a hand pump and a ram, or use a small cylinder with check valves and a ram. Note the movement of the ram relative to the mechanical input. Insert a pressure gauge from which you can read the pressures directly.

1-2 Note the change in speed of drop of the jack with varying openings of the needle valve, and with equal openings and varying loads.

1-3 Take oil samples from various machines and check the condition of the oil. Allow the sample to settle for 24 hr and check for foreign material and separation from the oil of water or other contaminants. Mix 1 cm^3 of the oil with equal parts of clean solvent and pass it through filter paper. Examine the residue with a microscope. Attempt to classify the type of residue, its size, and its potential effect on machine elements.

1-4 Remove filters from various machines and note the degree of contamination and type of material trapped by each filter. Check the fluid beyond the filter in the same manner as in Experiment 1-3.

1-5 Make a simple viscosity tester with a fixed orifice. Note the difference in readings when the oil is at room temperature, 110°F, and 125°F. Repeat with various oils. Repeat the test with standard automotive-type hydraulic brake fluid. Compare the viscosity readings obtained.

1-6 Fabricate a clear plastic test fixture similar to Fig. 1-5. Pass fluid through the unit and measure the pressure with a manometer at points x, y, and z.

2
Hydraulic Symbols

Graphic communications, like languages, are easiest to learn if they can be directly identified with the associated subjects. This chapter is intended to acquaint you with the basic symbols, terminology, and rules associated with the use of graphic symbology. The specific use of the symbols will be expanded as you continue your study of fluid-power-transmission devices and systems. A good knowledge of the material in this chapter will simplify your continuing studies.

Hydraulic symbols are used throughout the world in the design, operation, and maintenance of fluid-power systems. A knowledge of hydraulic symbols will enable you to read and understand circuit diagrams and other important drawings of fluid-power systems. This knowledge is a valuable aid in your work with all types of hydraulic equipment. Some of the symbols we will study in this chapter are called *simplified symbols*. Other, more complex symbols will be explained as you study specific components (valves, controls, etc.) in later chapters. Simplified symbols are widely used in commercial circuit diagrams. More complex symbols may be used when a basic valve or control is modified to make it perform a new or different function.

Graphical symbols used in this book are in accord with the International Standards for Fluid Power Systems and Components—Graphical Symbols—First Edition—1976-08-01, ISO Standard 1219-1976 (E-F), which supersedes former Joint Industrial Council (JIC) and American National Standards Institute (ANSI) standards for previous issues. Standards for fluid-power devices and systems for production machinery can be procured from the Joint Industrial Council, 7901 Westpark Drive, McLean, VA 22101.

American National Standards for graphical diagrams and American

National Standard Y14.17-1966 Drafting Standard for Fluid Power Diagrams can be procured from the American National Standards Institute, 1430 Broadway, New York, NY 10018.

Specialized Fluid Power Standards can be procured from the National Fluid Power Association, 3333 N. Mayfair Road, Milwaukee, WI 53222. The National Fluid Power Association also publishes an annual two-volume compilation of all current standards used within the fluid-power industry.

2-1 CIRCUIT ELEMENTS

A *hydraulic circuit* consists of one or more fluid-power devices connected by piping or tubing. In a diagram or drawing of a hydraulic circuit, a solid line [Fig. 2-1a] represents a *pipe,* a *tube,* or some *other conductor* capable of handling major flow in the circuit. This type of pipe is often termed a *major conductor;* it conveys fluid to various parts of the circuit.

Pilot lines are usually of much smaller fluid-carrying capacity (and diameter) than major lines. Figure 2-1b shows a pilot line. The length of each dash is at least 20 line-widths, with the space between dashes approximately 5 line-widths. The fluid pressure in pilot lines is usually the same as that within the major piping. Pilot-line pressure may, however, be any value needed in the circuit.

Fluid-drain and *air-exhaust lines* [Fig. 2-1c] may be made of lighter-weight material and may not be able to withstand as high pressures as pilot or major pipes do. The length of the dashed lines and the space between them in the symbol are about equal, each being approximately 5 line-widths.

Pipes that cross each other (but are not interconnected) are drawn with a small loop [Fig. 2-1d]. Note that Fig. 2-1d shows two *major lines* (or pipes) crossing, but not interconnecting. Figure 2-1e shows a major line being crossed by a pilot line (or pipe). (In fluid-power terminology the words *pipe, tubing, line,* and *conductor* are often used interchangeably to designate the structure that conveys the fluid from one device to another.) Figure 2-1f shows two drain lines crossing each other without physical interconnection.

When pipes are joined so that there is a connecting path for the fluid to each of the adjoining pipes, a *connector dot* (about 5 widths of associated symbol lines) is placed on the diagram so that it clearly shows the point of junction. Figure 2-1h shows two major lines joined at the point designated by the dot. In an actual circuit, this is usually a pipe or tubing tee. Figure 2-1i shows the point at which a pilot line is connected to a major line. The pilot line is usually smaller in diameter, and the connection may be a tee with a reduced-diameter outlet to the pilot line. Figure 2-1j shows two major lines intersecting. The dot indicates that these lines are interconnected. A fitting in the form of a *cross* with four connections or two tees in

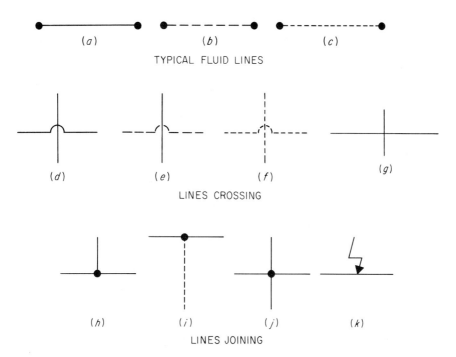

Fig. 2-1 Line symbols. (*a*) Working line. (*b*) Pilot line. (*c*) Drain line. (*d*) Crossing work lines. (*e*) Crossing work and pilot line. (*f*) Crossing drain lines. (*g*) Alternate crossing work lines. (*h*) Joined working lines. (*i*) Pilot line joining working line. (*j*) Intersecting working lines. (*k*) Electric line.

close proximity may be used for this connection. Note that these symbols do not indicate whether the piping is fabricated from *tube and fittings* or from other types of material. Instead, the symbols show the path of the fluid through the circuit. Electric lines are shown by Fig. 2-1*k*.

Figure 2-2*a* shows how a flexible connection is indicated in a circuit diagram. This symbol indicates some type of hose or similar flexible device.

Most fluid-power systems have a tank or reservoir in which to store fluid during some part of the cycle. Pipes connected to the reservoir can affect the operation of the entire hydraulic system. Because of the importance of the location of the end of the pipe (either above or below the surface of the fluid in the tank), symbols to indicate the pipe location have been devised. A U-shaped symbol [Fig. 2-2*b*] indicates a hydraulic-fluid reservoir or tank. The U is twice as wide at the base as the height of the uprights. In Fig. 2-2*c* the main fluid line touches the bottom of the U, indicating that the pipe terminates *below* the fluid level in the tank. Drain lines from many pumps and fluid motors are connected below the fluid level so that air will not enter the device through the drain. In Fig. 2-2*d* a major pipe is shown terminating *above* the fluid level in the tank. This

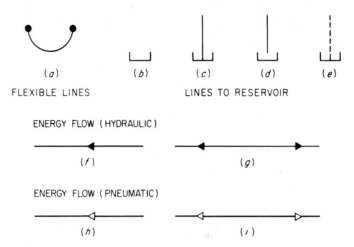

Fig. 2-2 (*a*) Flexible line. (*b*) Vented tank or reservoir. (*c*) Pipe terminated below fluid level. (*d*) Pipe terminating above fluid level. (*e*) Drain terminating below fluid level. (*f*) Energy flow (hydraulic) in one direction only. (*g*) Energy flow (hydraulic) in both directions. (*h*) Energy flow (pneumatic) in one direction only. (*i*) Energy flow (pneumatic) in both directions.

arrangement is used when it is desired to drain a fluid-power machine during its idle periods.

An equilateral triangle [Fig. 2-2*f*] is placed on a fluid line to show the direction in which the fluid is expected to travel. If a solid equilateral triangle is used, it indicates that the fluid is a liquid. An outline of a equilateral triangle, such as Fig. 2-2*h*, indicates that the fluid is compressed air or a gas. If the fluid can travel in both directions at different times, two equilateral triangles facing in opposite directions [Fig. 2-2*g* and *i*] are used. Flow-direction triangles can be omitted from circuit diagrams unless there is a definite need for an indication of the flow direction or the fluid medium.

The symbol for a *plugged terminal* (or end) point of a pipe is indicated by an X [Fig. 2-3*a*]. Note in Fig. 2-3*b* how the X is located in a *composite* (more than one connection) symbol. This symbol indicates that the connection to the hydraulic device is provided with a *plug* instead of a connecting line. Nonfunctional plugs used as construction devices are not shown on hydraulic symbols and can be disregarded. Even though construction plugs may be used as an aid in cleaning a device, they are not shown on the circuit diagram.

A functional plug may be located at the surface of a manifold block, which can be used to house many control devices. The terminating cross of Fig. 2-3*c* may be on a valve symbol as in Fig. 2-3*b* or on an enclosure (*envelope*) such as that indicated by the centerline. All parts that are either within a common housing or manifolded together as a unit are indicated by the enclosure formed by the centerline [line broken by an occasional dash in Fig. 2-3d]. A test station may be used for a gauge connection or for connecting other test devices. Plugs may also be provided at high points in a conducting line to serve the dual purpose of bleeding air from the system on start-up and providing a gauge opening. A needle-type shutoff valve may be permanently installed in place of a plug if air must be removed frequently or if a gauge must be used for checking the pressures when machine tooling is changed.

Figure 2-3d shows an envelope containing a *filter cartridge* and a *spring-loaded check valve,* which in this installation serves as a low-pressure relief valve. The enclosing dashed line indicates that all these devices are within this common enclosure or envelope. The enclosure is usually a metal body or housing. The square with the bisecting dashed line indicates the *filter.* Connecting lines are terminated at opposite corners. A wedge-shaped arrowhead with a ball between the arms indicates a *check valve.* This check-valve symbol represents a ball in a seat and is not intended as an arrow. Fluid flows from point A to point B through the filter or check valve (serving as a low-pressure relief valve) or both, depending on the state of cleanliness of the filter cartridge.

The test station on the left side of the envelope can be used to obtain a pressure reading of the fluid at that point. It may also be used to drain a series of test specimens of the hydraulic fluid before the fluid reaches the

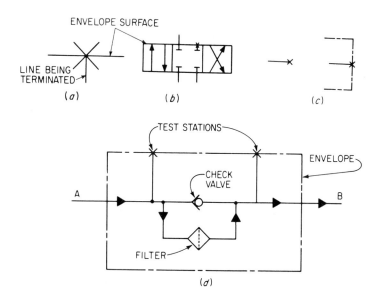

Fig. 2-3 (*a*) Plugged terminal point. (*b*) Plugged on a four-way-valve symbol. (*c*) Test-station symbol. (*d*) Typical test-station installation.

filter. At the test station, to the right of the filter, samples of the fluid can be collected after passage through the filter. This station can also be used for a gauge connection to check the fluid-pressure drop across the filter and check-valve assembly. Test stations can be located in various parts of a fluid-power circuit to permit checking of pressure or the condition of the fluid without major piping alterations.

A fixed-size, nonadjustable *pipe restriction* is indicated by the symbol in Fig. 2-4a. Dimensions of the hole in the restriction can be shown in parentheses, as in Fig. 2-4b, for convenience in identifying the expected degree of control. Hole dimensions assist in identifying the various orifices in a complex hydraulic circuit. An idea of the rate of fluid flow and the fluid pressures to be expected in various parts of the circuit can be obtained by studying the orifice sizes. The length of the orifice and its diameter may also be given on the symbol. These dimensions are important in pressure-drop calculations.

Explanatory notes about the orifice may be desirable if the orifice is created with pins or wire in the hole to establish the desired net flow area. Use of a properly secured pin or wire may make cleaning of the orifice simpler. Changing the size of the pin or wire can provide an inexpensive method of varying the orifice area without machining the orifice itself. Such information is usually shown in explanatory notes. In some installations the explanatory notes may be given in a maintenance manual instead of on the circuit diagram.

A sharp-edged orifice is relatively unaffected by the viscosity of the fluid (resistance to flow of the fluid). A longer flow path through the

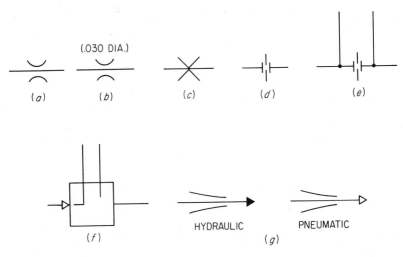

Fig. 2-4 (a) Fixed restriction (viscosity sensitive). (b) Fixed restriction with size indicated. (c) Fixed orifice unaffected by viscosity. (d) Orifice plate. (e) Orifice plate with sensing line. (f) Pitot tube. (g) Nozzles.

Fig. 2-5 (*a*) Quick-disconnect without check valves. (*b*) Quick-disconnect with check valves (disconnected). (*c*) Quick-disconnect with one check valve (connected). (*d*) Quick-disconnect with check valves in both sections.

orifice will be affected by the viscosity of the fluid as it passes through the restriction. The symbols in Fig. 2-4*a* and *b* anticipate a resistance to flow affected by fluid viscosity. Figure 2-4*c* shows the symbol for a sharp-edged orifice which would be relatively insensitive to flow restrictions resulting from changes in viscosity.

Sensing devices associated with flow rate in conductors may use orifice plates that can be changed to alter flow conditions. The symbol for an orifice plate is shown in Fig. 2-4*d* and *e*. Note the sensing line before and after the orifice created by the plate inserted in the line with a fixed hole size [Fig. 2-4*e*].

A pitot tube is inserted in a line to measure velocity of flow. Figure 2-4*f* illustrates the symbol used for a pitot tube. Hydraulic and pneumatic nozzles are indicated by the symbol shown in Fig. 2-4*g*.

When it is desired to disconnect two pipes or a pipe and a hydraulic unit quickly, *quick-disconnect couplings* may be used. These couplings use mechanical locks that are easily released and reassembled. Flexible hose of some type is generally used with the coupling. Tools are not needed to disconnect or reconnect. Some couplings are fitted with integral check valves that close when the coupling is disconnected. Closing of the check valve prevents loss of hydraulic fluid from the unit served by the coupling. Check valves are not needed in couplings serving a hydraulic unit that is self-draining because the fluid drains from the unit after it stops operating.

Figure 2-5*a* shows two quick-disconnects coupled together without integral check valves. One-half of a quick-disconnect coupling with a check valve to prevent fluid loss is shown in Fig. 2-5*b*. Figure 2-5*c* shows a connected coupling with half equipped with a check valve and the other half without a check valve; Fig. 2-5*d* shows the symbol for coupled connectors with check valves in both halves.

2-2 FLUID PUMPS

The basic envelope (or enclosure) symbol for a pump is shown in Fig. 2-6*a*. It is a circle. Lines outside the envelope are not part of the symbol but represent pipes connected to the pump [Fig. 2-6*b*]. Rotating shafts in pumps are shown by curved arrows on the side of the shaft nearest the viewer. Figure 2-6*c* shows a pump shaft that rotates in only one direction;

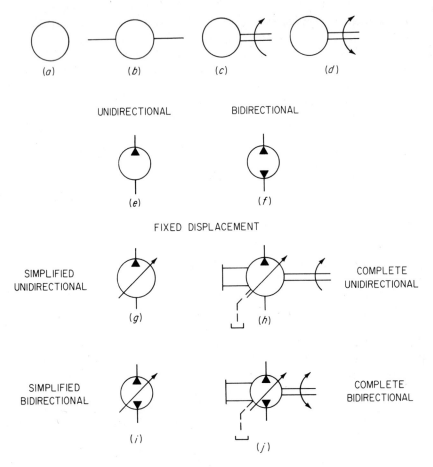

Fig. 2-6 (a) Basic pump symbol. (b) Flow lines connected to pump. (c) Shaft turns in one direction. (d) Shaft turns in either direction. (e) Fixed displacement, unidirectional. (f) Fixed displacement, bidirectional. (g) Variable displacement, simplified. (h) Variable displacement, complete. (i) Simplified, bidirectional. (j) Complete, bidirectional.

a shaft that turns in either direction (clockwise or counterclockwise) is shown in Fig. 2-6d.

The equilateral triangle used to show energy flow through a conducting line serves a similar function in a pump symbol. A single, solid, equilateral triangle is used to show the energy-flow pattern in a fixed-displacement unidirectional pump symbol. Figure 2-6e shows the direction of flow through the pump. The line opposite the equilateral triangle indicates the intake line to the pump. The discharge line from the pump adjacent to the equilateral triangle is assumed to be the supply to the immediate circuit. Bidirectional pumps, capable of delivering pressurized liquid from either port, are indicated by the symbol of Fig. 2-6f.

38 *Industrial Hydraulics*

A slash arrow across the symbol [Fig. 2-6g] indicates that the pump displacement can be varied. Figure 2-6g shows the simplified symbol, while Fig. 2-6h shows the complete symbol indicating manual control by the parallel lines at the left with the vertical terminating line. The drain, if included, is always shown in the complete symbol. Direction of rotation is also included, if pertinent.

Bidirectional, variable-displacement pumps are illustrated by the addition of the slash arrow and the second equilateral triangle, as in Fig. 2-6i. The complete symbol, shown in Fig. 2-6j, includes the drain and manual-control-device symbol.

2-3 FLUID MOTORS

The basic circle used for pump symbols is also used for rotary fluid motors. A solid equilateral triangle shows the direction of energy and fluid flow in the symbol, depicting a fluid motor in much the same manner as a pump. The primary difference is that fluid-energy flow is away from the pump. Energy flow is into the fluid motor, where it is converted into rotary force and motion. Consequently, the equilateral triangle points in, as shown in Fig. 2-7a or b. Bidirectional operation is shown with two triangles, as in Fig. 2-7c. Variable displacement of the rotating group is shown by the addition of the slash arrow, as in Fig. 2-7d through f.

The variable-displacement unit of Fig. 2-7g [simplified] or 2-7h [com-

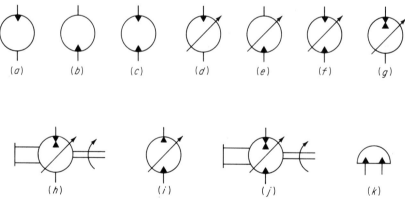

Fig. 2-7 (a) Basic symbol for a unidirectional motor. (b) Reverse rotation of unidirectional motor. (c) Fixed-displacement, bidirectional motor. (d) Variable-displacement, unidirectional motor. (e) Reverse-rotation, variable-displacement, unidirectional motor. (f) Bidirectional, variable-displacement motor. (g) Variable motor in one direction, pump in other direction—simplified symbol. (h) Complete symbol. (i) Variable unit functional as pump or motor in one direction only—simplified symbol. (j) Complete symbol. (k) Limited-rotation motor.

Hydraulic Symbols   **39**

plete] functions as a pump in one direction and as a motor in the other direction. The motor, driving the cable drum of a mobile crane, often encounters this type of service. Raising a load requires the motor function. Lowering the load requires that the motor function as a pump so that the output fluid can be restricted to provide the desired lowering speed.

The symbol of the variable unit in Fig. 2-7i [simplified] or 2-7j [complete] shows a device that can function as a motor and as a pump in the same direction. For example, a mixer drive requires that energy be applied to start the unit into motion. When the mass is to be stopped, the motor becomes a pump. The delivery of the pump is then restricted to provide the needed deceleration of the mass.

A motor or rotary actuator, which has a limited rotating pattern that does not permit continual rotation in one direction, is illustrated by Fig. 2-7k. Many of these devices rotate less than 360° prior ro reversal.

Linear motors, consisting of a piston and cylinder, are represented by the symbols shown in Fig. 2-8. *Diaphragms* and other linear *actuators* are represented by similar symbols. A *single-acting cylinder* is shown in Fig. 2-8a. Fluid enters the left end of the cylinder and forces the piston to the right. An external force, such as gravity or moving machine members, produces reverse movement of the piston, forcing the fluid back into the hydraulic circuit. Figure 2-8d shows a modification of the single-acting cylinder fitted with an *internal spring* to return the piston to its original position. The spring space or *cavity* has an external drain that terminates *above* the reservoir fluid level. Were this drain terminated below the fluid level, the piston might act as a pump during the reverse stroke and fill with fluid from the reservoir.

A *double-acting cylinder* with a single rod end is shown in Fig. 2-8b. Sometimes it is desirable to use both ends of the piston rod. To achieve this, the piston rod is extended to pass through both ends of the cylinder [Fig. 2-8c]. This arrangement is termed a *double-acting, double-end cylinder.*

Integral devices to decelerate the piston and its associated load at the ends of the stroke are illustrated by the rectangle shown attached to the piston in Fig. 2-8e. The addition of a slash arrow [Fig. 2-8f] shows that the cushion or deceleration device is adjustable.

If the relationship of rod diameter to cylinder bore is significant, it may be desirable to indicate this, as in in Fig. 2-8g.

Telescopic cylinders, as the name suggests, provide long strokes by the use of multiple sections which are nested within the concentric assembly as they retract. Figure 2-8h shows a single-acting assembly. Figure 2-8i shows a double-acting telescope cylinder. Combinations of two cylinders with different-size pistons are used to provide pressure intensification or, inversely, volume amplification. A single, common rod may be employed. The symbols shown in Fig. 2-8j through *l* are not intended to show construction. They indicate a pressure-intensification function or, under certain conditions, a volume amplification.

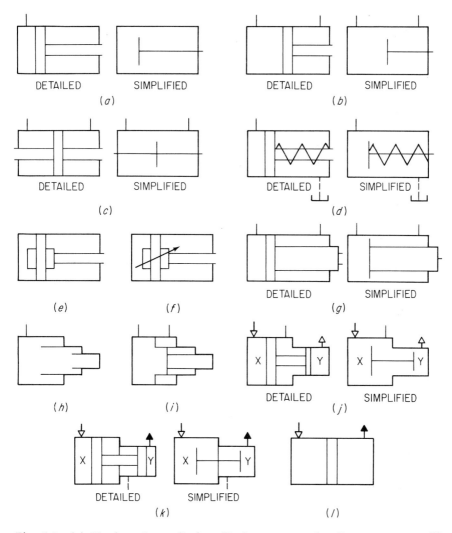

Fig. 2-8 (*a*) Single-acting cylinder, diaphragm, or other linear actuator. (*b*) Double-acting, single-rod cylinder. (*c*) Double-acting, double-rod-end cylinder. (*d*) Double-acting cylinder, pressure in one direction, spring return, single-rod cylinder with drain above liquid level. (*e*) Cylinder assembly with fixed cushion, advance and retract. (*f*) Adjustable cushion, advance and retract. (*g*) Oversize rod. (*h*) Telescoping cylinder, single-acting. (*i*) Telescoping cylinder, double-acting. (*j*) Pressure intensifier or volume amplifier, pneumatic. (*k*) Hydraulic and pneumatic intensifier. (*l*) Air-oil pressure transformer.

The assembly in Fig. 2-8*j* shows that a pneumatic pressure x is transformed into a higher pneumatic pressure y. Figure 2-8*k* shows that a pneumatic pressure x is transferred into a higher *hydraulic* pressure y.

Figure 2-8*l* shows an air-oil actuator. Equipment of this design transforms a pneumatic pressure into a substantially equal hydraulic pressure, or vice versa. Such equipment is most often used to obtain smooth hydraulic-flow control.

2-4 HYDRAULIC VALVES

The basic symbol for a hydraulic valve is a rectangle, termed the *valve envelope*. As with pumps and motors, the valve envelope represents the valve enclosure or body. Lines within the envelope show the flow directions between the valve inlet and outlet openings, termed *ports*. To show the change in flow conditions when the valve is opened or closed (*actuated*), two systems of symbols are used: (1) single envelope and (2) multiple envelope.

Single-envelope symbols are used where only one flow path exists through the valve. Flow lines within the envelope indicate static conditions when the actuating signal is not applied. The arrow can be visualized as moving to show how the flow conditions and pressure are controlled as the valve is actuated.

Multiple-envelope symbols are used when more than one flow path exists through the valve. The envelopes can be visualized as being moved to show how flow paths change when the *valving element* in an envelope is shifted to its various positions.

Figure 2-9a shows a typical single envelope; multiple envelopes are shown in Fig. 2-9b. Flow lines to the valve ports are added to the two types of basic valve envelopes in Fig. 2-9c and 2-9d. Note in Fig. 2-9d that the ports are shown at the center envelope segment.

A variety of valve-port conditions are shown in Fig. 2-10. Thus, the double arrow in Fig. 2-10a denotes a blocked or closed port in a single envelope where flow may be expected to pass in either direction. The blocked condition is indicated by the fact that the arrows do not line up with the lines coming to the single square. Figure 2-10b shows the same condition in a multiple-envelope valve. The single envelope signifies that an infinite number of positions may be encountered between the closed (normal) and open positions, such as might be expected with a relief valve where the opening is a function of the orifice size that must be created to provide the needed pressure level in an automatic function.

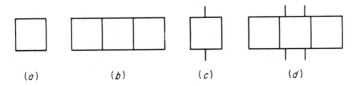

Fig. 2-9 (a) Basic envelope. (b) Multiple basic envelopes. (c) Ports attached to basic envelope. (d) Ports attached to normal multiple basic envelopes.

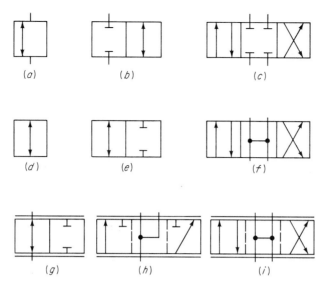

Fig. 2-10 (a) Blocked port. (b) Blocked port, finite positions. (c) Three-position, blocked ports in neutral, finite positions. (d) Open port, infinite positions. (e) Normally open, finite positions. (f) Open center, three finite positions. (g) Two-position, infinite crossover positions. (h) Two-position, infinite crossover position, three-way diversion valve. (i) Two-position, infinite crossover position, four-way valve with neutral fully open.

The multiple envelope of Fig. 2-10b and c indicates a finite number of positions which the directing elements can be normally expected to assume. The normal position of Fig. 2-10b is indicated by the left block. The normal position of the unit, shown in Fig. 2-10c, is the neutral position in which all the ports are blocked. This condition is referred to as a *closed center*. Shifted to conditions in the left block, the flow is in the direction indicated by the arrows. When shifted to the right, the flows are reversed. This is referred to as a *four-way function;* one input port can be directed to separate outlets or blocked, while a return line is directed to an opposite port in a predetermined pattern.

A through flow is indicated by the symbol of Fig. 2-10d. An infinite number of positions of the flow-control elements from full-open to closed can be anticipated with this design. The two-element symbol of Fig. 2-10e indicates two finite positions. The three-element symbol of Fig. 2-10f indicates three finite positions, with the neutral position indicating an interconnection of all associated lines.

The two-element symbol of Fig. 2-10g is provided with parallel lines at the top and bottom. This signifies that it represents a valve with two finite positions; however, the nature of the application is such that it may be shifted to infinite positions in the process of moving from one position to the other. An example is a cam-operated two-way valve used to decelerate flow; the cam may follow a random pattern in the functional movement from open to closed position.

The diversion valve of Fig. 2-10h indicates that all ports are interconnected as the flow-directing element is moved. The dashed lines between the blocks indicate that the neutral position is not a finite one. The information contained in the center block indicates only the conditions as the valve elements are shifted from one finite position to the other. The open-center condition of the symbol shown in Fig. 2-10i would not normally be shown in a two-block symbol. Addition of the third position between the dashed lines provides the needed data without indicating three finite positions.

Figure 2-11a and b summarizes the basic symbol, normal and actuated conditions for normally open and normally closed single-envelope valves. Figure 2-11c shows an alternate arrow design which indicates that one port is always connected to one flow path, whether normal or actuated.

The valve in Fig. 2-12a is a *spring-loaded relief valve*, piped so that the discharge flows to a reservoir having the outlet below the fluid level. The

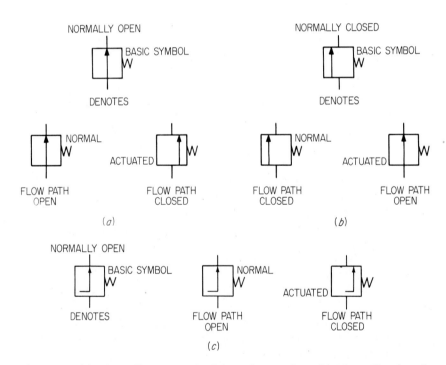

Fig. 2-11 (a) Normally open, single-envelope valve. (b) Normally closed, single-envelope valve. (c) Alternate arrow design to indicate that one port is always connected, whether actuated or at rest.

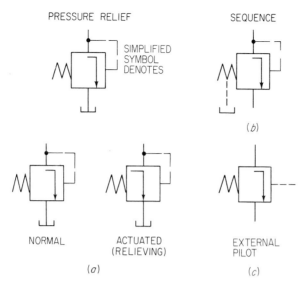

Fig. 2-12 (*a*) Relief valve. (*b*) Sequence valve. (*c*) Unloading valve.

signal source for a relief valve is ahead of the normally closed element. As the pressure rises to a value at which the valve opens and the flow passes through the valve, it can be imagined that the arrow is aligned as in Fig. 2-12*a*, where it is shown in the actuated position. A sequence valve, as shown in Fig. 2-12*b*, has a similar function to that of the relief valve. Because the outlet port may be directed to a pressurized area, it may be necessary to provide an external drain for the spring pocket or for draining a pilot mechanism if it is included in the design.

An unloading valve usually drains internally because the secondary port normally connects to the major return line. If pressure is expected on the valve outlet, a drain may be necessary. The operating signal, shown in Fig. 2-12*c*, is usually from an external source.

Two pressure-reducing valve symbols [Fig. 2-13*a* and *b*] show the basic symbol as a normally open two-way valve that may assume infinite positions during normal function. The *signal source* is at the valve outlet or *low-pressure secondary port*. Figure 2-13*b* shows the addition of a major port discharging to the tank. The flow-directing element indicates potential flow in either direction. Normal flow is directly through the valve. As a predetermined reduced-pressure value is attained, flow diminishes to a balance point. If external forces cause secondary pressures to increase, the control element functions beyond the usual levels and diverts excess fluid to the reservoir to ensure a maximum pressure level.

The symbol for a *check valve* is shown in Fig. 2-13*c*. Flow in this valve is from left to right. Do not confuse this symbol with an arrowhead. The circle represents the moving element of the check valve, usually a *ball*, and

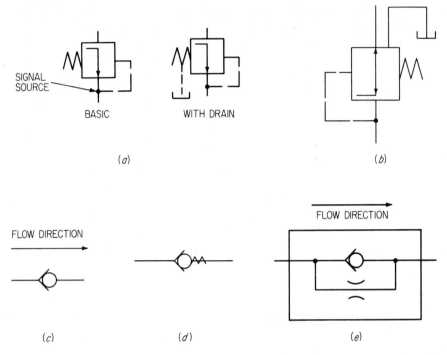

Fig. 2-13 (*a*) Nonrelieving reducing valve. (*b*) Relieving-type reducing valve. (*c*) Check valve—free. Opens if the inlet pressure is higher than the outlet pressure. (*d*) Spring loaded. Opens if the inlet pressure is greater than the outlet pressure plus the spring pressure. (*e*) Unit allowing free flow in one direction but restricted flow in the other.

the slant lines represent the *seat*, against which the ball is pressed when the valve shuts.

Figure 2-13*d* shows the check valve with a bias spring. The valve opens if the inlet pressure is greater than the outlet pressure plus the spring pressure.

In Fig. 2-13*e* the check valve is fitted with an orifice. The box symbol provides an indication of the composite nature of the device and shows the fluid-energy flow path. The fluid flows freely from left to right through the check valve. Should the check valve shut and the fluid flow reverse itself, the orifice would restrict the amount of fluid passing from right to left.

Two *pilot-operated check-valve* symbols are shown in Fig. 2-14*a* and *b*. This type of valve has an internal piston mechanism, which may either force the valve open or hold it closed, depending on the design. The dashed line in Fig. 2-14*a* represents the pilot line used to conduct fluid to the piston mechanism to hold the check valve open. A drain line may be shown leading from the lower part of the enclosure. A baffle device is designed for certain models. The cavity between the baffle and upper surface of the operating piston must be drained. Generally, this drain is terminated above the fluid level in the reservoir. In Fig. 2-14b the pilot line is arranged to hold the valve closed. When pilot pressure is *relaxed* (relieved) in this valve, fluid may pass in either direction through the valve in certain designs. In other designs the check will perform in a conventional manner until pilot pressure is applied. At that time the check is held closed so that flow is not possible in either direction. Releasing the pilot pressure then permits the usual function.

A manual shutoff valve of either the *gate* or *globe* type is represented by a symbol like that in Fig. 2-14c. This type of valve is *not* intended for control of rate of flow in an infinite adjustment pattern.

For several types of *variable-flow-rate-control valves,* a symbol similar to that for an orifice is used. A slash arrow is drawn across the symbol [Fig. 2-14d], indicating that flow can be varied by changing the orifice size. Pressure compensation is indicated by the symbols in Fig. 2-14e through h. Variations in inlet pressure do not affect the rate of flow.

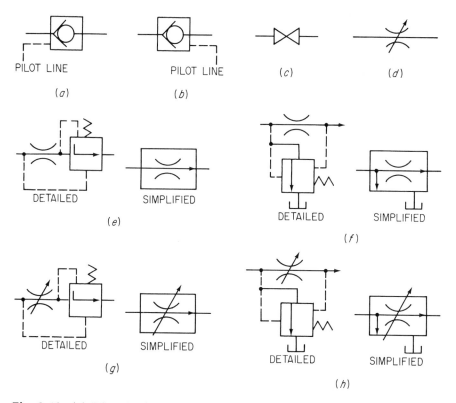

Fig. 2-14 (*a*) Pilot check valve—pressure applied to open valve. (*b*) Pressure relaxed to open. (*c*) Manual shutoff valve. (*d*) Variable-flow-control valve. (*e*) Pressure-compensated flow-control valve with fixed orifice. (*f*) With fixed output and relief port to reservoir. (*g*) With variable orifice. (*h*) With variable output and relief port to reservoir.

Hydraulic Symbols **47**

2-5 MULTIPLE-ENVELOPE VALVE

In this type of valve the basic symbol consists of (1) an envelope for each operating position of the valve, (2) internal flow paths for each valve position, (3) arrows indicating flow direction through the valve, and (4) external ports at the normal or neutral valve position.

Figure 2-15a shows a *two-position, three-connection valve* in its *neutral* or *rest* position. The external ports are always shown at the neutral or rest position in the valve symbol. Actuated, this valve is shown as in Fig. 2-15b. Note that the port is shown as connected to the flow path represented by the arrowhead in the right-hand box. Two-position, four-connection valves vary only in that the fourth connection is added, as shown in Fig. 2-15c and d.

Three-position, four-connection valves have three working positions. An additional two positions may be shown [Fig. 2-16d] with dashed dividers, indicating that the inner squares show a transitional condition. Thus, the symbol in Fig. 2-16d indicates that all valve ports are interconnected as the valve is shifted from neutral to an extreme position or from the completely shifted position back to neutral. Figure 2-16a shows this type of valve in the normal rest position where there is no need for showing the transitional conditions. In Fig. 2-16b the valve is in its actuated position with fluid flowing through the right-hand box. Flow is through the left-hand box in Fig. 2-16c. Common neutral conditions for a variety of three-position, four-connection valves are shown in Fig. 2-17. Identifying letters for the various ports may vary with different manufacturers. The existing standards do not cover the identification of the ports, other than the U-shaped symbol indicating return to the tank and the equilateral triangles indicating energy flow.

A complete symbol for a three-position, four-connection, spring-centered, solenoid controlled, four-way, directional-control valve with three ports interconnected and one port blocked in neutral is shown in Fig. 2-18.

Figure 2-19a shows a two-position, three-connection, spring-offset, solenoid-controlled valve. When the solenoid is deenergized, the spring urges the internal working mechanism to a position where there is a direct flow through the valve, as indicated by the box adjacent to the spring symbol. This is considered the normal position, and the external lines would be connected to the left-hand box. The two-position, four-connection, solenoid-controlled valve in Fig. 2-19b does not have a *rest position* (no-motion position) established by an energy source such as a spring or gravity. Because of this, the external connecting lines are usually shown connected to the box in what might be considered the rest position in the circuit in which the valve is used. The right-hand solenoid would be energized prior to the valve coming to a rest position; the external lines would then be connected to the right-hand box [Fig. 2-19b].

Pilot-operated, two-position, four-connection, spring-offset valves usually have some provision to drain the end cap containing the spring

48 *Industrial Hydraulics*

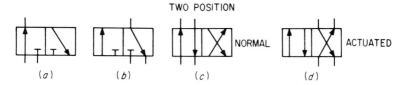

Fig. 2-15 (*a*) Normal position of a two-position, three-connection valve. (*b*) Actuated position of a two-position, three-connection valve. (*c*) Normal position of a two-position, four-connection valve. (*d*) Actuated position of a two-position, four-connection valve.

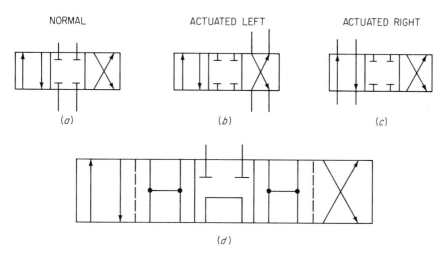

Fig. 2-16 (*a*) Normal position of a three-position, four-connection, four-way, directional-control valve. (*b*) Actuated position of valve in *a*; envelopes to left. (*c*) Actuated position of valve in *a*; envelopes to right. (*d*) Dash dividing line to indicate transitory conditions as valve is shifted from neutral to extreme positions or from extreme shifted positions back to neutral.

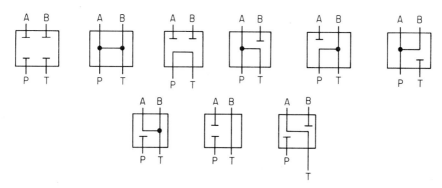

Fig. 2-17 Four-way, directional-control valve neutral configurations. Identifying letters vary with manufacturers.

Hydraulic Symbols **49**

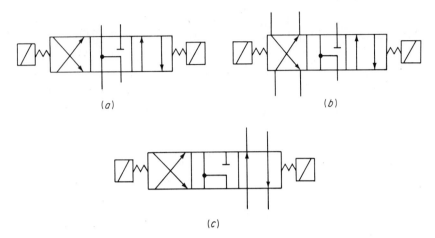

Fig. 2-18 Complete symbol for four-connection, three-position, spring-centered, solenoid-control, directional-control valve; three ports interconnected and one port-blocked in neutral. The three positions of the valve are: (a) Neutral, centered by springs when solenoids are deenergized. (b) Actuated, envelopes to right by energizing left solenoid (adjacent to block, showing energy flow). (c) Actuated, envelopes to left by energizing right solenoid.

mechanism. Figure 2-19c shows the drain, with the pilot connection on the opposite end. External lines are connected to the box adjacent to the spring.

Figure 2-20a shows a three-position, four-connection, spring-offset, mechanically controlled valve with one external port plugged. In this symbol the center or neutral position is illustrated, even though the unit does not stay in the neutral position for any fixed length of time. The symbol informs the viewer that all ports will be blocked on the neutral transitional crossover point until the reversal function is started. The dashed line between blocks indicates that the valve does not stay in neutral because the bias spring will urge the flow-directing elements to a predetermined position. The knob on the operator attached to the left end of the symbol in Fig. 2-20a indicates a manual lever. At each end of Fig. 2-20c the two parallel lines with the vertical bars at the ends indicate a

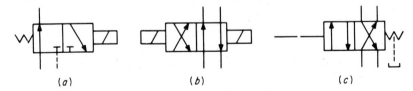

Fig. 2-19 Directional-control valves. (a) Two-position, three-connection, spring-offset, solenoid-control valve. (b) Two-position, four-connection, solenoid-control valve. (c) Two-position, four-connection, spring-offset, pilot-operated valve.

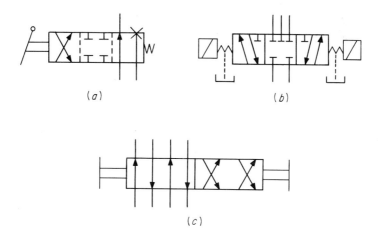

Fig. 2-20 Directional-control valves. (*a*) Three-position, four-connection, spring-offset, mechanical-control valve; one external port plugged. (*b*) Three-position, five-connection, spring-centered, solenoid-control valve. (*c*) Two-position, eight-connection, mechanical-control valve.

mechanical operator, which might be a stem, lever, cam, or knob. Five connections to a three-position valve can be made, as shown in Fig. 2-20*b*. The drain connections prevent full pressure being exerted on the shaft seals.

A multiplicity of connections can be incorporated within a two- or three-position valve. As an example, the two-position, eight-connection, mechanical-control valve in Fig. 2-20*c* can provide a predetermined flow pattern with a variety of external connections.

2-6 TYPES OF CONTROLS

When a spring is used in a control function, it is indicated by a wavy zigzag line [Fig. 2-21*a* and *b*]. It is placed so that one end touches the envelope. Note in Fig. 2-20*b* that the spring is in line with the rectangle with the diagonal line, which represents a solenoid. This signifies that *either* the spring *or* the solenoid can be the energy source to position the flow-directing element in the valve. Pilot pressure may emanate from an internal or external source.

Figures 2-21*c* shows the hyphenated line coming from an external source. A pilot function can result from either an application or a release of pilot pressure. Figure 2-21*d* shows the equilateral triangle pointing away from the major symbol, indicating that the operation is by release of pressure. Pilot actuation by differential pressure is shown in Fig. 2-21*e*.

The general symbol for manual operation is shown in Fig. 2-21*f*, with the symbol for a push button in Fig. 2-21*g*, lever in Fig. 2-21*h*, and pedal

Hydraulic Symbols **51**

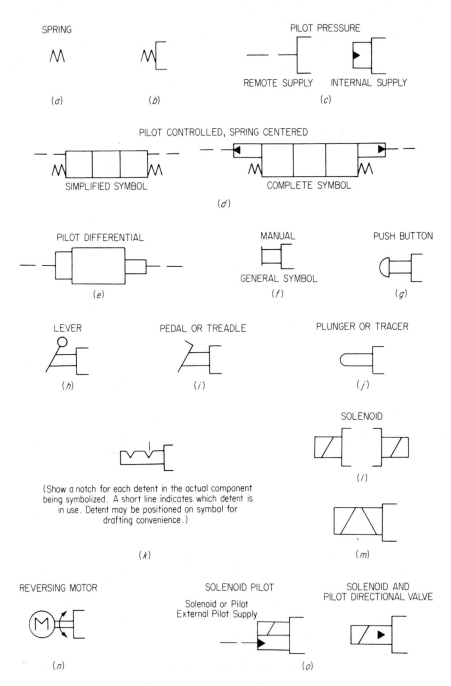

Fig. 2-21 (*a*) Spring actuator. (*b*) Attached to box. (*c*) Basic pilot symbol. (*d*) Spring-centered pilot assembly. (*e*) Pilot differential. In the symbol the larger rectangle represents the larger control area, i.e., the priority phase. (*f*) Manual operator used as general symbol without indication of specific type, i.e., foot, hand, leg, arm. (*g*) Push button. (*h*) Lever. (*i*) Pedal or treadle. (*j*) Plunger or tracer.

COMPOSITE ACTUATORS (AND, OR, AND/OR)

BASIC One signal only causes the device to operate

OR One signal OR the other signal causes the device to operate

AND One signal AND a second signal both cause the device to operate

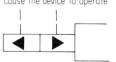

AND/OR The solenoid AND the pilot OR the manual override alone causes the device to operate

(p)

The solenoid AND the pilot OR the manual override AND the pilot

The solenoid AND the pilot OR a manual override AND the pilot OR a manual override alone

(q)

SINGLE-STAGE ELECTROHYDRAULIC SERVO VALVE

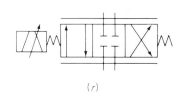

TWO-STAGE WITH MECHANICAL FEEDBACK

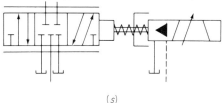

(r)

(s)

TWO-STAGE WITH HYDRAULIC FEEDBACK

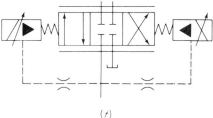

(t)

(k) Detent. (l) Solenoid. (m) Solenoid with two windings operating in opposite directions. (n) Reversing motor. (o) Solenoid OR pilot, solenoid AND pilot directional valve. (p) Composite actuators—AND, OR, AND/OR. (q) Pilot, manual, and solenoid combination actuators. (r) Electrohydraulic servo valve: a unit which accepts an analog electrical signal and provides a similar analog fluid-power output. Single stage. (s) Two-stage with mechanical feedback—with indirect pilot operation. (t) Two-stage with hydraulic feedback—with indirect pilot operation.

Hydraulic Symbols **53**

or treadle in Fig. 2-21*i*. Figure 2-21*j* indicates a plunger or tracer. A detent symbol [Fig. 2-21*k*] is most generally used with mechanically or pilot-operated valves. Some small-size pilot valves with dual-solenoid operation employ light detents to prevent spool drift because of machine vibration. The detent mechanically holds the flow-directing mechanism in the desired position until a specific force value is applied to cause movement from that position to another finite position. Individual solenoids [Fig. 2-21*l*] or a dual solenoid, with two windings operating in opposite directions [Fig. 2-21*m*] or reversing motor [Fig. 2-21*n*], provide electrical signal application. Combinations of actuators are common. Applications may dictate use of solenoid OR pilot or solenoid AND pilot actuation [Fig. 2-21*o*]. AND, OR, AND/OR symbols are shown in Fig. 2-21*p* and *q*. Electrohydraulic servo valves are shown in Fig. 2-21*r* through *t*.

2-7 RESERVOIRS FOR FLUIDS

There are two classes of hydraulic-system reservoirs. *Vented reservoirs* or tanks are indicated by a U-shaped symbol [Fig. 2-22*a*]. *Pressurized tanks* are indicated by Fig. 2-22*b*. A line from a manifold may be vented to permit free flow of fluid back to the reservoir, as in Fig. 2-22*c*.

2-8 MISCELLANEOUS UNITS

Figure 2-22*d* shows the symbol for a gauge for pressure or vacuum. Figure 2-22*e* shows the symbol for a temperature-indicating gauge. The sensing bulb may be shown at a remote point if it assists in clarifying the circuit details. Flow meters [Fig. 2-22*f*] are of two types. *Rate* is indicated by the left symbol. The symbol at the right indicates a *totalizing meter* such as those used to indicate water consumption in homes.

Rotating joints in fluid conducting lines are indicated by the symbol in Fig. 2-22*g* when only one line is within the assembly. Two or more lines are shown as in Fig. 2-22*h*, which also indicates a drain from the seal assembly.

In the pressure switch, fluid energy is transmitted as an electrical signal after the mechanism converts it from pressure as it enters the valve into motion which actuates an electrical switch. The symbol for the pressure switch shown in Fig. 2-22*i* shows the contacts involved.

The top accumulator symbol in Fig. 2-22*j* is used worldwide as a single symbol to designate any type of accumulator. American National Standards Institute graphical symbols also use the symbol of Fig. 2-22*j* as a general-purpose symbol. They, however, also recommend use of more explicit symbols such as those shown in Fig. 2-22*j* for spring-loaded, gas-charged, and weighted accumulators. The symbol for a spring-loaded accumulator encloses a zigzag and parting line to indicate use of a spring

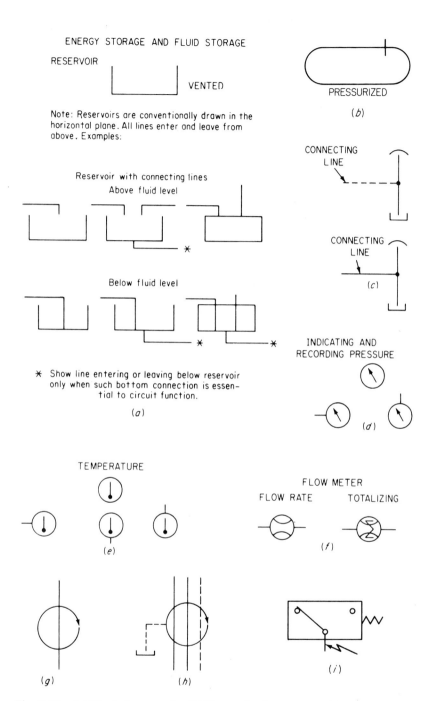

Fig. 2-22 (*a*) Vented reservoir. (*b*) Pressurized reservoir. (*c*) Vented manifold. (*d*) Pressure or vacuum gauge. (*e*) Temperature gauge. (*f*) Flow meters, flow rate, and totalizing. (*g*) Rotating connection with one flow path. (*h*) Rotating connection with two major working lines, pilot line and drain. (*i*) Pressure switch.

Hydraulic Symbols

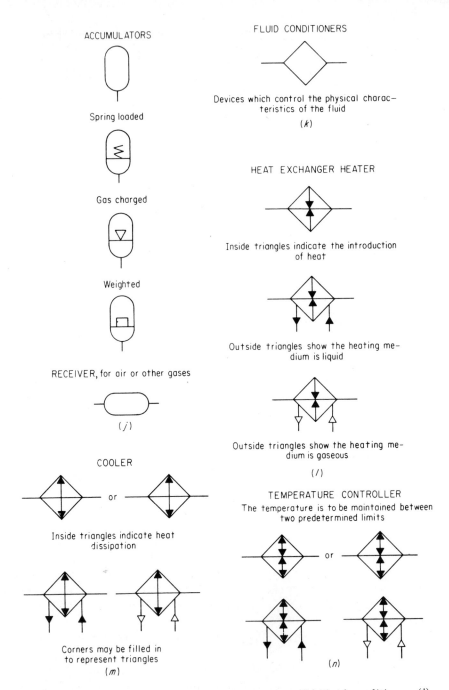

Fig. 2-22 *(Cont.)* (*j*) Accumulators and air receiver. (*k*) Fluid conditioner. (*l*) Heater. (*m*) Cooler. (*n*) Temperature controller.

56 *Industrial Hydraulics*

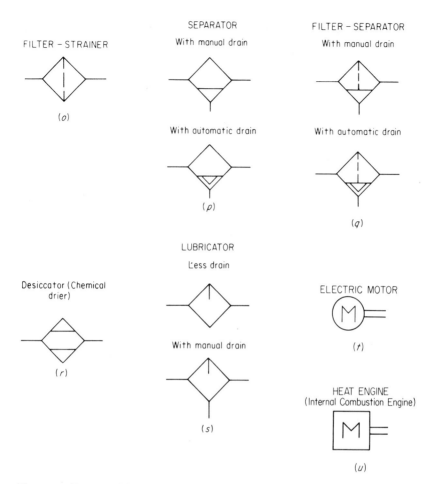

Fig. 2-22 (*Cont.*) (*o*) Filter strainer. (*p*) Separator. (*q*) Filter separator. (*r*) Desiccator. (*s*) Air-line lubricator. (*t*) Electric motor. (*u*) Heat engine.

as an energy-storage medium. The symbol for a pressurized-gas accumulator (hollow equilateral triangle plus dividing line) usually indicates an inert gas as the energy-storage medium. A small rectangle on the dividing line is used to indicate a weighted-type accumulator. The symbol for a receiver in Fig. 2-22*j* is usually associated with compressed air. It may also be used to show an auxiliary tank connected to a gas-charged accumulator to provide a larger quantity of gas under pressure to increase the energy-storage capacity.

Symbols for fluid conditioners employ a square resting on a corner with the associated major lines connected at the corners on a horizontal plane, as in Fig. 2-22*k*.

The symbol for a heater [Fig. 2-22*l*] shows the introduction of energy into the major line. The cooler symbol [Fig. 2-22*m*] indicates heat dissipation or energy flowing from the major line. Dual arrows [Fig. 2-22*n*] indicate that energy can flow in either direction. A heater, cooler, or temperature-controller symbol can be provided with external lines, indicating whether transfer is by means of gas or liquid. A solid equilateral triangle in a line, terminating at the midpoint on the diagonal of a square, indicates the medium is a liquid; a hollow equilateral triangle indicates a gas or compressed-air or a radiator-type heat exchanger, such as that used on conventional, internal-combustion, water-cooled engines.

The symbol for a filter or strainer employs a dashed line bisecting the square [Fig. 2-22o]. No attempt is made to differentiate between the various types of filters or strainers. But the position of the symbol in the circuit will usually provide a clue as to the type. Strainers are found most often on pump-suction lines, where introduction of major, large-size contaminants into the pump is prevented by the strainer. Filters may be found in any part of a circuit. However, machine tool circuits often employ high-pressure, in-line filters just ahead of critical components. Mobile equipment may use return-line filters most effectively because of the expected motion of the tank and resulting agitation of the fluid. There are no fixed rules as to strainer and filter usage, so the material list on a drawing may be needed to indicate the specific information.

Much pneumatic equipment is used in association with hydraulic systems. Because of this, it is usual to find pneumatic conditioning device symbols on hydraulic drawings. Pneumatic directional-control valves, used in circuitry, employ symbols similar to the equivalent hydraulic function. Hollow, equilateral, energy-flow triangles provide the clue as to the fluid medium. Figure 2-22p shows a separator used to remove moisture from pneumatic lines. The symbol of Fig. 2-22q illustrates a combination filter and separator for pneumatic lines. A desiccator, or chemical drier, is shown in Fig. 2-22r.

Pneumatic lines usually require the introduction of a lubricant into the line to minimize rust and corrosion. The oil may also serve to assist in lubricating operating devices in the circuit, such as air-piloted hydraulic valves. Figure 2-22s shows the units with and without a manual drain.

Figure 2-22t shows the symbol for an electric-drive motor. The symbol of Fig. 2-22u illustrates a heat engine, such as a steam, diesel, or gasoline engine.

2-9 COMPOSITE SYMBOLS

A *component enclosure* [Fig. 2-23a] may surround a composite symbol or a group of symbols to represent an *assembly*. It is used to convey more information about component connections and functions. The enclosure indicates the limits of the component or assembly. External ports are assumed to be on the enclosure line and indicate connections to the components. Flow lines cross the enclosure line without loops or dots.

In a typical component enclosure, one dash approximately 20 line-

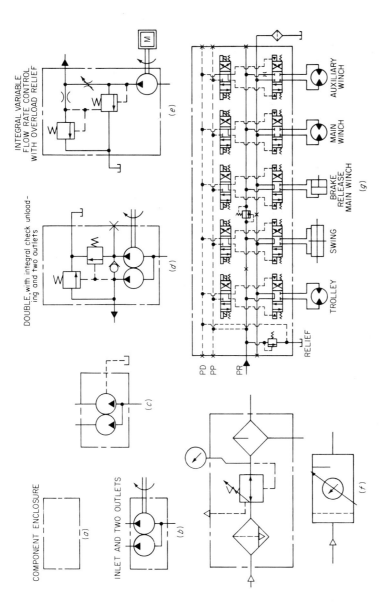

Fig. 2-23 (*a*) Component enclosure. (*b*) Double fixed-displacement pump with one inlet and separate outlets. (*c*) Double fixed-displacement motor with one inlet and separate outlets. (*d*) Double pump with integral check valve, unloading and relief valves. (*e*) Single pump with integral variable-flow-rate control and overload relief valve. (*f*) Control assembly for air supply to machine circuit. (Upper figure complete symbol, lower figure simplified symbol.) (*g*) Mobile stack valve assembly.

Hydraulic Symbols **59**

widths long is separated from the continuing line by a space of approximately five line-widths. The solid connecting line running lengthwise or around corners should be approximately 10 dash-lengths long. This can be varied but should include at least one dash to each side, top, and bottom. This enclosure is often referred to as a *centerline-type enclosure.*

An enclosure containing two pumps on a common shaft with a common suction line and independent discharge lines is shown in Fig. 2-23b. An enclosure containing two motors of equal size with a common shaft used to equalize or split the flow of fluid can be shown in a similar manner [Fig. 2-23c]. The drain may be included in some designs if the motor case is not designed for pressures equal to the working pressure.

Composite symbols, such as those in Fig. 2-23d, can be easily interpreted if a logical study pattern is followed. The best way to do this is to trace each line from its source to its endpoint. To start, follow the suction line from the tank to the pump. Starting from the pump opposite the rotating shaft (left pump), note that there is a working line passing through the enclosure to the circuit. Immediately beyond the pump discharge is a line connected to a check valve. The position of the check valve indicates that the flow from this pump can never pass through the check valve. Thus, this point can be considered a dead end as far as the delivery of the pump is concerned. Directly above the left pump is the symbol for a relief valve. The discharge of this relief valve is piped through the enclosure to the tank. Below the relief valve is a pilot line going to a valve in the right-hand pump circuit. This pilot line usually provides only a signal, and movement of fluid in it is relatively small. The pilot line will also be a dead end when the signal to this valve is completed. The left-hand pump, opposite the drive shaft, is provided with a circuit that can normally supply all machine pressure requirements, but it may not provide sufficient flow for the required speed. The extra pump is included in most circuits of this type to provide additional speed through added volume of fluid.

The discharge of the right-hand pump is connected to the input of a normally closed, two-way valve. Between the pump discharge and the inlet of the two-way valve a connection is provided to the inlet side of a check valve. This check valve permits discharge of the right-hand pump to connect with the delivery of the left-hand pump, combining the volume of both pumps for delivery to the circuit. When the pressure signal to the left-hand pump becomes high enough, it causes the pilot-operated two-way valve to open, permitting the right-hand pump discharge to pass freely through this valve back to the tank. This return flow combines with the discharge from the relief valve and passes through the enclosure to the tank.

The symbol in Fig. 2-23d does not necessarily indicate precise values of the fluid pressure in the various parts of the circuit, but certain relationships can be assumed. The highest fluid pressure will be developed by the left-hand pump. The piloted two-way valve must be adjusted to open at a pressure *less* than the relief-valve opening pressure. If this adjustment is incorrect and the pressure required to open the piloted two-way valve exceeds that of the relief valve, then the delivery of both pumps will combine and pass through the relief valve to the tank when the circuit is not accepting fluid under pressure. The testing station shown provides a place to install a gauge so that the point at which the right-hand pump starts to divert fluid to the reservoir can be checked and adjusted if necessary.

By tracing each line in this way and by knowing the basic symbols used, a complete understanding of all functional operations of a hydraulic circuit can be obtained. Figure 2-23*e* shows a flow-and-relief circuit enclosure. Trace its various parts in the same way as described above.

Figure 2-23*f* illustrates the composite and simplified symbols for filters, regulators, and lubricators in an assembly that serves to clean, lubricate, and regulate air diverted from a central system to be integrated into a fluid-*power* circuit.

Stackable valves, such as those used widely in mobile hydraulic circuits, may include many functional units in one manifolded assembly. Note the components included in the circuit of Fig. 2-23*g*. The first two directional-control valves employed for major flow directing are in series. The last three are in a parallel arrangement. All pilot valves are in a parallel supply and return to the tank circuit. Filtering is accomplished within the return to the tank line.

International standards provide best communications. However, five types of hydraulic diagrams have been used in commercial practice. These are (1) graphical diagrams, (2) circuit diagrams, (3) cutaway diagrams, (4) pictorial diagrams, and (5) combination diagrams.

Graphical diagrams [Fig. 2-24] show each piece of hydraulic apparatus, including all interconnecting pipes, by means of approved standard symbols. Diagrams of this type provide a simplified method of showing the function and control of each component.

Graphical diagrams are incorporated into specific *circuit diagrams.* Many large manufacturing groups have special forms designed to show special features that are useful to their designers and maintenance people. Several examples are included in the standards issued by ANSI.

Cutaway diagrams require significant time to produce. Generally, this type of diagram is used by manufacturers of fluid-power components to illustrate features within their product structure.

Pictorial diagrams are rarely used in modern fluid-power communications. Prior to the development of the ISO International Standards symbols and circuit diagrams, the pictorial diagram with much verbiage served as the communications medium.

Combination diagrams, like the pictorial diagram, have become obsolete and can be found most often in old machine tool instruction books.

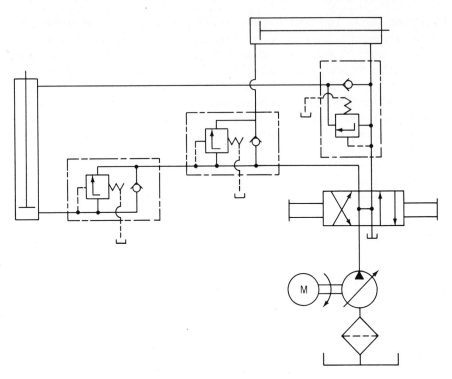

Fig. 2-24 Typical graphical diagram of hydraulic equipment.

SUMMARY

Graphical symbols for fluid-power circuits are international in nature. They provide clear-cut circuit information, regardless of language barriers. In a well-prepared circuit diagram, every part of a hydraulic circuit must be clearly shown, with a minimum need for auxiliary notes. Symbols are available for most commercial components. By following basic rules, it is possible to create easily understood symbols for special applications.

Hydraulic symbols show connections, flow paths, and the function of the components represented.

Symbols can indicate conditions occurring during transition from one flow-path arrangement to another. Symbols can also show if a device is capable of infinite positions or if it contains only finite working positions. A direct relationship is indicated between operator devices and the resulting flow paths. Symbols do not usually indicate construction or circuit values, such as pressure, flow rate, and temperature. Symbols do not indicate the location of ports, the direction of spool movement, or the position of control elements in actual components. (Modified symbols may show some of these features in special units.)

Hydraulic symbols may be rotated or reversed without altering their meaning except in the case of vented lines and lines to the reservoir, where the symbol must indicate whether the pipe terminates above or below the fluid level. Line-width does not alter the meaning of symbols. A symbol may be drawn any suitable size. The symbol size may be varied for emphasis or clarity.

Where flow lines cross, a loop is used, except within a symbol envelope. The loop may be used in this case if clarity is improved.

In multiple-envelope symbols, the flow condition, shown nearest a control symbol, takes place when that control is caused or permitted to actuate. Each symbol is drawn to show the normal or neutral condition of a component unless multiple circuit diagrams are furnished showing various phases of circuit operation. Arrows are used within symbol envelopes to show the direction of flow through the component in the application represented. Double-end arrows indicate reversing flow.

External ports are located where flow lines connect to the basic symbol, except where a component enclosure is used. External ports are at the intersections of flow lines and the component enclosure symbol, where the enclosure is used.

REVIEW QUESTIONS

2-1 What is meant by a normally open valve?

2-2 What symbol in a flow-control-valve symbol indicates that the valve can be adjusted?

2-3 On which side of the shaft is the indicating arrow for rotation of a pump or motor placed?

2-4 What does the diagonal arrow across an orifice or spring symbol signify?

2-5 What is a testing station? List two purposes for which a testing station is used.

2-6 Do hydraulic symbols indicate the location of ports in a component?

2-7 Must symbols be drawn to a fixed scale?

2-8 How is flow rate indicated on symbols?

2-9 Why are manifolds sometimes vented?

2-10 What is a component enclosure? Draw an enclosure having four components in it. Show the piping connections.

2-11 What is an envelope? For what types of symbols is an envelope used?

2-12 Sketch the following symbols: (a) working line, (b) lines crossing, (c) lines to reservoirs, (d) plugged connection in a four-way valve, (e) test station, (f) fixed restriction, (g) quick-disconnect with one check valve, (h) fixed-displacement single pump, (i) fixed-displacement motor with a drain, (j) double-end-rod cylinder, (k) remotely piloted unloading valve, (l) nonrelieving reducing valve, (m) pressure-compensated, flow-control valve, (n) pressurized reservoir, (o) pressure or vacuum gauge, (*p*) rotating connection with three flow paths and drain, (*q*) component enclosure.

2-13 An equilateral triangle indicates energy flow. What does a hollow triangle signify? A solid equilateral triangle?

2-14 What do the parallel lines adjacent to the three-position, directional-control valve signify? What is meant by finite positioning?

2-15 What is the major difference between a liquid-heater and a liquid-cooler symbol?

2-16 What is meant by solenoid AND pilot operation? By solenoid OR manual operation?

2-17 Why is there a double-headed arrow in the body of the symbol of Fig. 2-13*b*?

2-18 Which cylinder in Fig. 2-24 will move first? Why? Describe the action in each direction.

2-19 How is a temperature gauge indicated?

2-20 Can some of the physical characteristics of an accumulator be indicated in its symbol?

3
Fluid-Power Pumps and Motors

3-1 PURPOSE OF PUMPS

Every fluid-power system uses one or more *pumps* to pressurize the hydraulic fluid. The fluid under pressure, in turn, performs work in the output section of the fluid-power system. Thus, the pressurized fluid may be used to move a piston in a cylinder [Fig. 1-2] or to turn the shaft of a *hydraulic motor* [Fig. 3-16c].

The purpose of a pump in a fluid-power system is to pressurize the fluid so that work may be performed. Some fluid-power systems use low pressures—100 psi or less—to do work. Where a large work output is required, high pressures—10 000 psi or more—may be used. So we find that every modern fluid-power system uses at least one pump to pressurize the fluid.

3-2 TYPES OF PUMPS

Three types of pumps find use in fluid-power systems: (1) *rotary*, (2) *reciprocating*, and (3) *centrifugal* pumps.

Simple hydraulic systems may use but one type of pump. The trend is to use pumps with the most satisfactory characteristics for the specific tasks involved. In matching the characteristics of the pump to the requirements of the hydraulic system, it is not unusual to find two types of pumps in series. For example, a centrifugal pump may be used to supercharge a reciprocating pump, or a rotary pump may be used to supply pressurized oil for the controls associated with a reversing variable-displacement reciprocating pump.

3-3 ROTARY PUMPS

These are built in many different designs and are extremely popular in modern fluid-power systems. The most common rotary-pump designs used today are *spur-gear, internal-gear, generated-rotor, sliding-vane,* and *screw* pumps. Each type has advantages that make it most suitable for a given application.

SPUR-GEAR PUMPS

These pumps [Fig. 3-1] have two mating gears that are turned in a closely fitted casing. Rotation of one gear, the *driver,* causes the second, or *follower* gear, to turn. The driving shaft is usually connected to the upper gear of the pump.

When the pump is first started [Fig. 3-1a], rotation of the gears forces air out of the casing and into the discharge pipe. This removal of air from the pump casing produces a partial vacuum on the suction side of the pump. Fluid from an external reservoir is forced by atmospheric pressure into the pump inlet. Here the fluid is trapped between the teeth of the upper and lower gears and the pump casing [Fig. 3-1b]. Continued rotation of the gears forces the fluid out of the pump discharge [Fig. 3-1c].

Pressure rise in a spur-gear pump is produced by the squeezing action on the fluid as it is expelled from between the meshing gear teeth and the casing. A vacuum is formed in the cavity between the teeth as they unmesh, causing more fluid to be drawn into the pump. A spur-gear pump is a *constant*-displacement unit; its discharge is constant at a given shaft speed. The only way the quantity of fluid discharged by a spur-gear pump of the type in Fig. 3-1 can be regulated is by varying the shaft speed. Modern gear pumps used in fluid-power systems develop pressures up to about 3000 psi.

Figure 3-2 shows the typical *characteristic curves* of a spur-gear rotary pump. These curves show the capacity and power input for a spur-gear pump at various speeds. At any given speed the capacity characteristic is

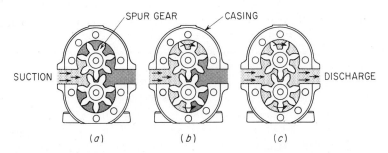

Fig. 3-1 Rotary spur-gear pump. (a) Vacuum draws fluid into pump. (b) Teeth carry fluid through pump. (c) Fluid is discharged under pressure.

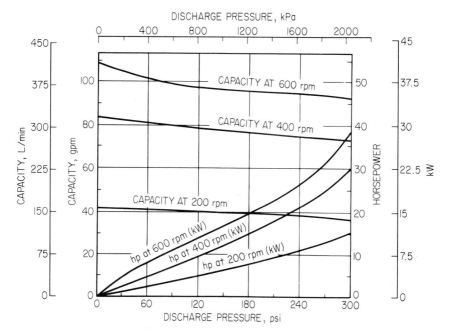

Fig. 3-2 Typical characteristic curves for a spur-gear rotary pump.

nearly a flat line. The slight decrease in capacity with rise in discharge pressure is caused by increased leakage across the gears, from the discharge to the suction side of the pump. Leakage in gear pumps is sometimes termed *slip*. Slip also increases with a rise in pump discharge pressure [Fig. 3-2]. The curve showing the relation between pump discharge pressure and pump capacity is often termed the *head-capacity* or *HQ* curve. The relation between power input and pump capacity is the *power-capacity* or *PQ* curve.

Power input to a spur-gear pump increases with both the operating speed and discharge pressure [Fig. 3-2]. As the speed of a gear pump is increased, its discharge rate in gallons per minute also rises. Thus, in Fig. 3-2, the horsepower input at a discharge pressure of 120 psi is 5 hp at 200 rpm and about 13 hp at 600 rpm. The corresponding capacities at these speeds and this pressure are 40 and 95 gpm, respectively, read on the 120-psi ordinate where it crosses the 200- and 600-rpm *HQ* curves.

Figure 3-2 is based on a spur-gear pump handling a fluid of constant viscosity. As the viscosity of the fluid handled increases (i.e., the fluid becomes thicker and has more resistance to flow), the capacity of a gear pump decreases. Thick, viscous fluids may limit pump capacity at higher speeds because the fluid cannot flow into the casing rapidly enough to fill it completely. Figure 3-3 shows the effect of increased fluid viscosity on the performance of a rotary pump in a fluid-power system. At 80-psi discharge pressure the pump has a capacity of 220 gpm when handling

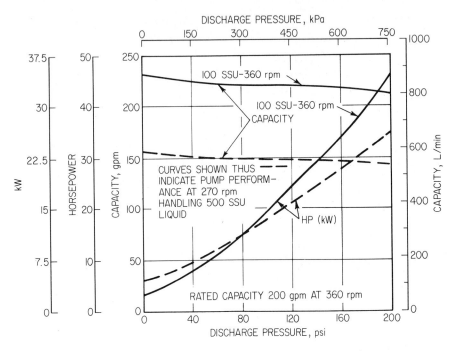

Fig. 3-3 Effect of increased fluid viscosity on the performance of a rotary pump.

fluid of 100 SSU viscosity. Capacity of this pump decreases to 150 gpm when handling fluid having a viscosity of 500 SSU. The power input to the pump also rises, as shown by the power characteristics in Fig. 3-3.

Capacity of a rotary pump is often expressed in gallons per revolution of the gear or other internal element. If the outlet of a positive-displacement rotary pump is completely closed, the discharge pressure will increase to the point where the pump driving motor stalls or some part of the pump casing or discharge pipe ruptures. Because this danger of rupture exists with every positive-displacement rotary pump, almost all fluid-power systems are fitted with a *pressure-relief valve*. This relief valve may be built as part of the pump or it may be mounted in the discharge piping.

The outlet of the relief valve for a fluid-power pump is always piped back to the reservoir in a location as far from the pump intake pipe as practical. Baffles shield the return pipe from the intake pipe; they are discussed in a later chapter. The flow-guiding baffle elements in a reservoir ensure maximum circulation of the fluid prior to its entering the pump intake. This circulation assists in removal of heat, entrained air, and foreign matter to permit maximum pumping efficiency. The outlet of the relief valve for a *fluid-transfer pump* [Fig. 3-4c] may be directed to the suction line of the pump. This may be essential when the transfer pump is not associated with a reservoir such as would be found in a fluid-power system. Relief valves are discussed in greater detail in Chap. 10.

68 Industrial Hydraulics

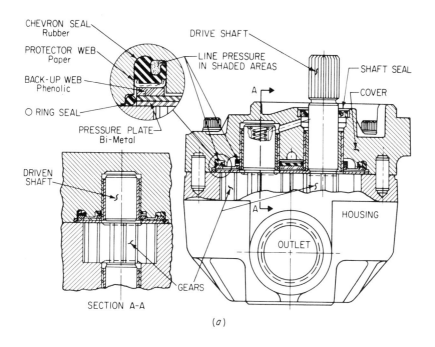

(a)

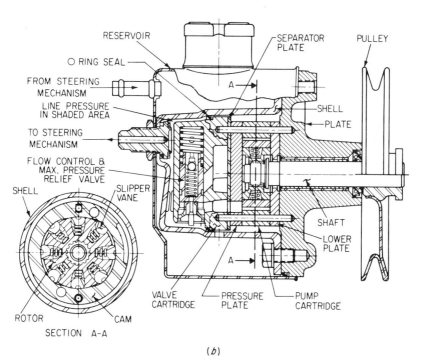

(b)

Fig. 3-4 (a) High-pressure gear pump with pressure loading. *(Michigan Division TRW, Inc.)* (b) High-pressure power-steering pump package. *(Michigan Division TRW, Inc.)*

Fluid-Power Pumps and Motors 69

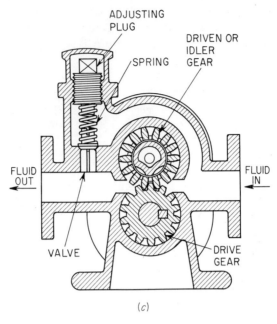

Fig. 3-4 (c) Fluid-transfer gear pump with integral relief valve directing fluid back to input port.

Care in adjusting relief valves is an important factor in the servicing of hydraulic-power-transmission systems. With standard rated loads, the relief valve can limit fluid pressure to a level that is compatible with the mechanical strength of the components in the power-transmitting system. Careless adjustments beyond known safe limits can result in rupture of the weakest link in the circuit [Fig. 3-4d].

High-pressure gear pumps, such as that in Fig. 3-4a, could overheat if the relief discharge were not permitted to circulate in a suitable manner. Internal leakage in this pump is reduced by the addition of pressurized plates at the side face of the gear. Many different devices are employed to direct pressurized fluid to these mechanisms for controlling clearances between the moving parts and the adjacent side walls. The unit illustrated employs a contoured seal which backs up the wear plate. Areas are carefully calculated to provide a minimum leak path as the pressure level is varied. By sensing the pump discharge pressure, the clamping action is automatically proportioned to the localized need.

Where the pump driver or driving element is controllable, a relief valve may not be needed. For example, some automobile fluid-power-steering systems have a pump driven by a lightweight V-belt. When the car steering wheel is turned to its full right or left travel, pressure in the pump will build to a level where the belt will slip, giving a characteristic squeal. This squeal warns the automobile driver that he has turned the steering wheel as far as it will go. Pump pressure builds up in the power-steering system at the maximum travel of the wheel because the piston in the actuating cylinder cannot move any further. The pressure at which the belt begins to slip is low enough to prevent damage to the fluid-power-steering-system components.

Higher road speeds, heavier loads, and more sensitive control dictate more efficient use of power-steering devices. Because of this, high-pressure units, such as Fig. 3-4b, were developed. A hydraulically balanced slipper-vane assembly provides pressures beyond 2000 psi. Because of the higher pressures, a flow-control and maximum-pressure relief valve is incorporated into the assembly. Flow is carefully calculated in assemblies such as this power-steering package, so that the maximum heat dissipation is provided, fluid is kept clean, and the noise level is maintained at an extremely low value. Belt slippage is not anticipated in the heavy-duty unit of Fig. 3-4b. Power-steering units are also incorporated into other special industrial drives where relief-valve protection is needed.

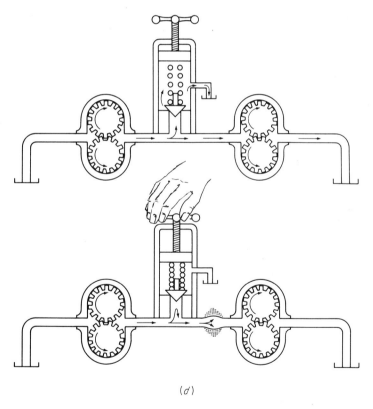

(d)

Fig. 3-4 (d) Turning relief valve adjustment without suitable reference to gauge or circuit information can cause rupture of weakest portion of power-transmission-circuit components.

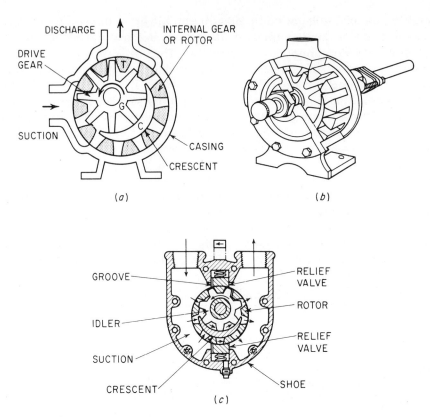

Fig. 3-5 (*a*) Internal-gear pump. (*b*) Cutaway view of an internal-gear pump. (*c*) Internal-gear pump fitted with a relief valve.

INTERNAL-GEAR PUMPS

These pumps [Fig. 3-5] have a *rotating member*, or *rotor*, on their periphery, which might be considered a gear. The teeth T of this gear are supported at one end of the rotor disk [Fig. 3-5a]. This rotor revolves concentrically in the casing. Supported off-center with the rotor is a free-running gear G [Fig. 3-5a] which meshes with and is driven by the rotor gear teeth. A crescent C supported from the *end cover* of the pump acts to ensure a seal between the suction and discharge.

With the rotor turning in a counterclockwise direction past the intake, the teeth of drive gear G are withdrawn from between the rotor teeth. This action produces a vacuum in the inlet, and fluid flows in to fill the spaces between the teeth. As the rotor continues to turn, fluid is trapped between the teeth, crescent, and casing. From here it is carried around and forced out the discharge by the internal-gear teeth meshing with those on the rotor. Figure 3-5b shows a cutaway view of a medium-pressure internal-gear pump.

Internal-gear pumps are also built with an internal relief valve [Fig. 3-

5c]. The suction is separated from the discharge by *hydraulically balanced shoes,* instead of the casing bore [Fig. 3-5a]. These shoes perform three functions: (1) They act as a relief valve by lifting off the rotor when the spring pressure is exceeded. (This is a low-pressure pump, and the spring loading is set to relieve the pressure at between 10 and 20 psi.) (2) The shoes permit chips and small particles of grit to pass through without damaging the housing or rotor. (3) They also compensate for wear, preventing leakage or slip from the discharge to the suction. Characteristics of the internal-gear pump resemble those of the spur-gear pump.

GENERATED-ROTOR PUMPS

These pumps [Fig. 3-6] have a pair of gear-shaped elements, one within the other, mounted in the pump casing. The inner rotor element is connected to the drive shaft and drives the outer generated rotor.

The generated-rotor device is often built into other machines in a cartridge form to provide lubrication, fluid transfer, or hydraulic power. The ring gear rather than the pinion can be driven when this facilitates the design of the machine structure.

Operation of the rotor elements is shown in Fig. 3-7. The inner element has one tooth fewer than the outer element. The tooth form of each element is related to that of the other in such a way that each tooth of the inner element is always in sliding contact with the surface of the outer element. Each meshing pair of teeth of the two elements engages at just one place in the pump. In Fig. 3-7 this is at X. Thus, tooth 1 will mesh with tooth 2, tooth 3 with tooth 4, and so on, when each pair reaches position X.

At the right-hand side of the point of mesh, *pockets of increasing size* [Fig. 3-7] are formed as the gears rotate, while on the other side the pockets decrease in size. The pockets of increasing size are *suction pockets,* and those of decreasing size are *discharge pockets.* In Fig. 3-7 the pockets

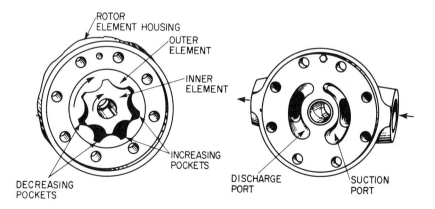

Fig. 3-6 Generated-rotor pump.

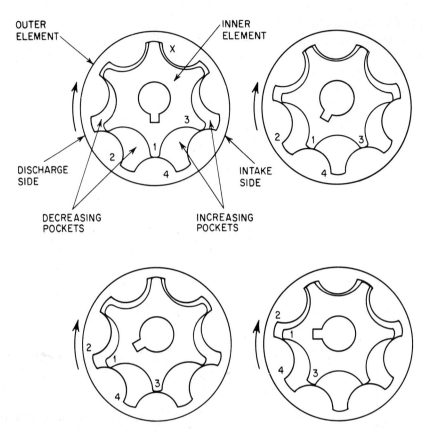

Fig. 3-7 Operation of the rotor elements in a generated-rotor pump.

on the right-hand side of the drawings are increasing in size as one moves down the page, while those on the left-hand side are decreasing in size. The intake side of the pump is therefore to the right, and the discharge is to the left. Fluid is forced out the discharge when the gears mesh at X [Fig. 3-7].

The inner and outer elements travel slowly with respect to each other, even when the drive shaft is rotating rapidly. Thus in one pump the generated rotor elements revolve at 200 rpm with respect to each other when the pump shaft is rotating at 1800 rpm. These pumps can deliver from fractions of a gallon per minute to more than 100 gpm. Pressures per set of rotors (called a *stage*) are usually limited to 2000 psi for continuous duty. This type of pump may be connected in a series of stages for higher pressures.

The generated-rotor element is also used in a modified form as a hydraulic motor displacement element, which is discussed later in this chapter.

SLIDING-VANE PUMPS

These pumps [Fig. 3-8] have a number of vanes which are free to slide into or out of slots in the pump rotor. When the rotor is turned by the pump driver, centrifugal force, springs, or pressurized fluid causes the vanes to move outward in their slots and bear against the inner bore of the pump casing or against a cam ring. As the rotor revolves, fluid flows in between the vanes when they pass the suction port. This fluid is carried around the

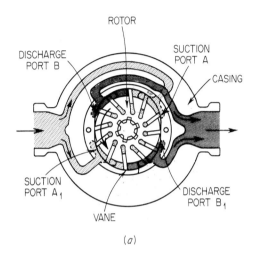

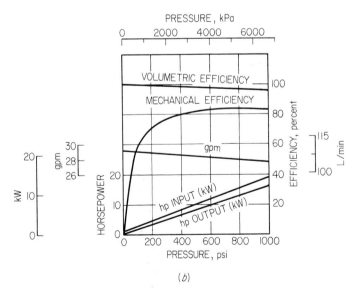

Fig. 3-8 (*a*) Sliding-vane rotary pump. (*b*) Characteristic curves for a sliding-vane pump.

pump casing until the discharge port is reached. Here the fluid is forced out of the casing and into the discharge pipe.

In the sliding-vane pump in Fig. 3-8 the vanes run in an oval-shaped bore. Centrifugal force starts the vanes out of their slots when the rotor begins turning. The vanes are held out by pressure which is bled into the cavities behind the vanes from a *distributing ring* at the end of the vane slots. Suction is through two ports A and A_1, placed diametrically opposite each other. Two discharge ports are similarly placed. This arrangement of ports keeps the rotor in hydraulic balance, relieving the bearings of heavy loads. When the rotor turns counterclockwise, fluid from the suction pipe comes into ports A and A_1, is trapped between the vanes, and is carried around and discharged through ports B and B_1. Pumps of this design are built for pressures up to 2500 psi. Earlier models required staging to attain pressures approximating those currently available in one stage. Valving, used to equalize flow and pressure loads as rotor sets are operated in series to attain high pressures, is shown in Chap. 10. Speed of rotation is usually limited to less than 2500 rpm because of centrifugal forces and subsequent wear at the contact point of the vanes against the cam-ring surface. Figure 3-8b shows the characteristic curves of the pump in Fig. 3-8a when operating at 1200 rpm and handling oil having a viscosity of 150 SSU at 100°F.

Two vanes may be used in each slot [Fig. 3-9] to control the force against the interior of the casing, or the *cam ring*. Dual vanes also provide a tighter seal, reducing the leakage from the discharge side to the suction side of the pump. The opposed inlet and discharge ports in this design provide hydraulic balance in the same way as the pump in Fig. 3-8. Both these pumps are constant-displacement units.

The *delivery* or capacity of a vane-type pump in gallons per minute

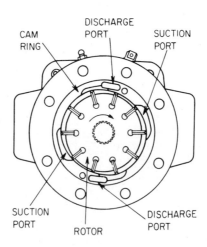

Fig. 3-9 Sliding-vane pump fitted with dual vanes.

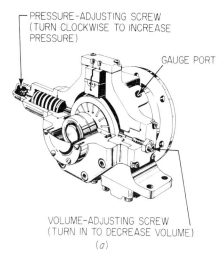
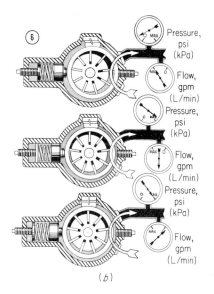

Fig. 3-10 (a) Variable-delivery sliding-vane pump. (*Racine Hydraulics, Inc.*)

Fig. 3-10 (b) Flow characteristics of variable-delivery vane pump.

cannot be changed without changing the speed of rotation unless a special design is used. Figure 3-10a shows a *variable-capacity* sliding-vane pump. It does not use dual suction and discharge ports. The rotor runs in the *pressure-chamber ring*, which can be adjusted so that it is off-center to the rotor. As the degree of off-center or *eccentricity* is changed, a variable volume of fluid is discharged. Figure 3-10a shows that the vanes create a vacuum so that oil enters through 180° of shaft rotation. Discharge also takes place through 180° of shaft rotation. There is a slight overlapping of the beginning of the fluid intake function and the beginning of the fluid discharge.

Figure 3-10b shows how maximum flow is available at minimum working pressure. As the pressure rises, flow diminishes in a predetermined pattern. As the flow decreases to a minimum value, the pressure increases to the maximum. The pump delivers only that fluid needed to replace clearance flows resulting from the usual slide fit in circuit components.

A relief valve is not essential with a variable-displacement-type pump of this design to protect the pumping mechanism. Other conditions within the circuit may dictate the use of a safety or relief valve to prevent localized pressure buildup beyond the usual working levels.

For automatic control of the discharge, an adjustable spring-loaded *governor* is used. This governor is arranged so that the pump discharge acts on a piston or the inner surface of the ring whose movement is opposed by the spring. If the pump discharge pressure rises above that for which the governor spring is set, the spring is compressed. This allows the pressure-chamber ring to move and take a position that is less off-

TEMPERATURE--Data plotted is at oil temperature of 120° F., using oil with 200 SSU viscosity at 100° F.

FLOW RATE--Maximum pump delivery is plotted. For volume and pressure settings other than shown, refer to "Interpolation Data."

DRIVE SPEED--1760 rpm. Performance for other drive speeds will be proportionate. Refer to "Interpolation Data."

HORSEPOWER--Horsepower data is power required at the pump input shaft.

INTERPOLATION DATA--To obtain specific input horsepower for volume and pressure settings other than shown, the following formula should be used: Proper values of E_f must be inserted for accurate results. Interpolate E_f values to obtain those not listed.

$$HP = \left(\frac{GPM \times PSI}{1714}\right) + E_f$$

To obtain corrected horsepower for drive speeds other than 1760 rpm use the following formula:

$$HP = \left(\frac{.00386 \times RPM \times PSI}{1714}\right) + E_f$$

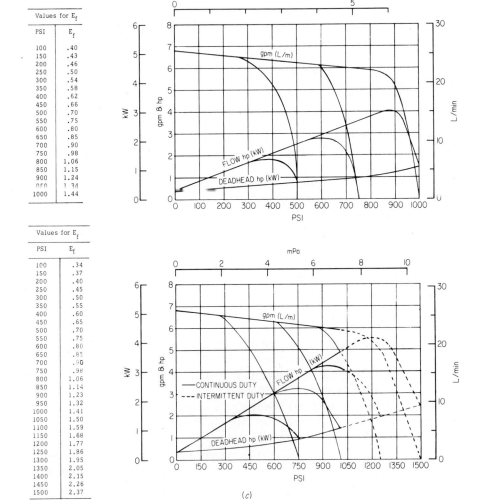

Fig. 3-10 (c) Variable-displacement-pump compensator characteristics.

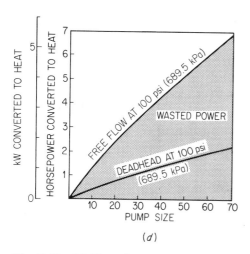

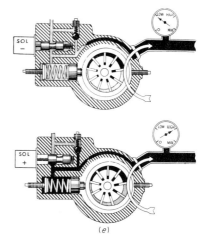

Fig. 3-10 (d) Use of two-pressure pump to save horsepower.

Fig. 3-10 (e) Solenoid-controlled, two-pressure governor.

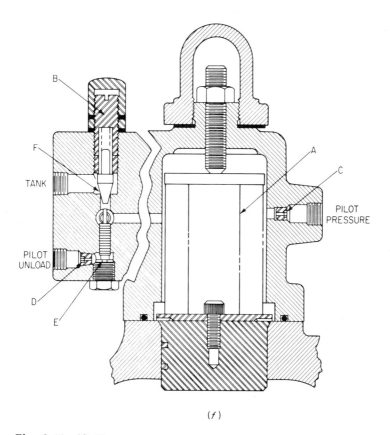

Fig. 3-10 (f) Two-pressure governor employing remote-control signals. *(Racine Hydraulics, Inc.)*

Fluid-Power Pumps and Motors 79

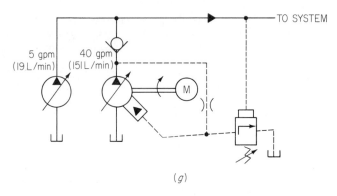

Fig. 3-10 (g) Circuit for two-pressure governor.

center with respect to the rotor. The pump then delivers less fluid, and the pressure is established at the desired level. The discharge pressure for units of this design varies between 100 and 2500 psi.

The characteristics of a variable-displacement-pump compensator are shown in Fig. 3-10c. Horsepower input values are also shown so that the power input requirements can be accurately computed.

Variable-volume vane pumps are capable of multiple-pressure levels in a predetermined pattern.

Two-pressure pump controls can provide an efficient method of unloading a circuit and still hold sufficient pressure available for pilot circuits.

The black area of the graph of Fig. 3-10d shows a variable-volume pump maintaining a pressure of 100 psi against a closed circuit. Wasted power is the result of pumping oil at 100 psi through an unloading or relief valve to maintain a source of positive pilot pressures. Two-pressure-type controls include hydraulic, pilot-operated types and solenoid-controlled, pilot-operated types. The upper view of Fig. 3-10e shows the solenoid deenergized so that the pilot oil is diverted to the tank. Thus, the pilot oil obtained from the pump discharge cannot assist the governor spring. Minimum pressure will result. The lower part of Fig. 3-10e shows the solenoid energized so that pilot oil assists the compensator spring. The amount of assistance is determined by the small ball and spring, acting as a simple relief valve. This provides the predetermined maximum operating pressure.

Another type of two-pressure system employs what is termed a *differential unloading governor*. It is applied in a high-low or *two-pump* circuit. The governor automatically, through pressure sensing, unloads the large-volume pump to a minimum *deadhead pressure* setting. *Deadhead pressure* refers to a specific pressure *level* established as a resulting action of the variable-displacement-pump control mechanism. The pumping action and the resulting fluid flow at deadhead condition are equal to the leakage in the system and pilot-control flow requirements. No major power movement occurs at this time, even though the hydraulic system may be providing a clamping or holding action while the pump is in deadhead position.

The governor is basically a hydraulically operated, two-pressure control with a differential piston that allows complete unloading when sufficient external pilot pressure is applied to the pilot unload port.

The minimum deadhead pressure setting is controlled by the main governor spring A [Fig. 3-10f]. The maximum pressure is controlled by the relief-valve adjustment B. The operating pressure for the governor is generated by the large-volume pump and enters through orifice C.

To use this device let us assume that the circuit requires a maximum pressure of 1000 psi, which will be supplied by a 5-gpm pump. It also needs a large flow (40 gpm) at pressures up to 500 psi; it continues to 1000 psi at the reduced flow rate. A two-pump system with an unloading governor on the 40-gpm pump will provide the needs.

We can unload the 40-gpm pump at 500 psi to a minimum pressure setting of 200 psi (or another desired value), while the 5-gpm pump takes the circuit up to 1000 psi or more.

Note in Fig. 3-10g that two sources of pilot pressure are required. One, the 40-gpm pump, provides pressure within the housing so that a maximum pressure setting can be obtained. The setting of the spring, plus the pressure within the governor housing, determines the maximum pressure capability of the 40-gpm pump. The second pilot source is the circuit proper, which will go to 1000 psi. This pilot line enters the governor through orifice D and acts on the unloading piston E. The area of piston E is 15 percent greater than the effective area of the relief poppet F. The unloading differential built into this governor control is, therefore, approximately 15 percent. The governor will unload at 500 psi and be activated at 15 percent below 500 psi, or 425 psi. By unloading, we mean zero flow output of the 40-gpm pump.

As pressure in the circuit increases from zero to 500 psi, the pressure within the governor housing also increases until the relief-valve setting is reached, at which time the relief valve cracks open, allowing flow to the tank.

The pressure drop in the housing is a maximum additive value, allowing the pump to deadhead. Meanwhile, the system pressure continues to rise above 700 psi, resulting in a greater force on the bottom of piston E than on the top. The piston then completely unseats poppet F, which results in a further pressure drop within the governor housing to zero pressure because of the full-open position of the relief poppet F. Flow entering the housing through orifice C is directed to the tank past the relief poppet without increasing the pressure in the housing. The deadhead pressure of the 40-gpm pump then decreases to the lower set value. Thus, at the setting of the unloading governor, the 40-gpm pump goes to deadhead. The flow rate to the circuit decreases to 5 gpm as the pressure increases to 1000 psi. At 1000 psi, the 5-gpm pump is also at its deadhead setting, thus only holding system pressure.

The 40-gpm pump unloads its volume at 500 psi. It requires a system pressure of 600 psi to unload the 40-gpm pump to its minimum pressure of 200 psi. The 600-psi pilot supply enters through orifice D and acts on the differential piston E. The pump's volume is reduced to zero circuit-flow output at 500 psi. The additional 100-psi pilot pressure is required to open poppet F completely and allow the pressure within the housing to decrease to zero.

As circuit pressure decreases, both pumps come back into service in a similar pattern.

SCREW PUMPS

These pumps [Fig. 3-11a] have two or more meshing screws to develop the desired pressure. These screws mesh to form a fluid-tight seal between

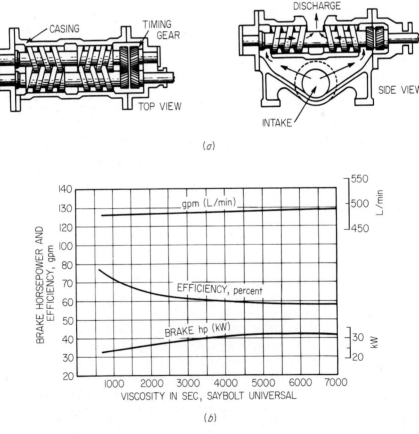

Fig. 3-11 (a) Screw-type rotary pump. (b) Characteristic curves for a screw-type pump.

the screws and between the screws and casing. The screws are arranged so that the threads of each run close to the bottom of the space between the threads of the opposite screw. Sides of the threads run with a small clearance where they intermesh. Timing gears on the screw shafts provide the turning force between the driven and driving screws. When the screws are turning, the space between the threads is divided into compartments which are separated from each other during complete convolution of the threads. When the screws turn in normal rotation, the fluid contained in these separate compartments advances toward the center of the pump, where the compartments discharge their contents.

The fluid is carried forward to the discharge by the screws very much as a nut on threads would work. Timing gears are usually of the herringbone type. The usual operating speed of screw pumps of this type is 1750 rpm. Figure 3-11*b* shows the characteristic curves of a 3.5-in-diameter screw pump handling oil at 325-psi discharge pressure while turning at 1150 rpm.

3-4 RECIPROCATING PUMPS

Popular types of reciprocating pumps for fluid-power systems include the *radial-piston, axial-piston, duplex,* and *triplex* pumps. Probably the most familiar reciprocating pump is the well-known farm lift pump [Fig. 3-12*a*]. In this pump the operator moves the handle up and down to produce a similar *reciprocating motion* of the *piston* in the pump *cylinder*. The piston slides the close-fitting cylinder or *barrel*. During the *suction stroke* of the piston, fluid is sucked from the reservoir into the cylinder of the pump [Fig. 3-12*b*]. The sucking action is produced by the closely fitted piston. During the piston upstroke, some air is removed from the pump cylinder. This reduces the pressure in the cylinder to less than the pressure of the surrounding atmosphere. Atmospheric pressure, acting on the surface of the fluid in the reservoir, forces fluid up the suction pipe and into the pump cylinder. The pump is thus said to produce a suction on the fluid. Note, however, that it is the difference in pressure between the atmosphere and the interior of the pump cylinder that produces the upward flow of the fluid.

At the inlet to the pump *suction pipe,* a *ball-type foot valve* is fitted. This valve is a solid rubber or metal ball that allows fluid to enter during the suction stroke but prevents fluid from leaving the pipe when the pump is idle. A similar valve is fitted at the inlet to the pump cylinder [Fig. 3-12*a*]. The valve in the cylinder is termed a *check valve* because it prevents fluid from draining out of the cylinder. A similar valve is mounted in the center of the hollow pump piston.

During the upstroke of the piston, the lower check valve is lifted off its seat and the fluid above the piston seats the valve in the piston [Fig. 3-

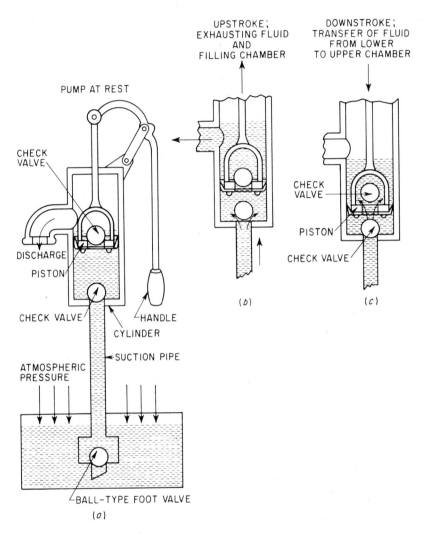

Fig. 3-12 Typical farm-type lift pump.

12b]. Fluid above the piston is discharged out the pump spout during this stroke.

When the piston reaches the top of its stroke, the movement stops for a moment. Then the piston begins its downstroke. During the downward travel, the piston check valve is lifted off its seat by the fluid trapped in the cylinder [Fig. 3-12c]. Pressure on the fluid trapped in the cylinder causes the cylinder check valve to close. This prevents fluid from leaving the cylinder. While the piston moves downward, the fluid moves from under the piston to above the piston. Thus, the pump is prepared for its next discharge stroke. Note that fluid is discharged by this pump *only* during the upstroke. Since the pump discharges during only one of its two strokes, the pump is termed *single-acting*. A reciprocating pump that discharges during both its strokes is termed *double-acting*. Since a double-acting pump discharges during both strokes, it has about twice the capacity of a single-acting pump.

Figure 3-13 shows a typical hand-operated reciprocating pump used as an auxiliary power source in fluid-power systems. Movement of the handle from 'eft to right produces a corresponding motion of the piston. During movement of the piston to the right, pressure exerted on the fluid by the piston forces discharge valve D off its seat. Suction valve S is forced onto its seat, preventing discharge of fluid to the reservoir. While valve D is off its seat, fluid is discharged out the right-hand side of the pump. During this same stroke, suction valve S_1 is lifted off its seat and fluid is sucked into the left-hand chamber of the pump. Valve D_1 is forced to its seat by the spring shown in Fig. 3-13, preventing fluid from the left-hand discharge line from entering the pump.

When the direction of the piston travel is reversed so that the piston travels from right to left, valve D is closed by its spring. Valve D_1 is opened by the increased pressure, while valve S_1 is closed by the pressure increase. Suction valve S is opened, and the piston draws fluid into the right-hand chamber while discharging from the left-hand one. The pump in Fig. 3-13 is double-acting because it discharges fluid on each stroke.

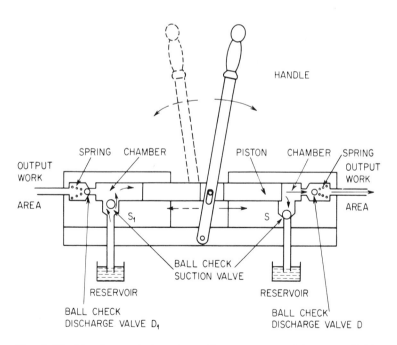

Fig. 3-13 Hand-operated reciprocating pump used in some fluid-power systems.

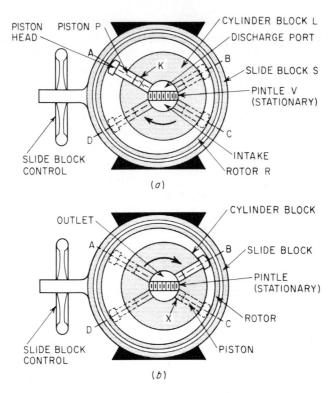

Fig. 3-14 Radial-piston reciprocating pump. (*a*) Fluid suction. (*b*) Fluid discharge.

RADIAL-PISTON RECIPROCATING PUMPS

These pumps [Fig. 3-14] operate on a principle similar to the simple lift pump [Fig. 3-12]. The radial-piston pump has a number of radial pistons P [Fig. 3-14a] in a cylinder block L which revolves around the stationary *pintle* or valve V. Rotor R houses the hardened-steel *reaction rings* against which the piston heads press. Slide block S is stationary; it houses and supports the rotor R, which revolves, being carried around by the friction set up by the sliding action between the piston heads and the reaction rings. A drive shaft is attached to the end of cylinder block L and provides the power needed to pump fluid.

In Fig. 3-14a, assume that there is fluid in space X of cylinder K when the piston is in position A. If the cylinder block L is rotated in a clockwise direction, piston P will be forced into its cylinder as the piston approaches position B. While moving into its cylinder, piston P will force fluid out of its cylinder and into the discharge port above the pintle. This pumping action is produced by the off-center position of the rotor in the slide block.

Figure 3-14b shows what happens after piston P reaches position B and discharges the fluid in its cylinders. The cylinder outlet is blocked as the piston moves from position B to C. Neither pumping nor suction will

86 *Industrial Hydraulics*

occur while the piston and cylinder are passing over the solid part or *land* of the pintle. After the piston moves to position C, however, it can begin to suck fluid into cylinder X. Suction continues as rotation of cylinder block L moves the piston to D, at which point the cylinder has taken in a full charge of fluid. This sucking action is produced by the tight-fitting piston being driven out of the cylinder by centrifugal force as the cylinder block revolves. Note that suction occurs only through the lower half of the pintle; discharge occurs only through the upper half of the pintle.

The piston once again passes the solid portion of the pintle—this time on the left-hand side of the pump. Once the piston passes this portion of the pintle and reaches position A, discharge again begins. Discharge continues until the piston reaches position B, where the cycle is repeated. Alternate suction and discharge continue as the rotor is revolved about its axis—suction on the underside of the pintle, discharge on the upper side. During this cycle the piston moves in and out as described above. The head of the piston is always held against the reaction ring of the rotor by centrifugal force. Figure 3-15 shows a cross section of a typical radial-piston pump used in fluid-power systems.

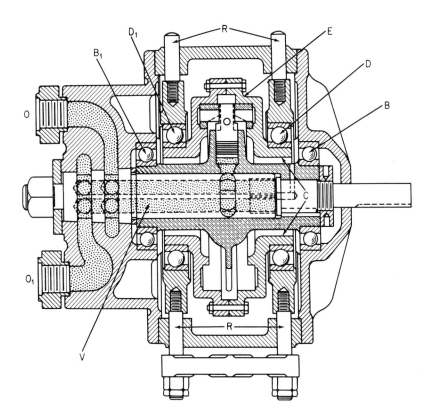

Fig. 3-15 Horizontal cross section through a radial-piston pump.

Note that in the discussion of the pumping action in the radial-piston pump, the center of rotor R was assumed to be different from the center of the cylinder block. This difference in center points produces the pumping action as the rotor is turned. If the rotor is moved so that its center coincides with that of the cylinder block, there will not be any pumping action. This is because the pistons will not reciprocate (move in and out) in their cylinders. With no pumping motion of the pistons, the pump does not discharge. The rotor, however, revolves in the same manner as described above.

Thus far, in the illustrations and discussion, it was assumed that the cylinder block was in an extreme offset position relative to the pistons. In this position the distance between the center of the rotor and the center of the cylinder block is a maximum. The pump discharges its maximum quantity when in this position because the piston stroke is longest. The quantity of fluid discharged by the pump can be varied by adjusting the distance between the two centers. Varying the center distance changes the length of the piston stroke and the quantity of fluid discharged.

The pump in Fig. 3-14a and b has only one piston, shown in different positions during its rotation. Single-piston pumps of this design are not practical. Usually, radial-piston pumps are fitted with an odd number of pistons—7, 9, 11, etc. With an odd number of pistons, no more than one piston will be blocked by the pintle at any given instant. Were an even number of pistons used, there would be times when two pistons would be blocked by the pintle. At other times, none would be blocked. This would mean that unequal numbers of pistons would discharge at various times, causing undesirable pulsation of the discharge flow. With an odd number of pistons, pulsation is reduced.

Figure 3-15 shows how the valve of pintle V works. This cylindrical valve has four holes parallel to its axis. Two of these holes are shown in Fig. 3-15. In the radial-piston pump, two holes in valve V connect with the intake ports, and two with the discharge ports at the left end of the valve. The left-hand cover of the pump rigidly supports valve V, and the ports connect to suction and discharge O and O_1. Either of these openings may be the discharge or suction, depending on the relative position of the center of the cylinder block and the slide block.

Ball bearings B and B_1 in the end covers of the pump support the cylinder block and hold it in a fixed position relative to the central cylindrical pintle or valve V. Bearings D and D_1 support the pistons and their rotor, which can be moved sidewise on guide rods R. Clearance C around the cylinder support permits displacing the center of rotation of the pistons from that of the cylinder assembly, which is fixed by the drive shaft. When the driving motor turns this shaft, the cylinder block, pistons, and enclosing case E turn with it to cause the pump to function as explained in Fig. 3-14a and b.

There are several designs of radial-piston pumps, but they all operate on the principle described in Fig. 3-14. They differ, however, in mechanical details, such as type of bearings, valve arrangement, and surface (curved shoes or slippers) on which piston thrust is applied. Piston rings are not normally used. Instead, the piston is accurately fitted into the cylinder, keeping leakage to a minimum.

Radial-piston motors [Fig. 3-16] are built in much the same way as radial-piston pumps. Flow through the radial-piston motor, however, is the reverse of that through the pump. A *hydraulic motor* produces power output to turn a shaft that performs an assigned task in a fluid-power system. This reversal of flow could be obtained in the pump [Fig. 3-14] by moving the slide block and rotor to the right. This would reverse the relation of the rotor and cylinder centers shown in Fig. 3-14. In this position the piston will draw in fluid as it travels from A to B and discharge fluid as it travels from C to D.

It is known from previous experience that if fluid is introduced into a cylinder under pressure, the piston will be pushed outward [Fig. 3-16a]. This principle is used in the radial-piston hydraulic motor in Fig. 3-16b. Three pistons are fitted in a cylinder block similar to that used in the radial-piston pump. The curved ends of the pistons bear on the rotor.

Assume that fluid under pressure is admitted to the cylinder containing piston 1. The piston must move outward in its cylinder, since the fluid is incompressible and two bodies (fluid and piston) cannot occupy the same space in the cylinder. For the piston to move out of its cylinder, the cylinder block must rotate in a clockwise direction. The piston moving out of the cylinder seeks the point of greatest distance between the cylinder block and rotor. Hydraulic fluid continues to enter cylinder 1 as the piston moves outward. As the cylinder block turns, piston 2 moves to the right, approaching the position formerly occupied by piston 3. Note that the distance between the cylinder block and rotor becomes progressively less as the piston moves from position 2 to position 1.

As piston 2 moves in a clockwise direction, it is pushed inward and it forces fluid out of the cylinder and through the pintle valve. Since there is little resistance to flow of fluid out of the cylinder, piston 2 is easily moved

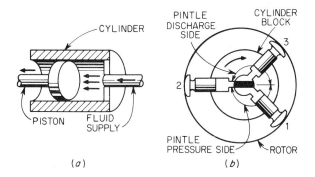

Fig. 3-16 (*a*) Piston and cylinder used in a hydraulic motor. (*b*) Simplified hydraulic motor.

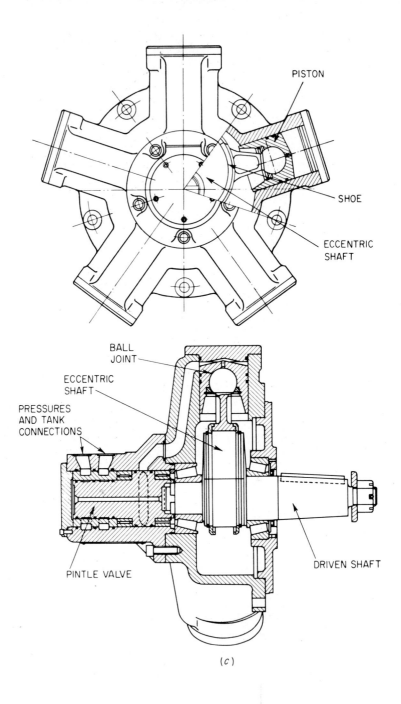

Fig. 3-16 (c) Radial-piston motor. (*Double A Products Co.*)

into its cylinder by the force exerted by the rotor. Fluid leaving the cylinder enters a reservoir or returns to the suction of the pump supplying the pressure. When piston 3 moves past the midpoint of the rotor, it reaches the pressure side of the pintle. Fluid is forced into the cylinder as it was in cylinder 1. Piston 3 then becomes the power piston and rotates the cylinder block. This action continues as long as fluid enters the various cylinders in the cylinder block and discharges through the upper half of the pintle.

The rate at which fluid is pumped into the cylinder determines the speed at which the cylinder block will turn. In a radial-piston pump the quantity of fluid discharged is varied by changing the length of the piston stroke. The length of the strokes is controlled by the amount the cylinder block is off-center with respect to the rotor. The speed of the radial-piston hydraulic motor is controlled by varying the quantity of fluid discharged by the pump supplying the motor.

Assume a pump-and-motor combination using radial-piston units. If the pump and motor are the same size (have the same number of pistons of equal diameter) and the rotor of each is offset the same distance, the hydraulic motor shaft will rotate at the same speed as the pump shaft, assuming that there are no frictional or leakage losses. Shortening the pump offset distance so that the discharge of four pump cylinders is required to fill one motor cylinder will produce a motor-shaft speed one-fourth that of the pump-shaft speed.

Since it is relatively easy to vary the pump-fluid output, the motor speed is also easy to control. To stop rotation of the motor shaft, the pump is adjusted so that its fluid output is zero. The direction of rotation of the motor shaft can be reversed by reversing the direction of fluid output from the pump. When the pump is reversed, fluid enters the motor on the top of the pintle valve, causing piston 3 in Fig. 3-16b to become the first to produce power output. This causes counterclockwise rotation of the cylinder block.

Operating parts of the motor, with the possible exception of the control devices, can be identical with those in the pump. Many radial-piston pump-and-motor combinations differ in only one respect—the rotor in the motor is fixed and cannot be shifted from side to side, as in the pump [Fig. 3-16c].

In some designs the stroke length of both the pump and motor pistons can be varied. With both units set at maximum stroke length, the motor speed equals the pump speed. If the motor stroke is adjusted to one-half the pump stroke, the motor shaft will revolve at twice the pump-shaft speed.

Some pumps and motors are lubricated by the hydraulic fluid. In these units it may be dangerous to operate at zero displacement of the pump for an extended period. For when there is no fluid flow, the pistons operate without lubrication. In other designs, an auxiliary lubricating supply is provided so that there is no danger during periods of zero displacement.

Hydraulic motors and pumps are widely used as close-coupled transmissions. They can produce an infinite number of variable speeds. The space requirement of these transmissions are small, and the operating efficiencies are high.

The reciprocating action within the double pump of Fig. 3-13 is associated with check valves to provide the needed flow relationships. Check valves provide a positive seal. High pressures can be retained by the check mechanism. Because of this, both radial- and axial-piston pumps are produced for high-pressure service employing check valves in the same manner as the simple dual pump unit of Fig. 3-13.

The radial-piston pump of Fig. 3-17 employs seven sets of check valves. These are located within the seven piston assemblies. An eccentric bearing assembly provides the reciprocating action of the pistons. Fluid from the reservoir is introduced into the pump housing. As the spring urges the

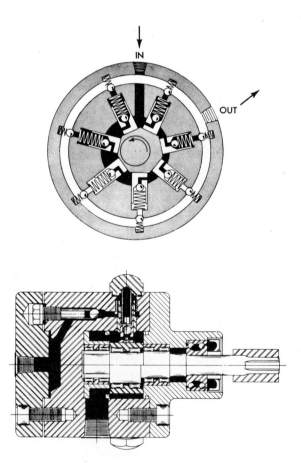

Fig. 3-17 Radial-check-valve-type piston pump. *(Racine Hydraulics, Inc.)*

piston against the cam, a free flow of fluid is permitted by the check-valve ball. As the cam pushes the piston against the spring, the check valve is seated. The contained fluid within the piston assembly is directed through the discharge check valves.

Two banks of seven pistons are employed in some models with dual cams 180° opposed to minimize bearing loads on the main drive shaft. Exhaust check assemblies can be segregated so that multiple flows can be obtained. Thus, the pump can supply independent flows to various circuit segments without the need for multiple-drive units.

Several modifications of the check-valve-type piston pump are available for variable-flow functions. The usual method of providing the variable flow is associated with the input characteristics. The piston accepts a full charge of fluid. On the compression stroke a movable sleeve is employed to control the point at which the fluid is pumped into the exhaust or high-pressure port. If the sleeve is in its rest or maximum-seal position, full volume is pumped into the discharge port. As the sleeve is retracted, the piston spills more fluid back into the housing prior to directing the pressurized remainder to the discharge port of the pump. If the sleeve is completely retracted, there is no pumping action. The sleeve, creating the delivery control, can be manually adjusted or moved as a result of a pressure signal and spring bias.

Direction of rotation of most check-valve-type radial- or axial-piston pumps is of no significance because of the basic need for simple piston reciprocation.

Generally, check-valve-type pumps require a head of oil or supercharging of the input line to provide satisfactory performance. In many installations this can be accomplished by locating the tank above the pump.

Because of the inherently rugged construction, these pumps can be produced to raise fluid pressures to 10 000 psi and above. Bearings and general construction factors dictate the actual ratings assigned by the manufacturer. Fluids must be compatible with the heavily loaded bearings at the elevated pressure loadings.

AXIAL-PISTON PUMPS

These pumps [Fig. 3-18] have a cylinder block with its pistons which is rotated on a shaft in such a way that the pistons are driven back and forth in their cylinders in a direction parallel to the shaft. This is called *axial motion*. In another axial-piston pump design, the cylinders and barrel remain stationary, but the drive plate rotates.

Fluid enters the cap of the pump shown in Fig. 3-18a. It is directed to a stationary valving plate. The cylinder barrel rides against the valve plate. A kidney-shaped port is provided so that the pistons can fill as they are retracted into the rotating barrel. A crossover condition exists in the valving element of the axial-piston pump similar to the action of the pintle in the radial-piston pump. A second kidney-shaped port directs the fluid being expelled to the opposite port in the rear cap. It will be pressurized to

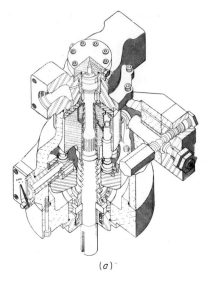 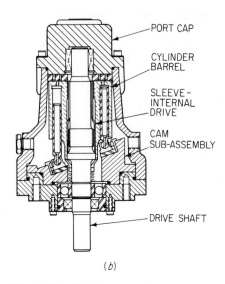

(a) (b)

Fig. 3-18 (a) Variable-displacement axial-piston pump. (*Racine Hydraulics, Inc.*)

Fig. 3-18 (b) Constant-displacement axial-fluid motor.

resistance or pressure level of the immediate circuit at the discharge port.

The major pistons are of a dual diameter. The cavity between the two diameters is pressurized by a small generated-rotor pump located in the rear cap. This provides positive piston positioning. Other pump designs employ springs, mechanical devices, or combinations to ensure piston return.

The plate, against which the head of the piston and its associated shoe ride, can be tilted. If it is at right angles with the drive shaft, there is no pumping action. Tilting to one side will establish one port in the rear cap as suction and the other as discharge. By tilting the plate in the opposite plane, the ports are reversed. This capability makes it possible to use the pump for both flow and directional control. The compensating mechanism can permit the pump to control the maximum pressure level. A pump with the ability to reverse flow through itself is referred to as an *overcenter pump*. Controls similar to those used for radial-piston pumps and variable-delivery vane pumps are also employed in axial-piston pumps. Thus the major control for the axial-piston pump is tilting of the plate which controls the length of piston stroke.

Fixed-angle pumps and motors, such as the motor shown in Fig. 3-18b, are available. Since these are motors, it is not always necessary to retract the pistons by pressure; therefore a single piston diameter is adequate for usual service. Mechanical piston return is provided for those services where the motor may act as a pump for a part of the cycle. At high rotating speeds it may be desirable to provide supercharge pressure for these high-pressure pumps. Generated-rotor pumps delivering fluid at 100 psi or more ensure proper piston fill and dependable action where the atmospheric pressure may be marginal. Pressures usually go to 5000 psi for this type of axial-piston pump.

In the *connecting-rod-type,* axial-piston pump [Fig. 3-19a] the entire cylinder barrel is carried in a yoke that can be set at various angles with the drive shaft. The block with its several pistons is rotated by the drive shaft through an intermediate shaft having a universal joint at each end. The pistons are connected to the driving flange by means of rods which have ball joints at each end. The drive shaft and cylinder block are supported by antifriction bearings. All working parts are submerged in oil that is continuously recirculated.

In Fig. 3-19b, with the cylinder block set at angle *a*, rotation of the structure through 180° produces the relatively short stroke of the piston shown. In Fig. 3-19c, with the block set at the larger angel *b*, the stroke and volume of fluid pumped are increased. With the block set at zero angle, in line with the drive shaft, there is no movement of the pistons in

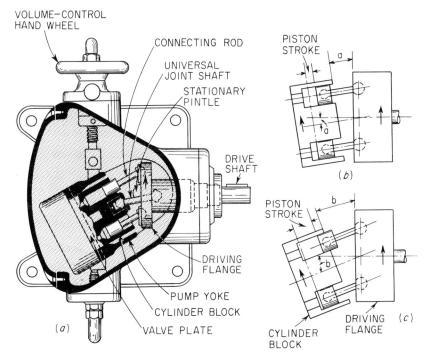

Fig. 3-19 (*a*) Connecting-rod type, axial-piston pump: cylinder barrel carried in yoke which can be set at various angles with drive shaft. (*b*) Cylinder block set at angle *a* produces short stroke of piston. (*c*) Block set at larger angle *b* increases stroke and volume of fluid pumped. *(Vickers Division of Sperry-Rand, Inc.)*

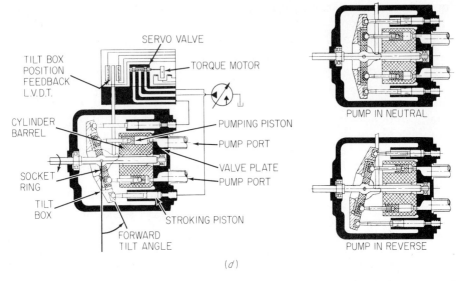

Fig. 3-19 (d) Electrohydraulic-pump control system. (*Racine and Vickers Armstrong, Inc.*)

or out of the cylinders. No fluid is pumped in this position. When the cylinder block is set at an angle above the drive shaft, the direction of fluid flow is reversed. Thus, flow can be adjusted in small steps from zero to maximum in either direction during operation. Suction and discharge take place through sausage-shaped ports in the valve plate. Passages in the pump yoke lead from the valve plate to the stationary pintles (on which the yoke pivots) and through these to external suction and discharge connections. In addition to manual control, a hydraulic cylinder or electric motor can be used to adjust the volume. Servo and automatic (pressure-compensated) controls are available. There are several makes of variable-volume, axial-piston pumps, and mechanisms for varying the pump stroke differ from one design to another.

Servo-valve control for axial-piston pumps offers excellent signal response. The pump of Fig. 3-19d is fitted with a *linear variable differential transformer* which senses the position of the tilt box and subsequently the piston stroke condition. This feedback device combined with the input information provides rapid response for control of direction, rate of acceleration or deceleration, running speed, and position. Further information relative to servo controls will be provided in a later chapter. Because of need for sensitivity, the pressure level in many servo systems is limited to less than 1000 psi, even though the pumps and motors may be capable of much higher pressures.

Axial-piston pumps and motors are widely used for close-coupled transmissions in much the same manner as are radial-piston units. The compact design of the axial pump and motor lends itself to closely coupled transmissions.

Axial-piston pumps require an extremely clean hydraulic system for completely satisfactory operation. Adequate mechanical cleaning of the fluid and system is necessary because of the closely fitted parts and the bearing loads involved. Pressures in these systems of 5000 psi are common, and under certain circumstances they may be higher.

Axial-piston and vane-type motors so closely resemble the corresponding pumps that a special discussion is not required. Applications of these motors will be discussed in a later chapter.

It should be noted, however, that fixed-displacement motors are often combined with variable-delivery pumps as hydrostatic transmission units. Generated-rotor, vane, and gear motors are often used to drive directly a speed reduction unit. The most compact units combine the speed reduction unit with the motor elements. A generated-rotor element can be timed with a commutator-type valve so that the natural geometry of the mechanism provides an integral speed reduction. This forms an extremely compact unit, providing large forces in a small package at relatively low speeds.

Planetary-gear reduction units built into a generated-rotor-type fluid motor can provide speed reduction up to about a 5:1 ratio in one gear set. Two or more stages are required for greater reduction.

The displacement mechanism, illustrated in Fig. 3-20a, provides a speed reduction of 7:1 because of the motion pattern. The rotor set, shown in Fig. 3-20b, is physically located within a housing, as shown in Fig. 3-20c.

Pressurized fluid, entering the rear cap, is directed by a disk-type valve assembly to the displacement element. Note in Fig. 3-20b how one circular

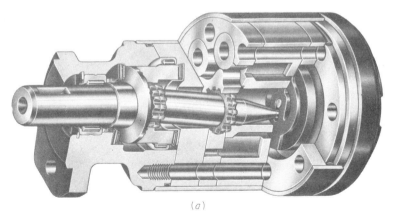

(a)

Fig. 3-20 (a) 7:1 speed reduction motor. (*Ross Gear Division, TRW, Inc.*)

Fluid-Power Pumps and Motors **97**

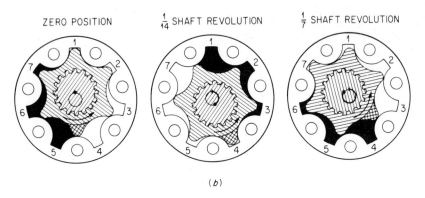

Fig. 3-20 (*b*) Rotor displacement pattern.

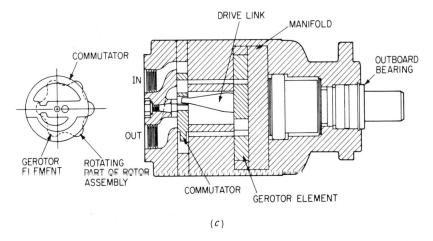

Fig. 3-20 (*c*) Motor assembly and rotor-timing element. (*Ross Gear Division, TRW, Inc.*)

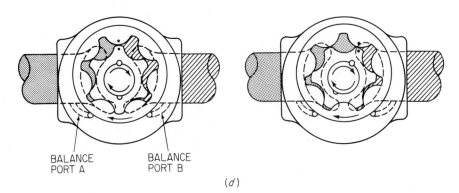

Fig. 3-20 (*d*) Fixed-eccentricity generated-rotor mechanism for hydraulic-motor use.

98 *Industrial Hydraulics*

movement of the commutator causes the movement of one rotor tooth one-seventh of a revolution of the output shaft. The valve drive link rotates the commutator to provide this desired motion pattern. The drive link also provides a universal joint between the displacement element and the output shaft. The output shaft is supported by bearings to accept heavy side loadings, which are commonly met in industrial applications and general-purpose agricultural drives. The universal-type drive element isolates the external loading from the displacement element and associated valving.

Figure 3-20d illustrates a generated-rotor unit specifically designed for hydraulic-fluid-motor use. The left view shows the rotor sealing from the pressure to exhaust port at the 12 and 6 o'clock positions. The right view shows how the lobes of the inner pinion gear seal against the arcs of the ring gear at two points at the upper and two points at the lower crossover point to create a seal from pressure to tank port.

The ring gear is eccentric to the pinion. Thus, there is a greater area at the top because of this eccentricity. Both the inner pinion and surrounding ring gear rotate together, rather than in the fixed and orbiting relationship shown in Fig. 3-20b, which is used for a speed reducing type of hydraulic motor.

Pressurized fluid is directed to the exterior surface of the ring gear so that it can be lubricated with a pressurized oil film. The same pressurized oil film serves to balance the ring gear so that friction drag is minimized.

Because of the balanced structure, the generated-rotor fixed-eccentricity motor can carry heavy loads at very slow speeds. It can also start heavy loads with pressures only slightly higher than running pressures. Wide speed range is possible because of the long arc of the kidney-shaped ports at the fluid-supply and fluid-discharge areas.

DUPLEX RECIPROCATING PUMPS

These pumps [Fig. 3-21a] find some limited use in industrial hydraulic systems. *Simplex* pumps have one fluid-handling cylinder. The duplex pump consists of two pumping elements mounted side by side. Single or double acting, the pump is arranged so that the discharge of the pumping elements overlaps. This arrangement reduces the pulsation of the fluid in the discharge pipe.

TRIPLEX RECIPROCATING PUMPS

These pumps [Fig. 3-21b] have three fluid-handling cylinders. These pumps are built as either single- or double-acting units. Triplex pumps are normally arranged so that each cylinder operates one-third of a revolution apart from the other two. This arrangement reduces pulsation in the discharge line to the minimum.

Pumps with more than three cylinders find occasional use in fluid-power systems. These are, however, specialized units and will not be discussed in this book.

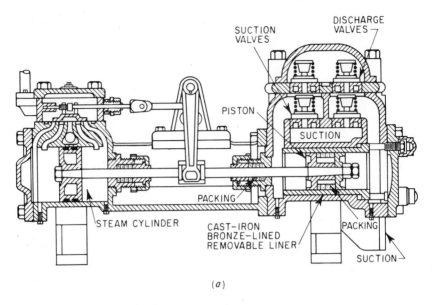

Fig. 3-21 (*a*) Typical duplex reciprocating pump.

3-5 CENTRIFUGAL PUMPS

Pumps of this type are often used in fluid-power applications where large quantities of fluid are needed for rapid movement of machine members. In some cases, centrifugal pumps are the only source of power—such as in elevators and some central-power systems operating at relatively low pressures. Centrifugal pumps are also widely used for auxiliary functions in hydraulic systems (such as circulating fluid through cooling and cleaning devices and supercharging larger piston or other rotary pumps). *Propeller pumps* find their widest use in fluid-power systems in the supercharging of other pumps.

To produce pressure, centrifugal pumps depend on the basic law of inertia: *A body at rest tends to remain at rest, and a body in motion tends to continue in motion with the same velocity and in the same direction.*

Figure 3-22 shows that if a pail of water is swung, it will remain in the pail if the rate is rapid enough to produce centrifugal force. Velocity of rotation has much to do with the water in the pail. If the bottom of the pail were to come loose at some point, the water would go off at a tangent to the circle of rotation. In Fig. 3-23 a place for *the bottom to come off*, or a discharge, has been provided. At the same time, a place has been provided for fluid to enter as the pressure is lowered by loss of fluid at the outer periphery.

Since velocity must have a direction, the direction in which the discharge takes place becomes important. To make maximum use of the velocity head developed, the discharge is generally in an approximately

100   Industrial Hydraulics

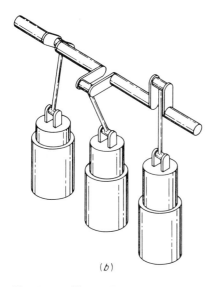

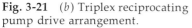

Fig. 3-21 (b) Triplex reciprocating pump drive arrangement.

Fig. 3-22 Centrifugal force holds the water in the pail.

tangential direction, but the exact size and shape of the chamber and the precise direction of discharge are matters which concern only the designer of centrifugal pumps. Note that fluid is thrown off the paddles or *impeller blades* all around the *volute* (circle). Fluid is thus added to space A at all points but escapes from it at only one point. To compensate for this, it is customary to increase the area of space A as the outlet is approached. By this means fluid velocity can be reduced at a more gradual rate, thus obtaining more efficient conversion of the velocity head to pressure head. For this reason it is customary to give the space A the volute shape shown in Fig. 3-23.

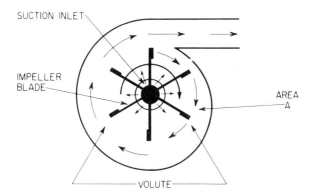

Fig. 3-23 Basic elements of a centrifugal pump.

Fluid-Power Pumps and Motors

The volute continuously expands at a definite ratio. The size and shape of the volute used are very important to the efficiency of the pump, as are the shape, size, and speed of rotation of the impeller. Impeller blades that curve backward with respect to their direction of rotation have higher efficiency than straight blades, but the reason for this and for other details of construction are matters which concern pump designers.

Figure 3-24 shows the basic parts of a centrifugal pump. Fluid enters the center or *eye* of the impeller and follows the volute passage to the nozzle. Energy imparted by the rotation of the impeller is transferred to the fluid. Modifications of the pump for special purposes include the use of diffusers, double intakes for balancing, and special impeller construction to increase efficiency. Increased discharge pressure may require use of more than one impeller or *stage*. In a multistage pump one impeller discharges into the inlet of the next impeller. The discharge pressure developed is approximately the sum of the pressure developed in each stage.

Peripheral turbine pumps and *propeller pumps* are quite similar to the basic centrifugal pump discussed above. Peripheral turbine pumps have different characteristics that make them suitable for higher suction lifts than the more common volute-type pump. Propeller pumps [Fig. 3-25] have little suction or lifting power, but they do have a fairly high pushing power that can be applied to large volumes of fluid.

3-6 JET PUMPS

In this type of pump [Fig. 3-26], a liquid or gas under pressure is introduced directly into the pump chamber to produce pumping action. Liquid or gas passing through the restricted nozzle undergoes a decrease in pressure and an increase in velocity. These changes produce a partial vacuum in the pump chamber. Atmospheric pressure acting on the fluid in the reservoir forces a column of fluid up into the pump chamber. Here the fluid combines with the input flow and is discharged from the pump.

Jet pumps find only restricted use in fluid-power systems. They are used in some heavy machinery as scavenger pumps to empty tanks located below reservoir level where drains from other equipment are terminated.

3-7 HEAD AND ENERGY IN PUMP SYSTEMS

No matter where or how pumps are used in fluid-power systems, all head and energy relations will conform to the laws of hydraulics that were discussed in Chap. 1. Like all other machines, pumps only transform energy; they do not create energy. The energy applied to a fluid by means of a pump goes either into the production of usable pressure or velocity in the fluid or into friction losses. The energy thereby released in the system precisely equals the input energy supplied to the system. A study of

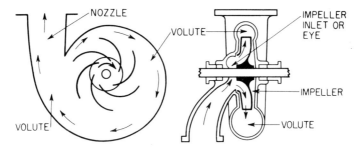

Fig. 3-24 Basic construction of a centrifugal pump.

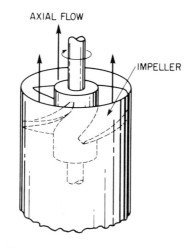

Fig. 3-25 Action of a propeller pump.

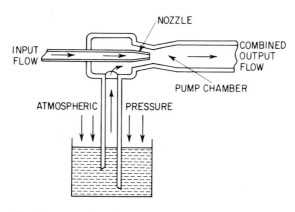

Fig. 3-26 Simple jet pump.

Fluid-Power Pumps and Motors **103**

energy relations in a pump system must take into consideration the factors involved in different types of pump installations. Three of these factors are:

1. The pump may stand above, at, or below the level of the fluid supply source. If the pump is at the same level as the free surface of the supply, a study of the system will show that there are no special problems, since the pump with its source of power acts on the fluid like any other applied force. If the pump stands below this level, as shown in Fig. 3-27, a certain amount of energy in the form of *gravity head* will be already available when the fluid enters the pump. If the pump stands above the fluid level, as shown in Fig. 3-28, a certain amount of energy is needed to suck the fluid into the pump.

2. By means of pumps, energy from a power source is often introduced into a system at some midpoint, as in closed-loop circuits. In some types of circuits the pump may even be used as a brake where gravity is a factor in a falling load, as in an elevator. The energy output may vary greatly in circuits of this type. Intermediate stages of a multistage pump pick up or accept the fluid at one pressure level and add more energy to the fluid to discharge the fluid at a higher pressure. When the pump becomes a brake, certain motor characteristics may be necessary that require modification of the basic design.

3. Many pumps operate in closed systems in which fluid from a reservoir is circulated through the system and back again to the reservoir by the pump. This is the case with cooling systems as well as with the most common machine tool hydraulic systems.

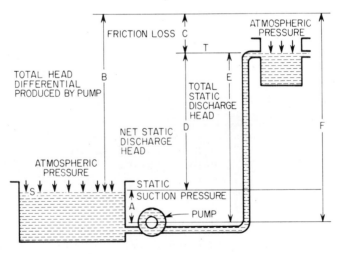

Fig. 3-27 Static suction head at pump inlet.

104 *Industrial Hydraulics*

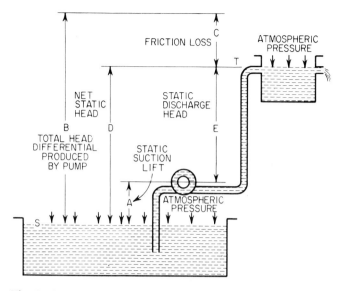

Fig. 3-28 Pump located above reservoir level.

When a pump is installed so that there is a *head* of fluid at the inlet, there are certain predictable relationships. If the pump stands at a vertical distance A below the free surface S of the source of supply, as in Fig. 3-27, there is a *static pressure head A* on the suction side of the pump. This head will be a part of the total input head necessary to produce the required output head F. The action of the pump produces the total head differential B, which can be divided into *friction loss C* and *net static discharge head D*. D is the vertical distance above S that the energy supplied through the pump has raised the fluid. E, the *total static discharge head*, is the vertical distance from the center of the pump to the surface of the fluid in the discharge reservoir and is equal to D plus A.

The two input factors are A and B, where A enters the system as initial *elevation* or *potential head* and B as a result of the energy supplied to the pump. On the outlet or output side are C and E, or *friction loss* and *total static discharge head*. Thus, the input and output are balanced and A plus B will equal C plus E, which in turn will equal F.

Velocity head is not mentioned because we are considering the overall problem of moving stationary fluid at level S to stationary fluid at level T. The fluid will, of course, have a velocity head at any point in the pump or pipes, but since the fluid ultimately comes to rest again within the system, all this velocity head will be converted into *potential head,* friction, or turbulent loss.

Atmospheric pressure will act upon the free surface of the fluid at S and at T, but since it creates equal heads on opposite sides of the pump, it cancels out.

Fluid-Power Pumps and Motors

Figure 3-28 shows the pump standing a vertical distance *A* above the free surface *S* of the source of supply. Energy must be applied to the fluid to get it into the pump. In addition, enough energy must be applied to produce static discharge head *E*, plus *C*, which is lost in friction, if the fluid is to be raised to the level *T*.

The total *differential head* produced by the pump action, *B*, is the total energy input. This is divided into *A* on the input or suction side of the pump, which represents the head consumed in drawing the fluid into the pump, plus *C* and *E* on the discharge side, which represent friction head and static discharge head. As a formula, *B* equals *C* plus *E* plus *A*.

Velocity head again does not appear because the fluid comes to rest at the end of the cycle. Atmospheric pressure cancels out as before; only in this case the atmospheric pressure at *S* was necessary to force the fluid up to the pump.

Head relations in a closed system (such as that commonly used in machine tools and other systems using a reservoir) are illustrated in Fig. 3-29. In this circuit a pump is driving a work piston back and forth in a cylinder. To do this against the resistance offered by the work, assume that the piston must develop a pressure equivalent to head *H*. Under this assumption, the total head differential *B* must be produced by the pump after the system has begun to operate. This is divided into friction head *C*, consumed in moving the fluid through the system, and static discharge head *D*, which is equal to *H* and therefore produces the pressure required to do the work. *B* equals *C* plus *D*, and *D* equals *H*, which was assumed necessary for the work operation.

Since the fluid returns to its original level and the system is closed throughout, there will be a *siphon* or *suction effect* in the return pipe which will exactly balance the *static suction lift A*. Atmospheric pressure will play a part in the operation of the system only during start-up and until the entire active portion of the system is filled with the fluid.

The various components need not be located as shown in Fig. 3-29. The cylinder, fluid motor, or other work device may not always be located higher than the pump. In some machine systems the pump may be located in the reservoir or at a position below the reservoir fluid level. But no matter at what levels the different components may be located, the levels will be in balance with the system in operation. Since operation starts and finishes with stationary fluid, whatever velocity head may exist at various points along the way will ultimately disappear into equivalent head, either as static pressure or as friction.

It has been pointed out that energy or work is the product of a force multiplied by the distance through which it acts. This principle applied to fluids gives the equivalent rule that energy is equal to the volume of fluid moved, multiplied by the head against which it is pumped. In each of the foregoing illustrations for a given volume of fluid, the energy relations are the same as the head relations.

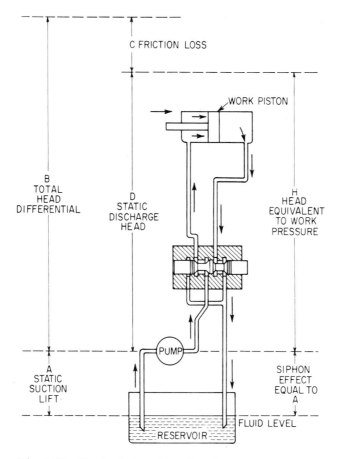

Fig. 3-29 Head relations in a closed system.

3-8 FACTORS DETERMINING SUCTION AND DISCHARGE HEADS

In *open systems,* suction head is limited by atmospheric pressure. This is also true of any siphon effect in a *closed system* if the reservoir is open to the atmosphere. The lift of pumps at sea level will be greater than at higher altitudes, since the lower atmospheric pressure at higher altitudes will be balanced by a shorter column of fluid.

If an attempt is made to lift a fluid more than the distance atmospheric pressure is capable of raising it, the fluid will vaporize. This is because at the reduced pressure above the fluid, the temperature at which it will vaporize is lower. At a certain suction lift, the fluid vaporizes even if cool and nothing but vapor reaches the pump.

This also explains why the maximum suction lift for hot fluids is considerably less than for cold. Volatile fluids are also difficult to pump

Fluid-Power Pumps and Motors **107**

since even a slight pumping action tends to vaporize them. Suction lift depends on the density of the fluid. It takes a longer column of light fluid than of a heavy one to balance atmospheric pressure at a given altitude. Thus, for oil with a specific gravity of 0.881, the theoretical limit of lift at sea level would be 34 divided by 0.881, or about 38.5 ft. In other words, it would take a 38.5-ft column of this oil to balance atmospheric pressure at sea level, as compared with a 34-ft column of water, since volume for volume this oil weighs less than water in the ratio of 0.881:1.0.

Applications using fire-resistant hydraulic fluids must be carefully analyzed to make certain that fluids having weight and volatility characteristics closely approaching water are compatible with the existing installation. Pump suction characteristics may be just as critical as the other factors (such as seals and certain types of paint) when using fluids with a higher specific gravity than oil. Most fire-resistant fluids have a higher specific gravity than oil. Pumps may require larger suction lines or relocation when fluids are changed and the specific gravity is different.

Liquid- or gas-operated reciprocating and rotary pumps are able to lift fluid if they are started empty, whereas centrifugal pumps must be primed before they will draw fluid. This is because ordinary centrifugal pumps are not sufficiently tight to pump air. Therefore the pressure above the fluid in the suction line will not decrease enough to allow atmospheric pressure to force the fluid up into the pump.

The pump may be primed by pouring fluid into the casing or intake area by hand, by the use of a bypass connection from the discharge side of the pump, by putting a check valve in the suction line below fluid level to prevent escape of the fluid from the intake line, or by exhausting the pump of air with a vacuum pump. Certain centrifugal pumps must be primed if the fluid level is permitted to drop below a point where air can enter the housing even though the pump may be below the fluid level when additional fluid is added. Other types may require that an air bleed at a high point be purged after the fluid level has been below a certain point or during the initial start. These factors vary with different pump designs. The manufacturer's recommendations should be closely followed for maximum efficiency of operation.

Suction lift is important in preventing *cavitation* in a pump. Cavitation is a mechanical wearing away of a metal surface by the action of the fluid, occurring usually in the form of pitting of the surface. It is caused by the fluid leaving the metal guiding surface, such as in the impeller of a pump. Where the fluid leaves the surface, a vacuum is created, into which air and fluid vapor are liberated from the fluid to form a bubble. This bubble is carried along with the fluid into an area of high pressure where the bubble suddenly collapses and may develop a pressure of several thousand pounds per square inch to compress the air in the bubble. This pressure may be sufficiently high to drive the particle of air into the metal. When this action is repeated in rapid succession, the metal is worn away and other mechanical effects may be produced, such as noisy operation and vibration due to the repeated blows of the collapsing bubbles.

It has been found that for a given head and capacity a low-speed pump will operate safely with a greater suction lift than a high-speed pump. If the suction lift is high (say over 15 ft), it is often necessary to use a lower-speed pump. If the suction lift is low, or if there is a positive head on the pump, the speed of the impeller may often be increased. A set of curves prepared by the Hydraulic Institute shows the upper limits of speed to prevent cavitation in various centrifugal pumps. These curves are important to engineers who select pumps for hydraulic systems.

On the discharge side, heads of almost any magnitude can be obtained. Pumps can be arranged to act in stages, so that each pump adds more pressure to that already supplied by the others. With centrifugal pumps the volume of fluid pumped will vary with the pressure, even with constant-speed operation. There is a maximum pressure attainable for any given speed of operation and density of fluid being pumped, beyond which no fluid will flow. For positive-displacement pumps this is not so; the volume depends only upon the speed of operation, regardless of pressure, within the capacity of the source of power and mechanical strength of the pump.

3-9 PUMP CHARACTERISTICS

The pressure developed by a pump or the volume of fluid it handles changes if the load against which it acts or the speed of operation is changed. In addition, pumps of different types will operate differently under similar conditions. The effect of discharge pressure on pump output differs, for example, between positive- and nonpositive-displacement pumps. The behavior of a pump under varying conditions is shown by curves known as the *characteristic curves* of the pump. These curves show the following facts about the specific pump: (1) the relation between head or pressure produced and the rate of discharge when the pump is running at constant speed (note how Fig. 3-30 shows the delivery in gallons per minute at different outlet pressures); (2) the horsepower which must be supplied to the pump to obtain different rates of discharge against different heads (note in Fig. 3-31 that a column to show horsepower input relative to pressure has been added); (3) the efficiency of the pump, or the usable energy divided by the energy input when the pump is run at different rates of discharge. Here the work that could be done by the fluid pumped is compared with the work required to pump the fluid. Figure 3-31 shows the performance chart of a widely used constant-delivery pump with both volumetric and mechanical efficiencies plotted. The manner in which pump efficiencies are calculated lies beyond the scope of this book.

Pumps have individual and definite characteristics that can be conve-

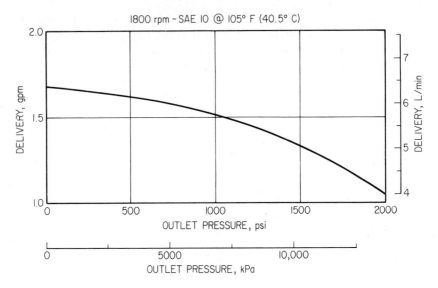

Fig. 3-30 Relation between pump discharge pressure and pump delivery.

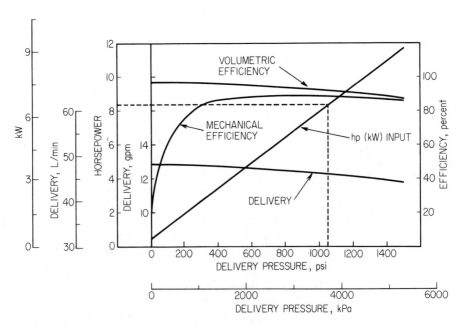

Fig. 3-31 Chart for determining pump horsepower requirements.

110 *Industrial Hydraulics*

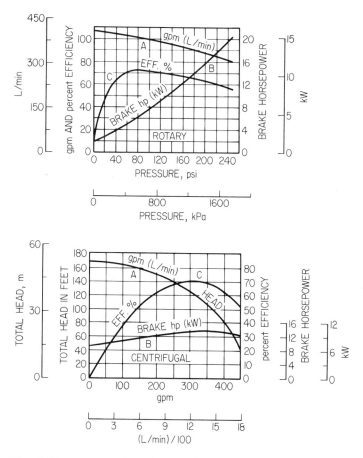

Fig. 3-32 Comparative characteristic curves for rotary and centrifugal pumps.

niently shown in graphical form. Figure 3-32 shows typical curves for a rotary and for a centrifugal pump with constant operational speeds. Rotary pumps usually run more efficiently at lower heads. Thus, in the rotary pump diagrammed, maximum efficiency is reached at a pressure of about 80 psi, curve C. The power necessary to run a rotary pump increases almost directly, as does the pressure produced by the pump at constant speed. This is shown in the diagram by the fact that the brake-horsepower curve B is almost a straight line. There is also relatively little difference in the rate of discharge for different pressures produced. This is shown by head-capacity curve A, which does not depart far from the horizontal. The departure shown is due solely to leakage and not to any inherent characteristics of the pump design.

The rate of discharge of most centrifugal pumps drops quickly as discharge pressure increases. The centrifugal pump diagram shows this

Fluid-Power Pumps and Motors **111**

by the fact that output equals about 250 gpm against a head of 140 ft, but drops to almost zero against a head of 170 ft, curve A. Centrifugal pumps usually run most efficiently at higher rates of discharge. The pump diagrammed is shown to run most effectively at a discharge of about 300 gpm, curve C, as compared with about 100 gpm for the rotary pump (combination of curves C and A). Relatively less input energy per output is needed for centrifugal pumps, up to a limit, as the rate of discharge is increased, curve B.

When centrifugal pumps are used in hydraulic circuits, the design varies according to the manner in which they are used. If the pump is used to supercharge rotary or reciprocating-piston-type pumps, the characteristics may be different from those of a pump used to move large quantities of fluid during rapid traverse functions. Multistage pump assemblies may also have considerably differing characteristics.

Sometimes these differences in centrifugal pumps are referred to as rising or falling characteristics. In a pump with a rising characteristic, the head increases up to a certain point as the pump is run at a constant speed and the discharge is increased from zero [Fig. 3-33]. In a pump with a falling characteristic, the head continuously decreases under constant speed as the discharge is increased from zero. A pump with a flat characteristic shows little variation, up to a certain point, as the rate of discharge is increased. These characteristics are obtained by design procedures outside the scope of this presentation.

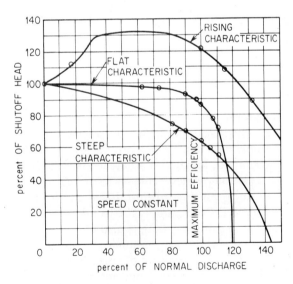

Fig. 3-33 Centrifugal pump discharge-head characteristics.

DESIGN CONSIDERATIONS FOR FLUID-POWER PUMPS AND MOTORS

Economic considerations usually dictate the size, type, and pressure level associated with a fluid-power pump or motor. Exceptions do exist. Hydraulic systems used to power aircraft-operating devices (landing-gear retraction, tail movement, etc.) require minimum total system weight. Highest possible pressures with minimum component weight are generally chosen. Significant premiums are paid to attain the specific goal of total weight reduction.

Higher operating pressures reduce the need for large components. The quantity of fluid in the system is reduced and premium quality components are incorporated in the total system. Maximum pressures of 3000 psi [20,685 kPa] are common.

Some heavy-duty construction machines (cranes, shovels, dozers, backhoes, etc.) are designed to function at pressures above 2000 psi [13,790 kPa]. Physical weight is not a prime consideration. Space to house the components, quantity of fluid, and improved response time dictate the higher pressures. Smaller conductors inherently tolerate higher pressures because of the reduced surface areas exposed to the pressurized fluid.

The least expensive systems generally function at pressures below 2000 psi [13,790 kPa]. The large number of components that are produced in this pressure range results in lowest unit cost. In the medium- (and low-pressure levels,) the cost of hose, pipes, fittings, and associated components is shared in many cases with other industries, such as chemical processing, lubrication, fluid transfer, etc. The availability of the lower pressure components is enhanced by the wide usage pattern. Hose becomes less flexible as working pressures increase. At the highest working pressures it is usually necessary to use rotary joints in various configurations to allow for needed movement of machine members.

Hydraulic motors are predominantly fixed-displacement devices that rotate at a speed directly related to the flow of pressurized fluid. Some hydrostatic transmissions employ motors with variable-displacement capabilities to permit an overdrive speed relationship, wherein a higher rotating speed is available at a reduced displacement value of the motor. Variable-displacement pumps require more complex working parts than do fixed-displacement types, resulting in higher unit cost. Hydrostatic transmissions for traction drives, powering winches, accelerating or decelerating loads, may require variable-displacement pumps and-or motors for most efficient use.

An intermittent use pattern may dictate use of fixed-displacement pumps and-or motors with appropriate valving to reduce initial cost. Operational savings may not justify initial higher costs. Reduced complexity of fixed-displacement pumps and motors may also be a determining factor in the choice of components and systems

SUMMARY

Every hydraulic system uses one or more pumps to develop pressure on the contained hydraulic fluid and move the fluid from one part of the system to another part.

Pressure is developed as resistance to flow is encountered. Pressure level, within a circuit, usually indicates resistance to flow caused by movement of a mass or friction. The required pressure level will be determined by the energy needed to perform the power-transmission function and the size relationship of the associated hydraulic system components.

The most common type of pump used in hydraulic systems is the rotary pump. Reciprocating pumps are used in specialized applications. Centrifugal pumps are used within a narrow group of installations.

The constant-displacement rotary pump delivers a fixed amount of fluid at each rotation. Its displacement changes with a change in the driving speed. The pressure produced equals the resistance to flow of the fluid being discharged by the pump. The discharge pressure must be limited by a control device which maintains a maximum pressure not exceeding the capabilities of the pump.

Variable-delivery pumps can change the rate of fluid discharge without changing the driving speed. The rate of fluid discharge can be changed by manual adjustment or by automatic devices. Variable-delivery pumps can be constructed so that the flow of fluid through the pump can be reversed without changing the rotation of the drive.

Higher pressures are usually produced by radial- or axial-piston pumps. Control valving may be check or pintle type. Mated rotating flat plates may alternately direct suction and discharge porting to and from the pistons in axial-piston pumps.

Radial-piston-type motors have the pistons within the rotary-drive portion [Fig. 3-16b] or are stationary, with connecting rods transmitting the force to the rotating-eccentric-drive rotor [Fig. 3-16c].

Axial- and radial-piston pumps and motors are widely used for close-coupled transmissions. Infinitely variable speeds are possible. Space requirements are minimized and efficiencies are high.

Gear-reduction units are provided in fluid motors to produce efficient, economical rotary force and motion at low speeds. Generated-rotor motors are available with an integral speed reduction function as a result of the motor design.

Centrifugal pumps find use where large flow volumes at relatively low pressure are required. Typical applications include providing large flows during rapid approach of large press cylinders, elevators, filling vessels for hydrostatic pressure tests, and pressurizing the suction of higher-pressure pumps in a multistage installation.

Suction characteristics of pumps vary widely. The location of the pump, type of plumbing installation, and supercharge or boost requirements must be considered to avoid cavitation within the pump.

Seals, bearings, and mechanical components can be affected by different types of fluids. The compatibility of components may change if fluid is changed.

Many pumps can be used as fluid motors with minor modifications. Hydraulic-fluid motors can be used safely in hazardous locations because of the complete absence of spark-generating components. Hydraulic motors can be fitted with seals so that they can be used under water and in contaminated areas. Lubrication is usually supplied by the fluid in the system so that a minimum of attention is required.

Input horsepower requirements can be calculated by using charts illustrating the *characteristic curves* of pumps and motors.

REVIEW QUESTIONS

3-1 What is meant by cavitation?

3-2 Why must the suction line to a pump be of a certain minimum diameter and length?

3-3 Describe the relationship between rotating speed and fluid delivery rate of a gear-type rotary pump.

3-4 Can a variable-delivery pump be used in a hydraulic system without an external relief valve? Explain.

3-5 What factors determine the pressure developed by a centrifugal pump?

3-6 Explain the difference between an axial-piston and a radial-piston pump.

3-7 What is meant by a multistage pump?

3-8 What is the function of a relief valve? Explain how a relief valve functions.

3-9 How many check valves are required to ensure satisfactory operation of a simple lift-type pump? Why? What is their function?

3-10 Do all piston-type pumps require check valves? Why?

3-11 What is a pintle valve?

3-12 What effect will excessive rotating speeds have on vane-type pumps? On piston-type pumps?

3-13 Why can generated-rotor or gear-type pumps be operated at high speeds? What is the limiting factor?

3-14 Why is fluid temperature important in the operation of a hydraulic pump? How does temperature affect cavitation?

3-15 What effect on the operation of a hydraulic pump can be expected when inlet filters are restricted by dirt and other foreign materials?

3-16 Explain how incorrect alignment will affect the performance of a hydraulic pump or motor.

3-17 Why is the suction line to a pump usually larger than the discharge line?

3-18 What is pulsating flow? What type of pumps deliver a pulsating fluid flow?

3-19 Why is it necessary to provide a supply of pressurized fluid to the intake port of some high-pressure pumps? What type of pump might be used for this purpose?

3-20 How does entrained air in hydraulic fluids affect the operation of pumps? Explain. How will this affect the other parts of a hydraulic circuit?

3-21 What is the purpose of the commutator in Fig. 3-20c?

3-22 How does the servo-type directional-control valve of Fig. 3-19d control the pump output?

3-23 What is the purpose of the small generated-rotor pump in the rear cap of the unit shown in Fig. 3-18a?

3-24 How does the solenoid of Fig. 3-10e affect the pump delivery?

3-25 What is meant by pressure loading in connection with a gear pump?

LABORATORY EXPERIMENTS

3-1 Make a piston pump using a simple hydraulic cylinder and two check valves. Note the power requirements at various discharge pressures.

3-2 Add two additional check valves to the piston pump of Experiment 3-1 at the opposite end of the cylinder. Connect the suction check valves to a common line and restrict the flow. Note the increased force necessary to move the handle or lever used to move the piston rod.

3-3 Completely close the suction line to the simple hand pump. Note the resistance to movement and the spongy action.

3-4 While operating a rotary pump, add a slight restriction to the intake pipe. Note the change in noise level.

3-5 Connect a relief valve to the discharge of a rotary hydraulic pump. Connect a tee pipe fitting in the suction line. Install a needle valve in this line. Direct the discharge of the relief valve to an open tank. Note the change in flow characteristics as air is admitted to the pump suction line through the needle valve. Note the change in noise level.

3-6 Connect a rotary pump to a rotary fluid motor. Use the fluid motor as a variable-speed-drive element. Immerse the suction line from the pump in an open tank. Connect the pump discharge to a relief valve. Direct the discharge of the relief valve back to the open tank in which the pump suction is located. Note the delivery characteristics from the minimum to the maximum rotating speed recommended by the pump builder. Observe the pressures that can be developed within the pressure rating of the equipment at various speeds. Note the point at which the fluid motor stalls. Check the stalled torque of the motor by the pressure developed by the pump.

3-7 Connect a relief valve to the line supplying the fluid motor in Experiment 3-6. Note the pressure necessary to start the pump rotating at various set values and comparable pressures necessary to maintain speed.

3-8 Connect an external drain to the pump and motor used in Experiments 3-6 and 3-7. Note the flow through the drain on both the pump and motor at various speeds and pressures.

3-9 Connect a hand crank to the shaft of a generated-rotor pump. Install a relief valve on the discharge and suction line to the pump from a small container. Discharge the relief valve into the small container. Note the physical effort necessary to rotate the pump. Observe the sponginess if air entrained in the fluid enters the pump.

3-10 Connect an external drain to the pump in Experiment 3-9. Note the drain characteristics as the unit is operated at various speeds and pressures.

3-11 Connect a fluid motor directly to a cable drum. Fasten dead weights to a cable rolled on the drum (through a pulley fastened to the ceiling if necessary). Note the pressure necessary to start the mass in motion. Observe the pressure needed to hold the weight in a fixed intermediary position.

3-12 Install gauges in the fluid lines to the motor in Experiment 3-11. Note the pressures when the weight is allowed to drop unrestricted for short distances and then is stopped by restricting the discharge port of the motor.

3-13 Connect relief valves to each cylinder or motor line in Experiment 3-11 and then repeat Experiment 3-12. Note the difference in pressure values with different relief-valve settings.

3-14 Connect a needle valve between the two motor lines in Experiment 3-11. Note the pressure in the cylinder lines when the weight is stopped quickly with different openings of the needle valve.

3-15 Fasten a bar to the shaft of a fluid motor. Note the starting torque with weights on the bar at varying distances from the shaft.

4
Hydraulic Cylinders and Rams

Hydraulic cylinders are widely used in industrial hydraulic systems. These cylinders are also called *linear motors* and *reciprocating motors*. The usual *hydraulic cylinder* consists of a circular tube, sealed at both ends, in which a *piston* and its *rod* move. The *piston rod* projects through either or both ends of the cylinder. Leakage of fluid out of the cylinder around the piston rod is controlled by a suitably designed *seal* usually containing *packing*. A hydraulic cylinder transforms the flow of pressurized fluid into a *push* or *pull* of the piston rod. Hydraulic cylinders are designed for a variety of services. In this chapter we shall study various types of cylinders and how they are used. A knowledge of hydraulic cylinders is a great help in your study of the uses of industrial hydraulics.

4-1 SINGLE-ACTING RAMS

A ram is a hydraulic cylinder designed to apply force in only one direction. Note in Fig. 4-1a that the piston, having but one diameter within the cylinder, is pushed upward by pressurized fluid to apply force. The piston moves in and out of the cylinder through the packed *gland* or seal. Fluid leakage out of the cylinder is prevented by the *packing* in the gland. The diameter of the piston determines the force developed when the fluid pressure remains constant. When the fluid pressure is reduced or *relaxes* in the *contained area*, the piston can be retraced by an external force, as shown in Fig. 4-1a. Fluid from the contained area must be drained to a tank or other low-pressure area. The retracting force can be gravity, a mechanical spring, or small-diameter auxiliary pistons.

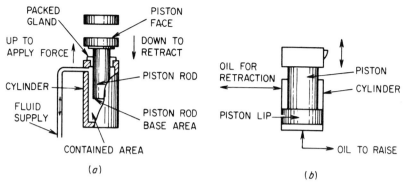

Fig. 4-1 (a) Single-acting ram. (b) Double-acting ram.

4-2 DOUBLE-ACTING RAMS

A *double-acting ram* is one in which both strokes (application and reduction) of force are produced by pressurized fluid. Thus, pressurized fluid raises the piston in Fig. 4-1b to apply force. To retract the piston and reduce the force, fluid is admitted to the top of the cylinder, forcing the piston downward. During force application, while the piston is rising, fluid above the *lip* may be allowed to drain out of the upper part of the cylinder. While the piston is descending, fluid must be allowed to drain from the lower part of the cylinder. The lip is machined on the ram during manufacture. Note that the area of the lip on which the fluid pressure acts during the downward stroke is considerably less than the piston-base area. This reduced area is satisfactory because the ram is not applying a significant force during its downward stroke.

4-3 DOUBLE-ACTING CYLINDERS

Figure 4-2a shows a *double-acting cylinder*. Like the double-acting ram, the double-acting cylinder is built so that the stroke in either direction is produced by the fluid pressure. Since the piston rod extends through the end of the cylinder, a packed gland is needed to prevent fluid leakage. If desired, a double-acting cylinder can be arranged so that it does its major work during either or both of its strokes.

Figure 4-2b shows another type of double-acting cylinder. This unit, a *double-rod-end cylinder*, has rods which extend through both ends of the cylinder. When the piston moves in either direction, it causes both ends of the rod to move because they are firmly attached to the piston. Work can be done by either rod or both rods. The speed of rod movement will be the same in either direction of travel when equal quantities of fluid are piped to either end of the cylinder.

Hydraulic Cylinders and Rams **119**

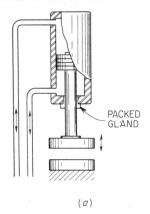

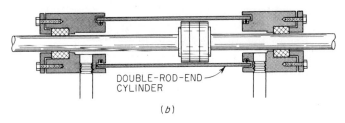

Fig. 4-2 (*a*) Double-acting cylinder. (*b*) Double-rod-end cylinder.

4-4 INTEGRAL BYPASS CIRCUIT

The cylinder in Fig. 4-3 has an *integral bypass circuit* at its upper stroke limit. When the piston rises to its point of maximum upward travel, fluid passes through the auxiliary port into the check valve. After passing through the check valve, the fluid enters the cylinder feed line. When the downstroke begins, the check valve prevents any fluid from entering the cylinder from pipe D [Fig. 4-3]. Pressurized fluid, supplied to the cylinder by pipe D, produces the downward stroke of the piston. Fluid supplied by pipe U produces upward movement of the piston. The *dashpot* at the bottom of the cylinder provides a cushion of fluid to prevent the piston from striking the bottom at the end of the downstroke.

Figure 4-4 shows a double-acting cylinder equipped with *secondary pockets* to *cushion* the piston at each end of its stroke. The pockets, one at each end of the cylinder, consist of a drilled passage fitted with an adjustable *needle valve*. When the piston approaches the end of its stroke in either direction, it forces fluid into the drilled passage. Flow of fluid past the needle valve is restricted so that the piston velocity is reduced.

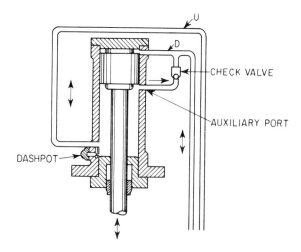

Fig. 4-3 Cylinder assembly with integral bypass port.

This prevents the piston from striking the cylinder end. The slowing of the piston motion is called *deceleration.* In this design, fluid enters the connecting port and the blind end, acts on the end of the piston rod, and passes over the lip of the isolation seal to start the piston on its reverse stroke. Fluid entering the connecting port at the rod end passes directly by the lip of the cushion seal and applies energy at the face of the piston. Other designs use a check valve to permit fluid to leave the cushion chamber until the piston movement provides a free flow path from the chamber. Cushion chambers are also called *captive areas* because the fluid is trapped in this area whenever the piston is at either end of its stroke.

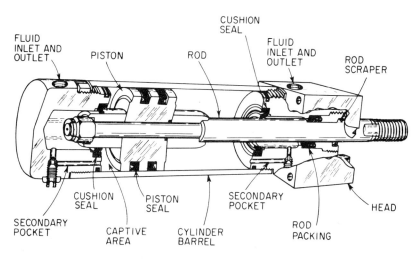

Fig. 4-4 Cushioned cylinder.

Hydraulic Cylinders and Rams **121**

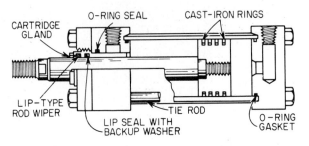

Fig. 4-5 Noncushioned cylinder with cast-iron rings.

Note that several seals are shown in Fig. 4-4. The type of seal chosen for a cylinder depends on the service for which cylinder is intended and on the fluid used. In machine tool service, cast-iron or other metal automotive-type piston rings [Fig. 4-5] may be used. The cylinder in Fig. 4-4 has synthetic *lip-type* seals. In other designs synthetic cups or leather cups of various types are used to provide a suitable sliding seal.

4-5 TELESCOPING-TYPE CYLINDERS

Figure 4-6a shows a series of rams nested in a telescoping assembly. This arrangement provides a relatively long stroke with good mechanical strength. An assembly of this type is used on dump trucks because of limited space and the need for a predetermined force pattern, as shown in Fig. 4-6b. Note that large forces can be developed when the load is the heaviest; the smaller diameters are employed as the load becomes lighter.

Diameter A [Fig. 4-6a] is relatively large; the ram in this cylinder provides a large force for the beginning of the lift of the load. (In any dump truck the greatest force is needed to begin lifting the load before dumping begins). As the required force decreases and ram A reaches the end of its stroke, ram B begins to move, providing the smaller force now needed to continue raising the load. When ram B reaches the end of its stroke, a smaller force is required to raise the load; ram C then moves outward to finish the lift-and-dump function. These three rams can be retracted by gravity acting on the dump or by pressurized fluid acting on the lip of each ram.

Outrigger-type stabilizing supports often use telescoping cylinders because of the usual small stowage space on the vehicle. Cranes, shovels, pile drivers, well-drilling rigs, and portable conveyors may support large loads by use of these devices. Because of the expected vulnerability to surrounding contaminants and mechanical damage, they are often shrouded. The shroud provides a dust and weather shield. Note in Fig. 4-6c that cylindrical shroud A, enclosing and attached to the body of the cylinder, slides within the load-carrying rectangular hollow column B welded to the frame of the rig. The rod is attached to the column head.

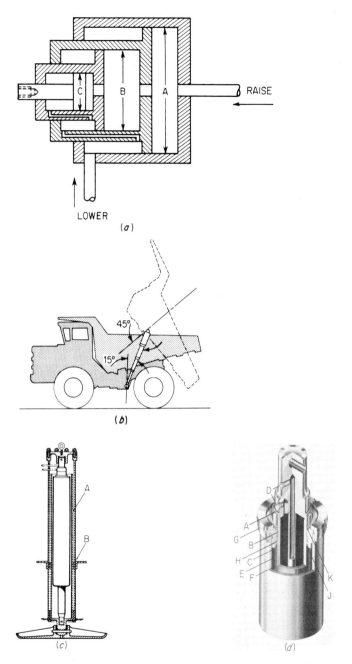

Fig. 4-6 (*a*) Telescoping cylinder. (*b*) Application of telescoping cylinder. (*c*) Mechanical shrouding to protect telescoping cylinder. (*d*) Telescoping cylinder with fluid passage in rod assembly.

Hydraulic Cylinders and Rams **123**

This arrangement shields the stuffing box and rod from the weather and from airborne cutting dust. It also safeguards the rig and jack against excessive bending stresses sometimes encountered on sloping terrain. Since the rod does not move linearly with respect to the housing and is ported, the hydraulic lines are short and fixed.

Figure 4-6d shows a cylinder assembly such as that used in Fig. 4-6c. The conducting line entering the bottom of the cylinder of Fig. 4-6a is directed through the center of the rod of the cylinder of Fig. 4-6d. The fixed element A contains both conducting ports D. Bronze glide bearings B support side loads while seamless tube C provides good column strength. Because of the need for long service life with minimum packing changes, the surface finish at E is usually about 8 rms, which is a polished surface. Surface F is often hard chrome plated by as much as 0.001 per side. The cartridge-type stuffing-box retainer G with lock retains the multilip packing H. Self-compensating packing gland J is further protected by self-contained wiper-scraper K.

4-6 DUAL RAMS

Simple construction and large forces are available with dual rams [Fig. 4-7]. These rams are used to turn a ship rudder having a large area projecting below the water. The rams are of equal size, producing equal speed and force in both directions of movement. By providing large rams and cylinders, the mechanism is made extremely stable.

4-7 INTEGRAL JACK RAMS

Cylinders and rams are often built as an integral part of a machine slide or press element. One such machine, a molding press with an upward acting ram, is shown in Fig. 4-8. The fluid reservoir is an integral part of the machine.

Pipe A connects to an area around the lip of the *main ram*. This area can

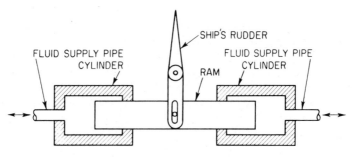

Fig. 4-7 Dual rams in ship's steering gear.

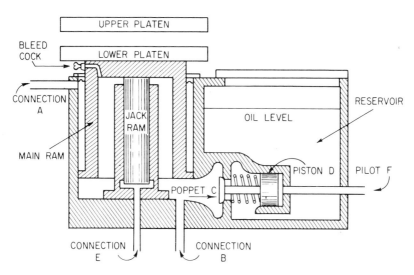

Fig. 4-8 Internal-jack-ram press with rising ram.

be pressurized to various levels. Some machines circuits provide a fluid pressure at this point equal to that in the remainder of the circuit. Equal pressure throughout the circuit provides a stabilizing influence on the main ram and assists the retraction of this ram.

To raise the main ram rapidly, fluid is pumped to connection E. The smaller inner ram, called a *jack ram,* has a greater area than the lip of the main ram. As the jack ram rises, it pushes the main ram upward. While rising, the main ram creates a vacuum at connection B. Fluid is sucked into the cylinder, but the relatively small size of connection B prevents sufficient fluid flow. The hole around poppet C provides a large flow area, and so the vacuum created by the main ram allows atmospheric pressure to force poppet C open, permitting the main ram to fill with fluid. At a predetermined point in the rise of the main ram, fluid under pressure is pumped through connections B and E. The main ram then no longer acts as a pump. As fluid is pumped to both the inner and outer rams, the loss of flow through poppet C permits the valve to close. The spring assists in closing the valve. As this occurs, the two rams travel together at a lower rate of speed until resistance is encountered. At this point the fluid pressure acts on the total ram area, minus the lip of the main ram.

In the retraction cycle, the pressure is relaxed at B and E. Connection A is still pressurized. Pressure is applied at piston D through pilot line F. This forces poppet C open so that fluid can return freely from the main ram to the tank. At the rest position, connections B and E are closed and piston D is relaxed so that the ram remains in a stationary position.

The bleed at the dome of the large ram is used to exhaust trapped air when the unit is pressurized. This reduces spongy action and ensures an immediate pressure rise when resistance is encountered. Note that the

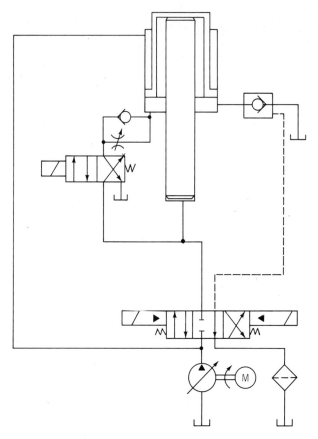

Fig. 4-9 Press circuit drawing using ISO symbols. Ram is the rising type.

entire ram assembly is at the same level as the fluid tank. Poppet assembly C and piston D are within the fluid reservoir.

Figure 4-9 shows the ISO symbolic-type circuit for this ram-type press mechanism. The variable-delivery pump provides flow and pressure in a pattern that permits rapid ram movement at low and medium pressures and a smaller fluid flow as the pressure exceeds medium values. At maximum pressure, as established by the control mechanism of the pump, the flow will be just sufficient to make up for leakage of the fluid and maintain the desired maximum pressure.

In this type of press, it is common to stop the drive motor between cycles, particularly when there is a long load cycle when the platens are open. The variable orifice in the line to the main ram permits controlled decompression if the following cycle is programmed.

At the end of the pressing function the main four-way valve is deenergized. The spool will return to neutral, as it is directly spring-centered. The small four-way valve in the main cylinder line will also return to the

126 *Industrial Hydraulics*

rest position because it is directly solenoid-operated with a bias return spring. In this position, the high-pressure fluid in the main ram can decompress by flowing through the adjustable orifice to the tank. Usually about 5 cc or less is adequate. After this time, the press return can be started by activating the pump motor and the solenoid on the main four-way valve used to direct pressurized fluid to the pilot of the check valve returning the major flow from the main ram to the tank. Actuating this right-hand end solenoid of the main four-way valve not only directs pilot oil to the pilot check valve but also provides a path back to the tank for the fluid entrapped in the jack or pilot ram. The filter on the return line will clean the fluid returning from the jack ram so that a fixed quantity of fluid will be filtered during each cycle.

4-8 SEALS, PACKINGS, AND ACCESSORIES

Figure 4-10a shows a rod scraper used to eliminate the possible penetration into the cylinder of foreign matter carried on the rod. In some corrosive atmospheres it is usual to provide a flexible-type boot to protect the rod. Figure 4-10b shows one method used to attach the boot to the

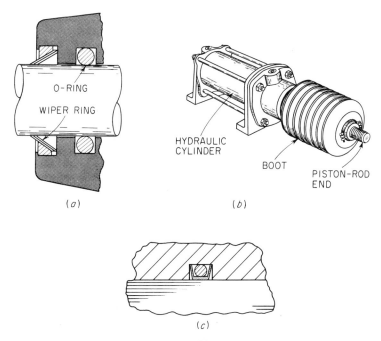

Fig. 4-10 (a) Piston-rod wiper ring and O ring. (b) Bellows-type boot to prevent dirt from entering a hydraulic system. (c) Plastics with low coefficient of friction backed up with O ring to provide resiliency and memory.

Hydraulic Cylinders and Rams

cylinder. Note how the wiper ring bears against the rod in Fig. 4-10a. The O ring is representative of many different types of packings that may be used. In some contaminated areas a protective coat of hard chrome over a hardened rod will provide satisfactory service without the use of of a boot. Carbon-steel rods are satisfactory in general-service usage and cost less.

Rod seals are generally made of resilient materials. The O ring and other molded shapes can seal for a long period of time at elevated pressures. Most of these seals are designed with an *initial squeeze* for low pressures and increased contact area on the rod as the pressures within the mechanism increase. At elevated pressures, usually about 1500 psi or above, the seal may function better with an adjacent backup ring of some less resilient material. The backup ring prevents the resilient seal from being squeezed out of the gland by the high pressure.

Plastics with an extremely low coefficient of friction can be used to bear against the metal-rod surfaces while the resilient O ring backs them up, as shown in Fig. 4-10c. A similar cap construction can be used on seals located at the surface of a piston, bearing against a cylinder bore. Provisions must be made to use relatively rigid plastic with minimum stretch and memory.

Cylinders and the machine members in a unit that is being actuated must be correctly aligned. Any misalignment will be reflected in wear of the piston-rod-guide bushings and packings. Rod scrapers [Fig. 4-10a] may be ineffective in cases of acute misalignment. Cylinder rods are rarely intended to carry an off-center loading without a special guide and adequate size rods.

4-9 LIMITED-ROTATION MOTORS

Figure 4-11a shows a *limited-rotation motor* or *actuator*. This *double-vane rotor* provides 100° rotation of its shaft. Single-vane designs provide 280° of shaft rotation. Figure 4-11b shows how cushion valves can decelerate the rotor and prevent impact at the end of the stroke. Snubbing occurs when the rotor blades close off the exhaust port near the end of the rotor's travel. When flow to the actuator is reversed, fluid flows through a check valve to start motor rotation in the opposite direction. A manifold-type body [Fig. 4-11-b] permits flange mounting of the *servo control valves* with a minimum of external piping.

High torque (twist) can be produced by large opposed pistons [Fig. 4-12a]. Gear racks, internally positioned by the pistons, engage a pinion gear in the center of the cylinder. As the pistons extend and retract, they rotate the pinion gear which is attached to the output shaft. This motor produces equal torque in either direction. Fluid powering the motor or actuator lubricates the rack and pinion.

Figure 4-12b shows how multiple vanes can give high torque in a small envelope. The rotor is pressure-balanced by ports drilled through the

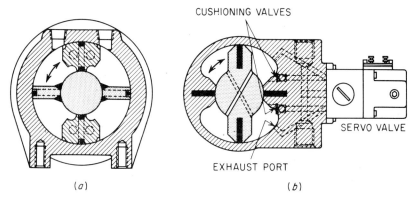

Fig. 4-11 (*a*) Limited-rotation, dual-vane motor. (*b*) Limited-rotation motor with dual valving to provide cushioning at end of stroke.

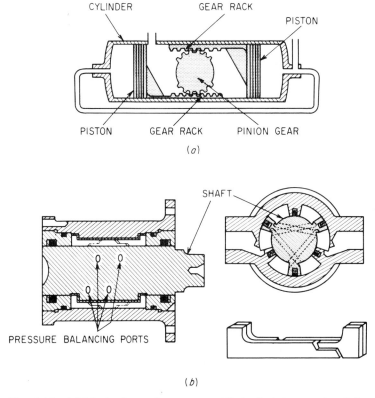

Fig. 4-12 (*a*) Limited-rotation motor with dual pistons and rack to provide increased rotation. (*b*) Limited-rotation motor with multiple vanes to increase torque potential in the small envelope.

Hydraulic Cylinders and Rams **129**

shaft. Seals on the rotor (shaft) and housing are laminated to reduce internal leakage and increase efficiency. The seals are made in sections that slide axially on one another with their joints staggered so that each joint is sealed by a continuous section of the adjacent seal. Fluid pressure in the actuator presses the seals axially and radially against their sealing surfaces.

The output shaft of the motor in Fig. 4-13a oscillates as the piston moves back and forth in the cylinder. An internal helix machined in the piston meshes with an external helix on the output shaft. Guide rods pass through holes in the piston to keep it from rotating. The actuator can be stopped at any point in its stroke. Its self-locking helix angle prevents rotation from external torque loads. This permits the actuator to be positioned with an *open-centered control valve*. Adjustable cushions can reduce shock at the ends of the stroke.

The *double rack-and-pinion design* [Fig. 4-13b] balances the forces rotating the output shaft. All moving parts operate in a oil-filled chamber sealed from the operating media. Suitable seals on the pistons permit the unit to hold loads in intermediate positions for indefinite periods of time. The unit in Fig. 4-13c is similar and incorporates a spring as a fail-safe feature in certain types of installations. Excellent acceleration and deceleration characteristics are possible when the limited-rotation motor or actuator is used over an arc of 180° or less with a crack-arm motion transmitting the force to a linear plane.

4-10 DIFFERENTIAL AREA CIRCUITS

Effective circuits can be built around the use of a cylinder with a piston-rod area approximately one-half the piston area. The circuit in Fig. 4-14a is called a *differential circuit.* Fluid from the head end returns to the tank when the cylinder is retracted. The piston and rod move (extend) at equal speeds because the fluid from the rod end is joined with that from the pump or fluid source. This causes the cylinder to extend at a relatively rapid rate. The displaced fluid volume is effectively equal to the volume of the rod. When maximum force is needed, the fluid from the rod end is directed to the tank. There are modifications of this circuit which can provide very fast operation with a minimum of energy input. Equal traverse speeds will result from connecting both ends of the cylinder together as shown, but the flow of fluid will be considerably different. A directional-control valve of adequate size must be used to ensure correct pressure drop in the circuit.

Differential cylinder action provides economical, rapid traverse of machine members. The variable-displacement pump, shown in Fig. 4-14b, for example, delivers 6 gpm at maximum capability. The rod area of the piston assembly is equal to one-half of the area of the piston face. By

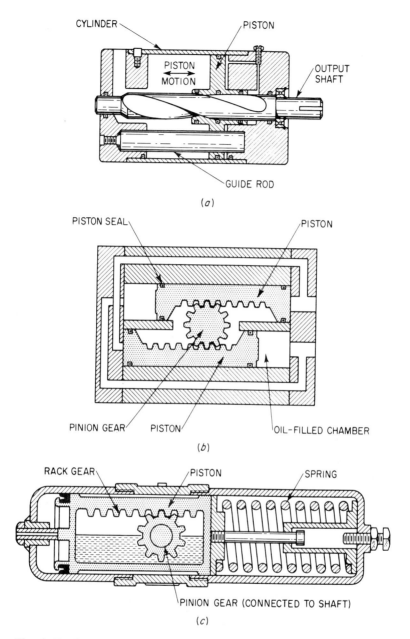

Fig. 4-13 Limited-rotation motors. (*a*) Helix-drive type has dowel to prevent rotation of drive piston. (*b*) Two pistons drive the pinion gear. (*c*) Integral-spring-return design.

Hydraulic Cylinders and Rams

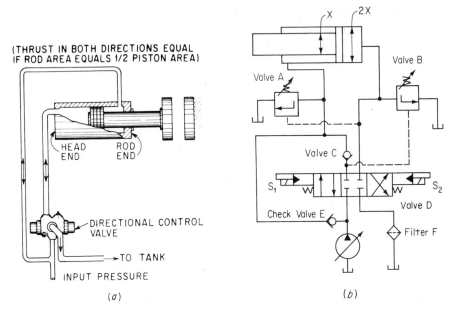

Fig. 4-14 (a) Differential circuit. (b) ISO symbolic circuit illustrating a regenerative hydraulic circuit.

energizing solenoid S_2, pump flow is directed to the blind end of the cylinder. Fluid from the rod end passes through check valve E, joining the pump flow. The resulting condition can be described as an effective 12-gpm flow through the directional-control valve to the blind end of the cylinder. It can also be thought of as merely a 6-gpm flow effective in displacing diameter X.

Friction, resisting the fluid flow, will limit some of the effectiveness of the circuit. As resistance is encountered by rod X, pressure at the blind end will increase. When this pressure equals the preset level of value A, the pilot fluid will actuate the flow-directing element so pressurized fluid in the rod end is relaxed as it flows through valve A to the tank. In this mode, the effective force is calculated by multiplying the pressure resulting from resistance by the full area of the piston face.

Check valve E prevents loss of pump delivery through valve A. By deenergizing solenoid S_2 and energizing solenoid S_1, fluid is directed through check valve C to the rod end of the cylinder. The blind end is directed to the tank. Valve A closes because of loss of pilot pressure. Valve B is then piloted open to divert part of the fluid from the head end back to the tank.

The 6-gpm flow to the rod end means that 12 gpm must be carried to the tank from the head end. Part of the fluid passes through filter F and valve D. The remainder passes through valve B. Circuit action is basically automatic. Desired pressure signal points can be established by adjusting valves A and B. Valve D can be actuated by solenoid OR pushpin AND internal hydraulic pilot pressure. It can be reversed by pushpin OR solenoid AND pressure OR centered by integral spring action.

4-11 FLUID-POWER INTENSIFIERS AND BOOSTERS

In an earlier chapter the driving of reciprocating piston pumps by steam or compressed air was discussed. This arrangement may be referred to as *boosting* an existing pressure (either the steam or air pressure). It is convenient to use this same principle to intensify the pressure on fluids by use of different *effective areas* in various parts of a circuit. Figure 4-15a shows how a 500-psi input pressure can be intensified to 5000-psi output pressure at one-tenth the fluid delivery rate. The intensifier unit has a high-pressure piston directly connected to a low-pressure piston with (in this unit) 10 times as much area. In this circuit the directional-control valve directs the low-pressure fluid to the tank when the pistons are in the neutral position. This holds the intensifier in neutral. If the E end of the four-way valve is actuated, fluid will be directed to the intensifier to cause the pistons to retract (move to the right), ready for a high-pressure stroke. When limit switch H is contacted by the contact on the piston, the unit is completely charged, ready for the forward stroke.

The low-pressure flow may also be directed through check valves D and C to fill the area which will be subjected to the highest pressure as well as to cause the return of the pistons A and B. The rod end of cylinder B is drained to tank (above the fluid level). The pilot line to valve R is pressurized when solenoid E is energized. This offers an additional flow path for the flow from the blind end of cylinder B, which reduces the resistance to flow as the unit is being readied for the next pressure stroke. High-pressure reverse flow is blocked at check valve D.

Relaxation of the high pressure may be done by a high-pressure pilot-operated check valve or through some other form of two-way valve that gently relieves the pressurized area without unnecessary shock. Limit switch G indicates when the full stroke has been completed. This may, if desired, start the piston back for another stroke. The intensifier in Fig. 4-15a can be used continuously for a high-pressure pumping action. Flow of the high-pressure fluid will, however, be pulsating. High-pressure check valve C may be omitted in certain applications where the reversal of piston B will gently diminish the pressure until a pilot check can easily be opened and the major actuation of the work cylinder can be continued for its return.

The pressure-booster circuit illustrated in Fig. 4-15b does not require external limit switches or solenoids for the reversal function. Cam valves E and F are actuated by both mechanical and pilot signal. The cross-sectional view in Fig. 4-15c shows how the components are physically located. Figure 4-15b shows the energy flow through the circuit.

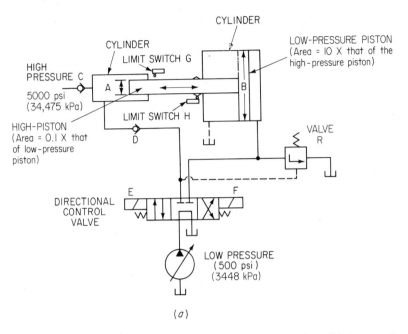

Fig. 4-15 (a) Electrically operated fluid-power intensifier. (b) Automatic reciprocating pressure intensifier. (c) Reciprocating intensifier with mechanical shift. (*Racine Hydraulics, Inc.*)

Low pressure (usually 1000 psi or less) is ported to the major flow-directing element and the two cam valves E and F.

To understand the operation of the pressure-booster circuit in Fig. 4-15b, assume movement of major piston to the right. Discharge passes freely through piloted valve G to the tank. Intensified fluid closes check valves D and A. High-pressure fluid is delivered to the outlet at a pressure level reflecting the resistance to flow encountered in that part of the circuit. If the ratio of areas of the piston-rod assembly is 5:1, the low pressure will be one-fifth that of the high-pressure outlet. Pressurized fluid in the left end causes the movement of the piston assembly. Fluid also passes through check valve C to the chamber surrounding the left, high-pressure piston. Pressurized fluid at the face of the spool in cam valve E prevents creep resulting from vibration, which may cause a premature shift to the spool. A larger-diameter piston at the end of the spool in cam valve F is pressurized to hold the pilot-valve spool shifted. The main spool is also held in the shifted position by live pressure.

As the main piston head contacts the cam surface at the end of valve F, it shifts the spool against the pressure load created by the bias piston. This directs pressurized fluid to the right end of the main spool. It also pressurizes the bias piston at the end of cam valve E. This causes valve E to shift. The shifting of pilot valve E releases pilot pressure from the left

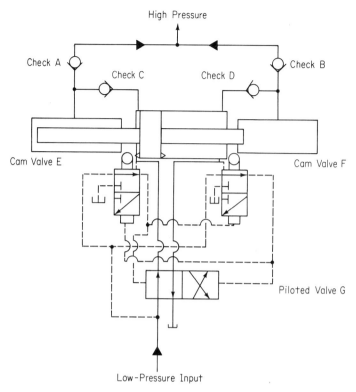

Fig. 4-15 (*Cont.*)   (*b*)

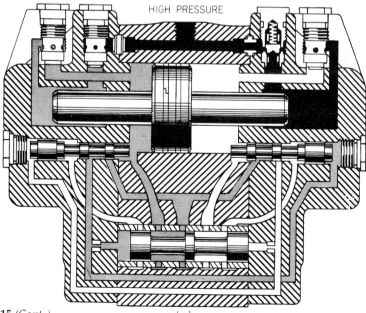

Fig. 4-15 (*Cont.*)   (*c*)

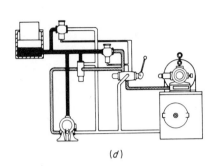
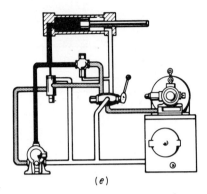
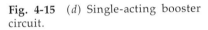

Fig. 4-15 (*d*) Single-acting booster circuit.

Fig. 4-15 (*e*) Double-acting booster circuit.

end of the main spool so that the main spool shifts, reversing the direction of the main piston assembly. The high-pressure discharge then closes check valves *B* and *C*; thus, flow is through check valve *A*.

This reciprocating flow continues as long as the output resistance is less than five times the pressure capability of the input flow.

Cataloged boosters of this basic design are available in many pressure-flow ratios. Common values are 3:1, 5:1, and 7:1.

Figure 4-15*d* illustrates a simple, hand-valve-actuated, single-action circuit. With the four-way-valve lever in the *forward* position, the ram will move forward with full volume from the low-pressure pump. When the movement of the ram is resisted by some type of load, pressure builds up and automatically opens the sequence valve. Full pump volume at low pressure is then diverted to the pressure booster. The booster, in turn, develops high pressure at low volume or movement rate of the ram for the final closing or squeezing action.

With the four-way-valve lever in the *reverse* position, the ram will return by gravity. In many instances a higher opening speed may be obtained by the addition of an auxiliary dump valve, as indicated. The dump valve is a pilot-operated, two-way valve that accepts a portion of the return flow so that the entire volume need not return through the four-way valve.

Figure 4-15*e* illustrates a double-acting booster circuit. with the lever-operated, four-way-valve lever in the forward position, the ram will move forward with full volume from the low-pressure pump in the same manner as the single-acting ram of Fig. 4-15*d*. When the movement of the ram is resisted by some type of load, pressure builds up and automatically opens the sequence valve. Full pump volume at low pressure is then diverted to the pressure booster.

The booster, in turn, develops high pressure at low volume on the ram for the final closing or squeezing action. With the four-way-valve lever in the reverse position, the ram will return with full volume from the low-pressure pump and the booster will not be in operation.

There are reasons, other than economic, that dictate use of an intensifier rather than higher-pressure pumps for certain applications. A list of these reasons is helpful in deciding when the use of an intensifier is justified. The reasons are:

1. Decompression of fluid under pressure can be at lower pressure values, permitting softer action with less expensive components.
2. Action and maintenance are simple.
3. High-pressure fluid is localized in what can normally be a low-pressure machine.
4. Pressure can be maintained over a period of time with low horsepower input.
5. There is a minimum number of high-pressure seals to maintain.
6. Input fluid to the intensifier or booster can be different from the working fluid in the circuit (i.e., steam, raw water, air).
7. Part of the piping can be low pressure, where stresses are minimized.
8. Fluid pressures can be quickly dissipated in the event of a line break.
9. Shock loadings from punch-and-shear operations can be more easily controlled.
10. High pressures can be varied with more sensitive low-pressure devices, according to the size ratio.

4-12 CYLINDER USES

Figure 4-16 shows a variety of uses for hydraulic cylinders. Study these diagrams now to learn some of the ways of using hydraulic cylinders in industry.

DESIGN CONSIDERATIONS FOR HYDRAULIC CYLINDERS AND RAMS

General-purpose hydraulic cylinders are rated for their ability to perform at a specific maximum pressure value. Hydraulic rams are designed for specific applications. The very nature of the device precludes the design of a general-purpose family of products. Many hydraulic cylinders and most hydraulic rams are thus designed for their anticipated use. General char-

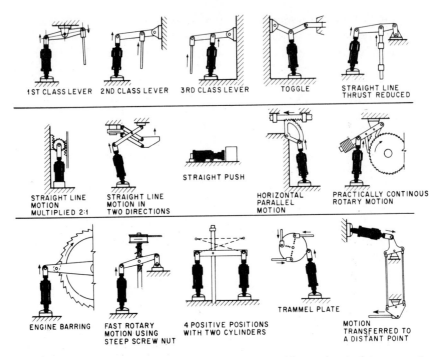

Fig. 4-16 Typical applications of cylinders to provide mechanical movements.

acteristics required for special applications generate a family of products. Some manufacturers gear their production output to a specialized industry. Some examples of the special-purpose products and their unique characteristics follow:

1. Steel mill and general heavy equipment
   Features: Fabricated from steel plate with heavy-duty thick-wall steel tubing. Oversize piston rods. Heavy-duty packing designed for quick, easy change. Extremely rugged.

2. Agricultural cylinders
   Features: Medium to light duty—often throw-away type with nonreplaceable packing. Construction carefully designed to match expected service life, which reflects seasonal activity with relatively short service life. Simple, easy replacement of a total unit may be featured in place of field repairs.

3. Machine Tool Cylinders
   Features: Precise construction designed for long service life. Easy seal and packing replacement. Rugged construction. Some mechanical protection is anticipated. Good maintenance can be expected.

4. Marine Applications
   Features: Rustproof construction. Oversize piston rods. Fail-safe structures.

5. Construction Machinery
   Features: Flange-type connectors for plumbing. Heavy-duty construction. Weatherproof design. Packing-gland protection. Long service life.

6. Aircraft cylinders
   Features: Maximum strength with lightweight materials. Careful design to reduce all excess physical weight. Anticipate maintenance on a carefully calculated schedule.

Machine-tool hydraulic cylinders are usually considered the general-purpose product. All cylinders follow a common pattern relative to operational characteristics.

The first decision that a designer must make relates to the type of cylinder as determined by the anticipated use. The second decision is governed by the available pressurized fluid. The force needed must be divided by available maximum pressure to determine the required area of the cylinder. If 30,000 lb [133,440 N] of thrust is needed, and 3000 psi [20,685 kPa] is available, the 30,000 [133,440 N] divided by 3000 [20,685 kPa] equals 10 in^2 [6.452 cm^2].

If this cylinder must move 24 in [60.96 cm] in 30 s, it will require 24 in [60.96cm] times 10 in^3 [64.52 cm^3], or 240 in^3 [3933.6 mL] per min. Thus, slightly more than 2 gpm [7.87 L/min] will provide the needed rate of movement and force at 3000 psi [20,685 kPa]. Two gal/min equals 2 x 231 equals 462 in^3/min.

To avoid bending under load, heavy-duty rods are often designed to be equal to one-half the area of the piston so that the return speed is twice the rate of the forward speed and the return force capability is one-half that of the forward stroke. Increased return speed and greater strength provide plus design values.

Available space may limit the size of a fluid-power cylinder. Thus, to provide the same force, the pressure must be increased as the number of square inches of area is decreased.

Pneumatic or light-duty hydraulic cylinders usually function at 1000 psi [6895 kPa], or less. Many machine-tool cylinders are rated at 2000 psi [13,790 kPa].

Agricultural cylinder design usually conforms to the 2000-psi [13,790-kPa] rating. Hydraulic pumps supplied as a standard option on tractors quite often have a 2000-psi [13,790-kPa] rating.

Mill cylinders, marine presses, and construction machinery use pressures in excess of 2000 psi [13,790 kPa]. Some presses and construction machines use pressures to 5000 psi [34,475 kPa], and above.

SUMMARY

Hydraulic cylinders are of two basic types: (1) those with piston rods of less than half the area of the piston, used for both push and pull functions, and (2) those with more than half as much rod area as piston area, used primarily for push functions and commonly called *rams*.

Cylinders with equal-diameter rods extending from each end move at the same rate of speed in each direction.

Limited-rotation motors can provide rotary action directly without the need for a rack and pinion or similar devices used to convert linear-cylinder action into rotary action.

Intensifier circuits can transform low pressures into high pressures by transmitting force through pistons with different diameters. The speed of fluid movement is directly proportional to the ratio of piston areas and the cubic content of the cylinders.

REVIEW QUESTIONS

4-1 Sketch a single-acting ram and describe its motion during a complete cycle.
4-2 What is the purpose of the piston lip in a double-acting ram?
4-3 Describe a double-rod-end cylinder and its motion.
4-4 What is the function of the check valve in Fig. 4-3?
4-5 What is the function of the cushion seal in Fig. 4-4?
4-6 What is a rod scraper? Discuss.
4-7 Sketch a telescoping cylinder and describe its motion.
4-8 What is an internal jack ram? Describe its motion.
4-9 Sketch one type of limited-rotation motor. Discuss its operation.
4-10 If the diameter of piston A is one-fifth that of piston B in Fig. 4-15a, what will the high pressure be if the low pressure is 250 psi?
4-11 Why are rods hardened on certain types of cylinders?
4-12 What does the term *intensification* mean as applied to hydraulic circuits?
4-13 What happens when a cylinder is incorrectly aligned during installation?
4-14 What parts of a cylinder are affected by a change of fluid in the hydraulic system?
4-15 Give an example of a machine that might use a double-rod cylinder.
4-16 What is a third-class lever? Draw a sketch of a hydraulic cylinder used with a third-class lever.
4-17 Sketch the symbol for a single-rod cylinder with adjustable cushions.
4-18 What is the purpose of the bleed valve in Fig. 4-8?
4-19 Why is the pilot line for valve B in Fig. 4-14b connected between valves D and C? What will be the maximum flow through check

valve E in Fig. 4-14b? What will be the maximum flow rate through valve D in Fig. 4-14d?

4-20 Why are there two pilot-operated check valves shown in the circuit of Fig. 4-15d?

LABORATORY EXPERIMENTS

4-1 Connect a cylinder to a four-way directional-control valve and source of pressure, as shown in Fig. 4-14a. Measure the operating speed.

4-2 Reconnect the cylinder in Experiment 4-1 in the conventional manner. Measure the difference in operating speed.

4-3 Face the rod of a second cylinder in alignment with that used in Experiments 4-1 and 4-2. Trap fluid in the blind end and observe the pressure on a dial gauge, using the connection of Experiment 4-1 and then Experiment 4-2. Why are the pressures different? Which trapped pressure is higher? Explain.

4-4 Build the circuit shown in Fig. 4-15a. Keep pressures within the rated capacity of the equipment. A manual directional-control valve can be substituted for the electrical valve for initial observation of pressures in each part of the circuit. Install a gauge in the line to the blind end of piston B and block the flow to the tank. Observe the pressure at B when full pressure is applied to piston A.

4-5 Obtain a cylinder similar to that pictured in Fig. 4-4. Drill a hole parallel to the needle-valve connecting hole in the blind end with a small pipe tap in the back of the blind cap. Connect a pressure gauge and note the reading when the appendage on the piston enters the cushion seal. Connect a small needle valve to this tapped hole, leading to the captive area. Drain the fluid into an open container. Note the speed and the resultant flow as the valve is adjusted. (Caution: Oil may splash because of high velocity.)

4-6 Connect a cylinder with piston rings, such as in Fig. 4-5, to a four-way directional-control valve. Install a high-pressure globe or needle valve in each cylinder line. Provide a tee connection between the shutoff valve and the cylinder. Install a pressure gauge in each line. Reciprocate the cylinder several times to remove entrapped air. Direct pressure to the blind end. Close the valve on the rod end. Note the pressure reading on the gauge. Direct pressure to the rod end. Close the valve on the blind end. Note the pressure reading on the gauge. Why are the pressure readings different? Will they be different on a double-rod cylinder?

4-7 Direct fluid to the blind end of the cylinder of Experiment 4-6. Permit piston movement halfway out and close the valve on the rod end. Attach a dial indicator to the rod end which can bear against a solid object and note the amount of movement at various pressure readings. Repeat the experiment with a cylinder similar to that of Fig. 4-4 equipped with resilient piston seals.

4-8 Open the shutoff valves of Experiment 4-6 and supply pressure to the directional-control valve at a pressure level of approximately 10 psi; check for piston breakaway friction. Shift the directional-control valve to extend the cylinder and retract in the following pattern. (The gauges in the cylinder lines should be capable of a maximum pressure reading of about 200 psi to provide the needed sensitivity.) Increase the pressure gradually on a cylinder like Fig. 4-4. Note the pressure at which movement commences in each direction of travel. Repeat the experiment with a cylinder such as Fig. 4-5. Note the pressure readings. Remove the packing from the rod gland and repeat with pressure at the blind end. Determine the effect of rod packing on cylinder movement. Calculate the efficiency of the cylinders with known frictional losses.

4-9 Connect a cylinder as in Fig. 4-16 as a second-class lever. Use a second cylinder as a load by filling it with oil and capping its ports with a pressure gauge after bleeding the air out. Note the pressure developed as the first cylinder is actuated.

4-10 Repeat Experiment 4-9 as a third-class lever. Note the difference in pressure values in the slave cylinder.

5
Fluid-Power Plumbing

Connecting tubes or pipes used to contain and conduct hydraulic fluid or gases from one part of a fluid-power circuit to another are termed *fluid-power plumbing*. Valves, strainers, filters, and other devices incorporated in the complete hydraulic installation are often referred to as part of the plumbing. In this chapter you will learn about the different types of plumbing used in modern industrial hydraulic systems of all types.

5-1 PURPOSE OF FLUID-POWER PLUMBING

The connections between a reservoir or supply tank and a pump, and from the pump through suitable valves to a linear or rotary motor, with lines to return the fluid to the reservoir, can constitute a fluid-power circuit. The plumbing is used both to contain and to act as a conductor for the fluid to the desired points within a circuit. Atmospheric pressure, acting upon the surface of the fluid within a reservoir, will cause fluid to flow into a pump that has created a vacuum in its suction pipe. The discharge of the pump will be connected to pipes and valves that can control pressure, rate of flow, and direction of flow. Pipes or tubes are also connected to the valves and linear or rotary motors to conduct the fluid back to the reservoir.

Figure 5-1 shows how a reservoir is fitted with an air vent to allow atmospheric pressure to act on the surface of the fluid in the reservoir. The piping is arranged to contain and conduct the fluid to the pump and then to the end-use area, which in this installation is the hydraulically operated platen press.

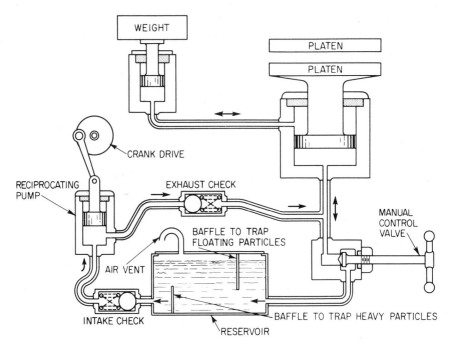

Fig. 5-1 Typical fluid-power plumbing.

5-2 REQUIREMENTS FOR FLUID-POWER PLUMBING

The conduit used to conduct fluid through a fluid-power system must be designed with several basic requirements in mind. Attention to the following points will ensure adequate performance:

1 The conductor must be of sufficient strength to contain the fluid at the desired working pressure. In addition, the conductor must have sufficient mechanical strength to contain the fluid under the highest possible shock pressures that may develop in the pipe during any portion of the machine cycle.

2 The conductor must have sufficient mechanical strength to support devices mounted in or on the pipe or tubing.

3 Mechanical strength of the conductor must be adequate to minimize or suppress shock waves generated by normal machine functions.

4 Terminal points (unions, flanges, etc.) must be provided at all junctions with parts or components that may require functional removal for maintenance or dismantling of parts of the machine.

5 Port seals or connections should be designed so that they are reusable with a minimum of time loss or maintenance work.

6 Pipe supports must be capable of damping shock waves and changes in direction and magnitude of fluid flow. This will ensure longer life for the system.

7 Lines must be kept clean and flushed regularly after the original installation. Any continuing source of contaminants in the lines (such as scale or oxidation) must be eliminated.

8 A smooth interior surface of the pipe or tube is usually of considerable importance in satisfactory operation of the system.

9 Conductors must be positioned so they will not be damaged by maintenance procedures, such as chip removal, machine loading or unloading, stockpiling, or moving of adjacent machine members.

10 The correct size of pipe to ensure the best flow conditions with the most economical use of material may require the use of a particular type of tube or pipe.

11 The physical position of the pump-suction line can be critical in some installations. (Note baffle ahead of suction line in Fig. 5-1).

12 A provision for air bleeds may be necessary at certain high points in the plumbing system.

13 Excessive use of elbows in place of tube bends and sharp reversals in flow direction will directly affect the pumping power needed to force the fluid through the conductors.

14 The length of pipe runs can be reduced by using manifolds and conductors that are in an integral part of the device being actuated by fluid.

15 Conductors must be so arranged that stresses are not induced in the terminating surfaces of the circuit components.

16 Taper threads on connecting devices may distort components. Straight-thread fittings and flanges are recommended to minimize potential damage.

5-3 MECHANICAL STRENGTH

Sufficient strength to contain the fluid under pressure is essential. However, mechanical strength may be one of the lesser factors in the overall requirements. Figure 5-2 shows the relative wall thicknesses of *standard, extra-strong,* and *double-extra-strong* pipe. Standard pipe is often referred to as *Schedule 40 pipe.* Schedule 40 pipe in sizes up to $1\frac{1}{2}$-in diameter (*nominal*) may provide an adequate safety factor in containing the fluid. Mechanical strength to support devices and positioning of the conductor may be of more importance than resistance to fluid pressure. These two

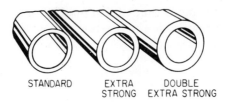

Fig. 5-2 Relative wall thickness of hydraulic piping.

conditions may dictate the use of *Schedule 80* (extra-heavy), or *Schedule 160* (not shown), double-extra-heavy. Resistance to flow created by the small passage through double-extra-heavy limits usage.

5-4 LINE MOUNTING OF COMPONENTS

Shutoff valves, gage cocks, and some control valves are "hung" on the pipe or tubing lines. Occasionally, filters, coolers, and other auxiliary items are positioned in a run of pipe. In some installations this not only is economically sound but also makes the components much easier to maintain. A rule of thumb, however, is to keep all these related components solidly mounted to something other than the pipeline and use the components as supports for the piping. Figure 5-3 shows how components are mounted in pipelines. When a heavy device must be mounted in the line, it may be necessary to use extra-strong pipe to provide sufficient mechanical strength. As can be seen from Fig. 5-3, it would be possible to use double-extra-strong pipe if greater mechanical and hydraulic strength were necessary to contain the fluid and support the component.

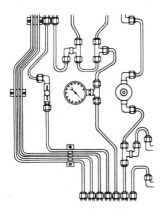

Fig. 5-3 Line mounting of hydraulic components.

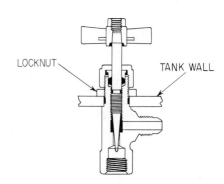

Fig. 5-4 Bulkhead mounting of needle valve.

Industrial Hydraulics

There are some installations where bulkhead mounting (through a wall or partition) of shutoff valves can be accomplished with slight additional first cost. Figure 5-4 shows how the operating handle is mounted on the opposite side of the wall from the piping connections in a bulkhead-mounted valve. However, most small valves lend themselves well to line mounting. When there are more than two connections to a valve, rigid mounting is often favored. Think of a line mount as a second choice, with rigid mounting of components as the preferred method.

5-5 SUPPRESSION OF SHOCK WAVES

Sufficient mechanical strength to contain and suppress shock waves is normally not very well understood by plumbers until they encounter a leaky hydraulic circuit where standard pipe, malleable fittings, and taper pipe threads have been used in jury-rig style to plumb-up a hydraulic system. This type of equipment has proved adequate for connecting laundry tubs, but it is not good enough for the simplest industrial hydraulic circuit. The high pressures at which hydraulic fluid is used imply that changes in direction of flow are certain to cause some transfer of energy to the lines in which the fluid is contained. These lines must be made of materials that are compatible with shock-resistant requirements. Shock waves can be minimized by careful layout of the piping. The suppression of shock may be a function of pipe strength.

5-6 TERMINAL POINTS

End or terminal points of piping cannot be overlooked in fluid-power circuits. The average journeyman starts a steam line at the boiler or a water line at a meter and provides unions only at points where expendable components such as water meters, heaters, or traps are located. In fluid-power systems, taper pipe threads and ground-joint union, have some limitations. Figure 5-5 shows a ground-joint union and other taper-pipe-thread fittings. The fittings may be manufactured from malleable iron, brass, steel, or any other metal that is compatible with the fluid to be handled. Experience in the fluid-power industry indicates that there are some newer preferred methods to terminate lines and provide union connections. The ground-joint union in Fig. 5-5 consists of a male threaded end with an internal thread to receive the pipe. The mating piece is provided with a collar that is threaded on the inside and which is fitted over a female end that is also treaded to receive a pipe. The true ground-joint design has a fitted socket between the two ends that have pipe fastened to them. They are matched together, and the collar is threaded over the mating area, providing the mechanical lock. The ground joint

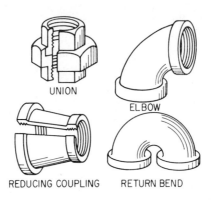

Fig. 5-5 Fittings having taper-pipe threads.

may be supplemented by a resilient gasket between the two mating surfaces.

Use of steel tubing in fluid-power systems is relatively new. Steel tubing has gained acceptance in most portions of the fluid-power industry. An important advantage of steel tubing is that every joint can be a union. Connections to the components can be with taper pipe threads or with straight threads and gaskets. Many other methods of connecting to machine ports have been devised for special applications.

A straight thread with locknut and gasket can do an excellent job in providing a rigid and tight connection between the tube and the terminal point in the component. The military forces established their tube fitting standard No. AND 10050 port connection several years ago, and this standard has been accepted throughout the military and by some commercial installations. The SAE has adopted a modification of this port seal design that can accommodate pipes in a much closer mechanical location without loss of efficiency. The SAE connectors incorporate a much narrower skirt on the retainer nut than the AND military design so that the AND fitting can be used in an SAE port, but the SAE fitting cannot be used in an AND port. The combination straight-thread port fittings and tubing has improved maintenance and extended service life of the plumbing system.

Figure 5-6 shows how the skirt of the connector in a straight thread acts as a cap over the O-ring-type seal. Other modifications of the fitting may contain a locknut with a back-up washer to hold the O-ring seal in position. The design in Fig. 5-6 serves as a swing joint that can be tightened in any direction at right angles to the straight-thread joint. The internal straight-thread boss is machined with a chamfer to contain the resilient seal.

In other modifications of this seal, a metal ring is used in place of the resilient ring for operation with higher temperatures and fluids other than regular hydraulic fluids.

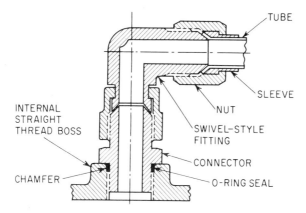

Fig. 5-6 Straight-thread type port connection and adapter.

5-7 EASE OF MAINTENANCE

Straight-thread adapters with tubing lines permit the use of easily reassembled components for simplicity of maintenance and long service life.

Taper pipe threads can be used in lines of more than ¼-in nominal pipe size if an auxiliary seal mechanism is fitted to the pipe. The lines, however, are not dependable without this mechanical seal. Further, the threading of the pipe removes metal from the pipe end—a critical point. Much care is necessary in the preparations of threaded pipe to ensure satisfactory service. Note the burrs in Fig. 5-7a resulting from pipe cutters used in normal threading. These burrs and all loose chips must be carefully and completely removed. The crushing action of the threads used

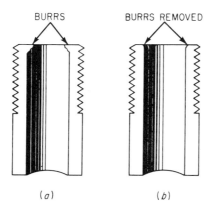

Fig. 5-7 Taper pipe thread used in hydraulic systems. (a) With burrs. (b) Burrs removed.

Fluid-Power Plumbing **149**

with a taper-thread connection directly affects reusability. Because of this and other reasons—such as the loss of strength caused by the metal removal and the scale found in the pipe—steel tubing has generally proved most satisfactory in conduit sizes from the smallest through 1¼-in outside diameter. Larger sizes of tube will require specialized handling equipment, and much care must be exercised to avoid stressing the metal. Manual handling of tube bending with portable benders is satisfactory in the smaller sizes. Above ½-in tube size it is usually desirable to have mechanical or power benders. Tube benders with established radii on the mandrels ensure the most satisfactory radius for optimum flow and minimum pressure loss.

Figure 5-8 shows how tubing should be installed so that it is easy to maintain and can be expected to provide maximum service life.

Welding has proved most satisfactory in pipe sizes from 1-in nominal diameter to the largest sizes of pipe used in hydraulic systems. This does not mean that certain jobs cannot use larger tubing with fittings or other jobs cannot use smaller weld sizes. However, this size range does indicate a trend that reflects satisfactory performances in a large number of installations over a long period. Some manufacturers have established this range of welded pipe as standard practice, and any deviations must be negotiated.

Welding flanges may be provided with pressure sealing rings made of resilient material. Usually they are provided with a basic squeeze just sufficient to prevent leakage at low pressure. Higher pressures merely push the seal into the corner of the groove, preventing loss of contained fluid. An O-ring-type seal is shown in the flange of Fig. 5-9a. The weld may be at the shoulder of the flange socket, as shown in Fig. 5-9b, or butt-welded, as shown in Fig. 5-9c. The face of the flange may also fit into a socket containing a metal gasket if temperature extremes are expected. Rings, such as shown in Fig. 5-9d, may be inserted inside the pipe to prevent weld spatter from lodging in the major flow path. Projections extend from the surface of the ring to position it between the pipe or

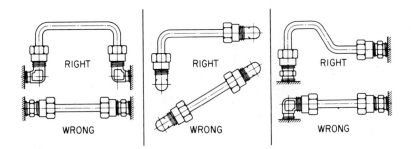

Fig. 5-8 Right and wrong methods of installing steel and other metallic tubing.

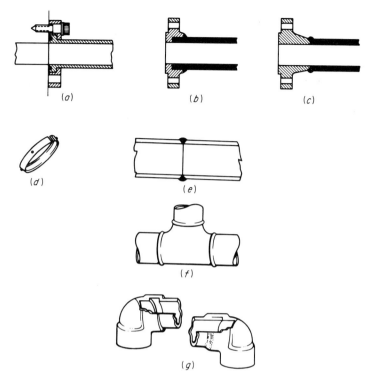

Fig. 5-9 Welding pipe joints.

fitting joints providing the most satisfactory positioning prior to welding. Beveled ends on a pipe, carefully positioned as shown in Fig. 5-9e, can provide the basic needs for a satisfactory welded joint. Proper welding provides a good, smooth interior surface with minimum resistance to flow. The exterior is also easy to keep clean because of minimum projections. Note in Fig. 5-9f the uniform diameter maintained by the use of butt-welding. The socket fittings shown in Fig. 5-9g add mechanical strength and further assist in preventing contamination of the inner surfaces.

When a weld is completed, it must be thoroughly cleaned. If acid cleaning is possible, it may be the most desirable. Otherwise, mechanical cleaning must be thorough. Much care is necessary with welding to minimize pipe distortion and contamination with weld spatter. When spatter occurs, small particles of metal may become temporarily attached to the pipe. Subsequently the particles come loose and jam the hydraulic components.

Welded pipe is probably the most rigid and when properly supported suppresses shock loadings well. Proper support may require a fastening bracket such as that shown in Fig. 5-10. Note that there is resilient

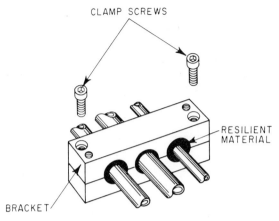

Fig. 5-10 Tubing and pipe support brackets for shock suppression and elimination of strain on the conductors.

material around the pipe and a suitable clamp to ensure adequate support without damaging the lines.

High-pressure pipe welding requires technically proficient personnel.

5-8 TAPER PIPE THREADS

When a joint using taper pipe threads is tightened, an *interference fit* is obtained. An interference fit occurs when the mating parts contact each other and interfere to the extent that some force must be used in assembling the joint. Two perfectly machined taper threads will carry the load over a wide bearing surface. During assembly a *crush fit* is said to be obtained because there is some crushing of the adjacent threads, caused by the metallic interference. If the tapers are not equal, a single point of contact results and the crush action is much less difficult. This also explains why it becomes more difficult to seal a joint each time it is taken apart and reassembled.

Imperfections in pipe threads also account for much of the tendency to *gall* or *tear* when the joint is disassembled. Pipe dope or joint compounds of various kinds will minimize the tendency to gall on disassembly and will prevent some of the abrasive action when assembling the joint. But pipe dope, when improperly applied, may also get inside the lines and harm pumps and control equipment. This has been such a problem that many manufacturers forbid the use of any compound when fabricating the piping for hydraulic circuits.

Use of straight threads or flanges that are welded to the pipe preclude the need for joint compounds. Thin plastic tapes have proved satisfactory as a combination sealer and gall preventive in taper-pipe-thread applications. In other installations a tinning of the male threads serves the same purpose. Electrolytically tinned pipe plugs of the taper variety will seal effectively with minimum distortion. They can also be easily removed for servicing. These electrolytically tinned plugs are often used in hydraulic equipment to seal construction holes. In some installations they serve as a convenient terminal point to connect gages to check pressure and flow values.

5-9 RAPID CHANGES IN FLUID-FLOW DIRECTION

Changes in the magnitude, pressure, or direction of flow can measurably affect the stability of the connecting lines. The resulting line shock is found in many fluid-power circuits. To say that all shock waves have been designed out of a hydraulic system is a relative statement. If *all* shock waves have been designed out of a machine, one can assume only that the machine is standing still. Realizing that it is impossible to eliminate all shock waves, the builder must support the piping so that it will be able to contain the expected changes in magnitude, pressure, or direction of flow. Relative changes in fluid movement will vary with the type of work being performed by the hydraulic system. There will be sudden changes in potential energy in the most simple accumulator circuit when a valve is suddenly opened and the stored fluid is directed into an empty pipeline. When the line fills with fluid, a sudden halting of the fluid velocity produces a shock wave closely akin to the hammer-on-anvil action in a forging shop.

It is possible to cushion the shock action of the fluid in the piping. But the piping must be constructed to withstand the most destructive action that can reasonably be expected in normal service. Designs anticipating every possible malfunction are uneconomical. Cushioning devices and other mechanisms to choke, throttle, or restrict flow are available, but even with these the fluid conduit must be heavy enough and constructed of suitable material to give long service life.

5-10 CLEANLINESS

Clean plumbing is necessary for dependable operation of industrial hydraulic systems. Some purchasers of fluid-power systems have minimized the use of welded and black pipe and demand that seamless tubing in the smaller sizes and mechanical tube in the larger sizes be incorporated in their systems. Welded joints in the larger sizes must be designed with provisions for mechanical cleaning or pickling with acid. This specification is understandable because fine performance levels are obtainable when welds are properly cleaned. Some systems using black pipe shed scale after 12 months of service.

5-11 RESISTANCE OF FLOW

Smooth interior surfaces in the piping system produce minimun turbulence. Some circuits have little or no problem with turbulence; other marginal operations must be carefully engineered to eliminate any trace of excess turbulence.

5-12 LOCATION OF CONNECTING LINES

Careless positioning of pipelines can lead to a continuing source of maintenance troubles. Lines located too close to chip-cleanout area can be subjected to continuous hammering until they fail physically and cause problems ranging from nuisance leakage to lost performance. An unprotected line may be gradually crushed by continual banging by stock handlers. This can result in slower movement of the machine member. The flow passage through the tube is reduced until such time as it prevents fluid flow and subsequent machine movement. This type of problem can be difficult to diagnose when the lines are not easily inspected.

Pump suction lines are particularly vulnerable. A restriction of the suction line, such as that shown in Fig. 5-11, prevents limited atmospheric

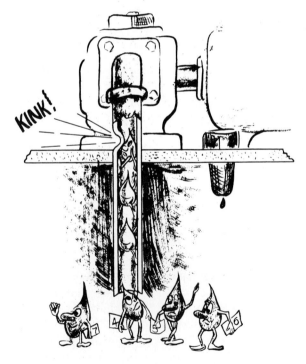

Fig. 5-11 Restricted pump suction line.

pressure from forcing fluid into the vacuum created by the pump mechanism. This can result in damage to the pump in addition to the unsatisfactory performance that is usually encountered with restricted suction pipes.

Careless installation of hoses can also cause some serious maintenance problems. Using the correct installation methods shown in Fig. 5-12 will ensure maximum service life from flexible conductors.

5-13 CHOICE OF FLUID CONDUCTOR SIZE

The choice of conductor sizes involves several factors. The trend in the fluid-power industry to standardize on certain pipe and tubing sizes dictates the following usage:

$1/16$-in nominal pipe size Used primarily for construction holes in fluid-power components, miniaturized systems, and bleed holes in plumbing fittings.

$1/8$-in nominal pipe size Used for construction holes, bleed ports, and some miniaturized installations.

$1/4$-in nominal pipe size Widely used for gage and test stations as well as auxiliary pilot connections. Used with $1/4$-in-OD tube for pilot lines, gauge lines, etc. Used less frequently with $5/16$-in-OD tube lines for similar functions. Used with $3/8$-in-OD tube for all the previous and for small power-transmission lines. Both tube and fittings are available in almost any industrial area. This is one of the most widely used sizes of hydraulic pilot lines.

$3/8$-in nominal pipe size Common in port connections in subplates for hydraulic valves handling flows less than 10 gpm. Use in threaded-connection valves is less common. Usually connects with $1/2$-in-OD tubes and fittings. Some manufacturers and users prefer the $3/8$-in pipe connection with $3/8$-in-OD tube because of the mechanical rigidity of the port connections.

$1/2$-in nominal pipe size Limited usage in fluid-power systems. Used with $1/2$-in-OD tube to provide mechanical strength at the taper-pipe-threaded port connection. Usage with $5/8$-in-OD tube provides flow to 15 gpm in usual hydraulic applications.

$3/4$-in nominal pipe size Probably the most widely used size in fluid-power systems. With $5/8$-in-OD tube it provides good mechanical port seal connections and capacities for 15 gpm or less. Used with $3/4$-in-OD tube it raises the nominal flow capacity to about 20 gpm.

Flows to 30 gpm are common, particularly in mobile hydraulic applications, using $7/8$-in-OD tubing and $3/4$-in nominal pipe size at the port

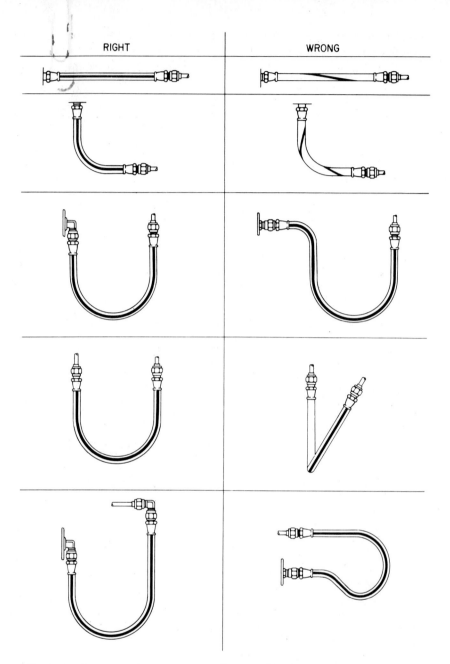

Fig. 5-12 Correct and incorrect installation of hydraulic hose.

156 *Industrial Hydraulics*

connections. The ⅞-in-OD tube size provides minimum pressure drop when used with ¾-in connectors.

- **1-in nominal pipe size** Widely used as pump suction lines with thin-wall pipe. Somewhat limited in pressure-line applications. Used with 1-in- or 1⅛-OD tube. Some installations combine 1-in low-pressure suction and return lines and ¾-in high-pressure lines in other portions of the circuit.

- **1¼-in nominal pipe size** High-pressure lines of this size form a team with 1½-in low-pressure suction and tank lines much in the same manner that 1-in and ¾-in lines are combined in hydraulic-circuit plumbing. The 1¼-in size is an industry standard for flows of 50 gpm or less in pressure lines.

- **1½-in nominal pipe size** Widely used as low-pressure suction lines and return-to-tank lines. Of secondary importance in large installations as pressure lines, generally welded when used with high pressures. Machine bending is necessary in these sizes.

- **2-in nominal pipe size** Tubing is rarely used; taper pipe threads are limited to low-pressure lines. Flanges and welding are the accepted method of connecting. The more critical flow characteristics in this size grouping and the following large sizes indicate that piping must be carefully sized. Piping drawings are desirable before systems are fabricated. A large number of valves in the 2-in pipe size are available for manifold or gasket mounting so that the pipe termination point is in a separate plate, thereby removing all stresses from the valving. Some installations are made on manifold plates that have drilled interconnecting lines with several components mounted to the plate. In some installations the valve is mounted on an independent plate and piped from this plate to the other components. These plates may have flange connections so that welded pipes can be removed for cleaning prior to being put into service. Angle check valves in the 2-in size are available with threaded connections for installation in tank lines to keep the lines full of fluid or to act as low-pressure safety valves when diverting return fluid through a cooling or filtering system.

5-14 PIPE CHOICE

Deviations from calculated fluid flows through large pipelines cannot be tolerated in hydraulic systems. All fittings and pressure drops must be carefully calculated. Support and pipe location can be critical. Grouting of pipelines in concrete is common practice and absorbs shock loadings in long lines. Flange connections and welding are used in both high- and low-pressure lines.

A decision whether to use pipe and pipe fittings or tube and tube fittings has many ramifications. Availability of maintenance materials may have much to do with the decision. If the hydraulic system is to be installed in an area where tubing is not common, it may be desirable to use pipe and flanged fittings even on small sizes to ensure adequate local maintenance supplies.

Some users of hydraulic equipment have saved so much maintenance time by using tube and tube fittings that they are willing to risk the inconvenience and lost time that may be caused by lack of immediate repair facilities. This decision is not one that can be broadened to an industry-wide recommendation. Each installation must be evaluated separately. Designers should consider that eventually tube and tube fittings will probably be universally used for smaller-size hydraulic lines and welded fabrication for larger-size fluid conductors.

The chart in Fig. 5-13 compares pipe and tubing sizes. The chart also includes a guide to usage in fluid-power systems. From this chart the designer can cross-reference the association between conductors and connectors. Recommended flow rates should be modified if a large number of turns are involved or exceptionally long lines are encountered. Suction lines for fire-resistant fluids are usually chosen one size larger than those indicated in this chart. At high altitudes, larger-size suction lines may be

Nominal pipe size	SAE straight thread	Tube identification	Size OD (tube)	Maximum flow, gpm suct.	Maximum flow, gpm pres.	Usage
1/16	5/16 – 24 3/8 – 24	-2 -3	1/8 3/16	— —	1/2 3/4	Lube – Air vent Lube – Air vent – Pilot
1/8	5/16 – 24 3/8 – 24 7/16 – 20 1/2 – 20	-2 -3 -4 -5	1/8 3/16 1/4 5/16	— — — —	1/2 3/4 1 2	Lube – Air vent – Instrument Air vent – Instrument – Pilot Pilot – Drain – Instrument Pilot – Drain – Instrument
1/4	7/16 – 20 1/2 – 20 9/16 – 18	-4 -5 -6	1/4 5/16 3/8	— — —	1 2 3	Pilot – Drain – Instrument Pilot – Drain – Instrument Pilot – Drain – Power
3/8	9/16 – 18 3/4 – 16	-6 -8	3/8 1/2	1 2	3 6	Pilot – Drain – Power – Suction Suction – Power
1/2	3/4 – 16 7/8 – 14	-8 -10	1/2 5/8	2 3	6 10	Suction – Power Suction – Power
3/4	7/8 – 14 1-1/16 – 12 1-3/16 – 12	-10 -12 -14	5/8 3/4 7/8	4 5 8	15 20 30	Suction – Power Suction – Power Suction – Power
1	1-3/16 – 12 1-5/16 – 12	-14 -16	7/8 1	10 12	35 40	Suction – Power Suction – Power
1-1/4	1-5/8 – 12	-20	1-1/4	18	50	Suction – Power
*1-1/2	1-7/8 – 12	-24	1-1/2	35	75	Suction – Power
2	2-1/2 – 12	-32	2	70	100	Suction – Power

* Flow in 1½ in. or larger pipe sizes should be calculated to ensure satisfactory service, particularly if wide temperature variations are expected in usual operation.

Fig. 5-13 Comparison chart—pipe vs. tube.

required because of the lower atmospheric pressure. Recommendations in this chart are based on 200 SSU oil viscosity at an average temperature of 100°F. The basic fluid rating is the same—200 SSU at 100°F. Thicker fluids may require larger lines to provide the most efficient system operation.

The physical position of suction lines and the need for air removal in hydraulic lines are well established. These are engineering design functions as well as plumbing problems. Using good fabrication practices with level lines or lines positioned to localize air traps in a desired position will minimize removal or control problems. Improperly positioned plumbing lines can cause malfunctioning of the pump and spongy operation in pressure lines because of the entrapped air.

Design problems relative to pressure drop and manifolded construction usually do not concern the plumber. But a note of warning is in order concerning excessive use of elbows and other fittings. *Changes in direction of flow are almost certain to be reflected in need for extra pumping energy.*

5-15 MANIFOLDS

Use of manifolded equipment has minimized many plumbing problems. As shown in Fig. 5-14, large numbers of connecting lines can be located in drilled or fabricated *manifold plates*. These manifold plates not only eliminate many fittings and potential leaks but also ensure a minimum number of errors in plumbing. The manifold may be drilled, cored and cast, or brazed in sandwich-type construction.

Manifold plates and other auxiliary equipment may be connected with high-pressure hose and quick-disconnect couplings. The quick-disconnect coupling in Fig. 5-15 is a device provided with suitable seals to prevent loss of fluid when the two coupling halves are being connected. The locking device is so constructed that a single mechanical movement will unlock the unit and allow the two halves to be parted.

There are several different types of quick-disconnects available. Some have integral valving to prevent loss of fluid when the coupling is severed. Other types have valves in only the female half and permit the other half to drain. The specific installation will dictate the type needed. Those used on earthmoving equipment and agricultural machines usually have protective covers as well as valves to prevent loss of fluid when the sections are parted. The protective covers prevent the entrance of contaminants. Many industrial applications require a check device in only one-half of the connector. Some test-rack facilities are constructed and used without any restriction to flow so that maximum flow values can be ensured in testing procedures. Some agricultural connectors are so designed that a heavy pull on the connector will cause the device to part without damaging the hydraulic lines in the event that they are forgotten when plows, etc., are mechanically disconnected or if they should be accidentally disconnected during the operating function.

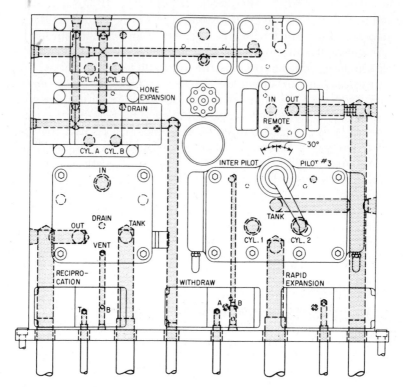

Fig. 5-14 Manifold mounted directly on machine with drilled interconnecting passages.

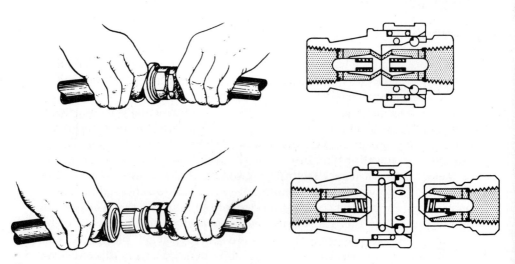

Fig. 5-15 Quick-disconnect coupling.

160 *Industrial Hydraulics*

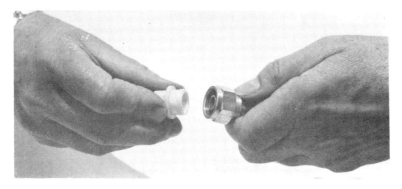

Fig. 5-16 Plastic caps or plugs protect fluid-power lines during service activities.

5-16 MECHANICAL PROTECTION OF CONNECTOR THREADS

Metal or plastic plugs and caps can be used effectively to prevent loss of fluid and protect threads when conductors are disconnected for service functions. Plastic closures [Fig. 5-16] provide a clean, lint-free plug or cap function. The relatively soft plastic will retain the fluid with little danger of damage to the threads. Metal plugs or caps are usually used if the line is to be pressurized during the service activity. Plugs and caps are available for most tube and pipe-thread configurations. Spare conductors and components should be fitted with closures of this type during a storage period to minimize contamination.

Brightly colored plastic closures will serve the primary function of protecting the disconnected components and also alert the maintenance personnel to the disconnected conductors. Most of the units are reusable; as such they are an important part of the service tools associated with fluid-conductor maintenance procedures.

DESIGN CONSIDERATIONS FOR FLUID-POWER PLUMBING

Most vendor catalogs include design information that applies specifically to their proprietary devices. In addition to furnishing the proprietary information, most catalogs also include charts, graphs, formulas, and fabrication data useful to both designers and maintenance personnel.

Several vendors make available handbooks with many charts listing pressure drops, flow considerations, and other data for various fluids under differing operational conditions plus easy-to-use instructions for fabricating pipe, tube, and hose assemblies. Standardized metric piping systems are being developed. Current information can be obtained from the National Fluid Power Association, 3333 N. Mayfair Road, Milwaukee, WI 53222. This association holds the secretariat for ISO (International

Standards Organization) Technical Committee 131. Subcommittee 4 is developing the standard for SI (International System of Units) fluid-power plumbing. Copies of available standards and procedures can be procured from NFPA as they become officially accepted and published.

SUMMARY

Steel tubing and straight-thread fittings have proved most satisfactory for industrial hydraulic plumbing in sizes up to and including ¾-in nominal pipe. Power bending and flaring equipment must be available to maintain larger sizes of tubing.

Larger lines are fabricated by bending, welding, and using flanged terminal points.

Threaded-connection valves, either taper- or straight-thread type, are generally limited to gage cocks, gate, globe, needle, and certain types of pressure-control valves. Line mounting of valves should be limited to small sizes.

Pressure-, flow-, and directional-control valves are available with flange connections and with all terminal points in a removable plate for connecting the piping. Manifolds may contain the major interconnecting lines, with the exception of the lines from the pump, the return lines to the reservoir, and the lines to the cylinder or motors. All remaining lines through sequence valves, check valves, reducing valves, etc., are contained within the manifold.

Maintenance of plumbing accessories, such as swing joints, packed slide joints, and swivels, must be done on a regular schedule dictated by the machine manufacturer or by past history or experience with the products involved.

A hydraulic system does not have to leak to be efficient. Good housekeeping, good maintenance, and proper choice of products are necessary for plumbing that will contain the working fluid and permit maximum efficiency of operation.

A review of possible causes after each malfunction, with suitable revisions to avoid repetition of the occurrence, can ensure trouble-free operation.

REVIEW QUESTIONS

5-1 For what purpose is the air vent provided in the reservoir in Fig. 5-1?

5-2 What is meant by the term *bulkhead mounting*?

5-3 Why is a taper-pipe-thread fitting difficult to reuse?

5-4 What is the purpose of the check valves built into quick-disconnect couplings?

5-5 What is the reason for using straight threads with tube fittings?
5-6 Can all shock be eliminated from hydraulic systems?
5-7 What type of malfunction can be anticipated with incorrectly positioned piping?
5-8 What is the reason for minimizing the number of bends in a pipeline?
5-9 What is the reason for interconnecting lines within a manifold block?
5-10 What is the purpose of integral valving in a quick-disconnect coupling?
5-11 How does a ground-joint union differ from a flange union?
5-12 What is a return bend?
5-13 Why is pipe dope prohibited in some installations?
5-14 What is the purpose of pipe dope?
5-15 How are burrs developed when fitting taper-threaded pipe? What malfunction may be expected if the burr is not removed?
5-16 What is the smallest size hydraulic piping usually joined by welding?
5-17 What is the purpose of the ring in Fig. 5-9*d*?
5-18 What process is generally used to clean welding flash and debris from pipe prior to use of the pipe in hydraulic systems?
5-19 Why is it necessary to calculate flow in larger-size conductors?
5-20 What effect does change of direction of fluid flow have on circuit efficiency?

LABORATORY EXPERIMENTS

5-1 Secure a ¾-in steel pipe nipple with taper-pipe thread in a vise. To the nipple fasten a malleable elbow, hand-tight. Note the position of the right-angle outlet of the elbow. Now tighten the elbow as much as possible with a 10-in pipe wrench. Note the position of the outlet of the elbow. Use an 18-in pipe wrench to tighten the elbow as much as possible. Note the position of the outlet of the elbow. Remove the elbow and repeat the experiment, noting the different position of the elbow at each retightening.

5-2 Attach a ¼-in elbow to the outlet of a pump with a tee in the line and a low-pressure gauge. Allow the discharge of the elbow to pass into an open tank. Pass approximately 5 gpm through the line and note the pressure drop. Add another 3-in long nipple and another elbow. Note the change in pressure required to pass the same flow of fluid. Repeat with additional nipples and elbows and note the increased resistance to flow.

5-3 Insert a quick-disconnect coupling in a test line and note the pressure necessary to pass an established flow. Disconnect the coupling and remove the check valves in one half. Retest and note the difference in flow resistance. Repeat with the check valves out of both halves. Note the resistance to flow.

5-4 Cut off a length of ¼-in black pipe with a commerical pipe cutter. Thread both ends and connect to a pump having at least 5 gpm delivery. Note the pressure required to pass hydraulic fluid through the pipe. Remove and ream the pipe ends to eliminate burrs and metal forced in during the cutting operation. Note any difference in pressure required to force the same flow through the pipe.

5-5 Partially crush the pipe in Experiment 5-4 to duplicate what might occur when a pipe is damaged on a machine. Note the pressure required to pass fluid through the pipe with varying degrees of restriction.

5-6 Measure the inside diameter of a ¾-in nominal pipe size flexible hose fitting. Compare this to the ID of standard-weight pipe and 0.072-in wall thickness steel tubing with a ⅞-in OD. Pass 20 gpm through a 36-in length of each with a tee at the upstream end. Note the difference in pressure drop on a sensitive gauge.

5-7 Fasten a ¾-in nominal pipe size taper-threaded malleable tee in a machinist's vise with the side outlet facing up. Insert an extra-heavy pipe nipple hand-tight. Measure the shoulder of the fitting with a micrometer caliper. Tighten one turn with an 18-in pipe wrench. Remeasure. Continue tightening the pipe and measure at each turn. Note the distortion of the fitting because of the wedging action.

5-8 Repeat Experiment 5-7 using plastic sealing tape as recommended by the tape manufacturer. Note the additional care needed to avoid component damage. (Less tightening is required with a suitable plastic tape seal to prevent leakage.) Permit a small portion of the tape to extend over the end of the pipe nipple. Note flaking at interior. Remember that this flaking could cause component malfunction as it comes loose in the fluid stream. Make certain the tape is properly positioned when making finished joints so that it will not penetrate into the flow path.

5-9 Connect a suitably rated hydraulic hose of at least 24-in length to a hand pump or pressurized oil source. Plug one end securely. Without pressure on the line, bend the hose 180°. Gradually apply pressure. Note the tendency for the hose to straighten. Caution: *Do not exceed the hose pressure rating or suddenly release the hose!*

5-10 Install a low-pressure gate valve in the suction line of a gear pump. Operate the pump with the valve full-open. Slowly close the valve to approximately half-open. Note the cavitation and the noise level in the pump. Do not run more than a few seconds with the suction line restricted. After listening to the noise, remove the valve from the suction line. (The pump may be destroyed if the valve is accidentally closed while in operation.)

6
Pressure Accumulators

Pressure accumulators are used in many fluid-power circuits to (1) store fluid under pressure, (2) cushion shock waves in the circuit piping, or (3) perform a combination of both functions. Storage of energy in the form of pressurized fluid and absorption of pressure surges are extremely important in most circuits. Proper performance of both these functions ensures efficient and reliable operation of industrial hydraulic equipment.

6-1 TYPES OF ACCUMULATORS

Three types of accumulators are used in modern fluid-power circuits: (1) *weighted,* (2) *spring-loaded,* and (3) *pneumatic* or *gas-charged* accumulators. Each type has characteristics which make it suitable for certain applications.

6-2 WEIGHTED ACCUMULATORS

This [Fig. 6-1a] is the oldest type of accumulator used. The accumulator consists of a ram X sliding in a packed cylinder C. Fluid under pressure, usually from the pump, enters the bottom of the cylinder after passing through check valve V. The pressurized fluid forces the ram upward. To prevent the ram from being forced out of the cylinder, a weight W is mounted on top of the ram. This weight is chosen to exert a predetermined pressure on the fluid entering the accumulator cylinder. *The pressure exerted by the weight and ram, in pounds per square inch, can be found by dividing the sum of the weights of the ram and weight, in pounds, by the area of*

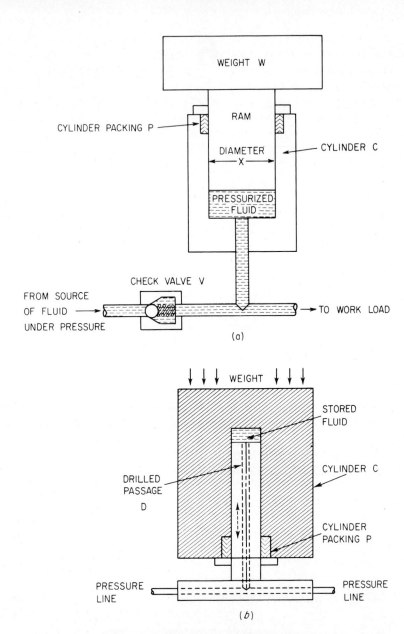

Fig. 6-1 (*a*) Storage of hydraulic fluid under pressure in a weighted accumulator. (*b*) Weighted accumulator with ram inverted.

the ram, in square inches, contacting the fluid. Thus, with a ram weighing 100 lb, carrying a 500-lb weight, and having a fluid contact area of 3 in^2, the pressure on the fluid is $(100 + 500) \div (3) = 200$ psi. This calculation neglects ram friction.

Friction does affect ram movement. The cylinder packing P tends to keep the weight from developing the maximum pressure on the fluid. The friction force exerted by the packing tends to slow the upward movement of the ram while the cylinder is being filled with fluid from the pressure source. During discharge from the accumulator, the friction force tends to slow the downward movement of the ram. Where the friction force restricting the ram movement is known, *subtract the friction force (pounds) from the gross weight of the ram (pounds) and divide the result by the ram area (square inches) to determine the pressure produced by the accumulator.*

The force exerted on the fluid by the ram is relatively constant. Regardless of the position of the ram, the fluid pressure in the cylinder is constant. Inertia of the ram assembly can, however, cause the fluid pressure to exceed that resulting from the gross weight and the ram area. Thus, if the ram is allowed to drop rapidly and is suddenly stopped by the fluid, the resulting pressure surge can be several times the normal accumulator pressure.

In some weighted accumulators the ram is fixed to a base [Fig. 6-1b] and the cylinder moves. Fluid passes upward or downward through a drilled passage D in the ram. When the fluid reaches the top of this passage, it enters the cylinder chamber, where it is stored for future use. Note that the cylinder in Fig. 6-1b has extremely thick walls. These heavy walls provide the weight needed to produce the desired pressure on the fluid. The thick walls also help contain the fluid in the cylinder, preventing rupture during periods of excessive pressure.

Note in Fig. 6-1a that a check valve is located between the source of fluid under pressure and the accumulator. This check valve prevents loss of fluid if the accumulator is expected to remain pressurized for an extended period. A *limit switch* may be provided at the extreme ends of the ram movement. This switch may stop the fluid pump or divert the fluid to another portion of the circuit when the accumulator is fully charged to its designed pressure. When the accumulator is discharged to a predetermined pressure, the switch mechanism can start the pump or divert the fluid under pressure back through the check valve into the accumulator so that the accumulator can be refilled. If electrical devices cannot be used in the circuit, it is possible to use pilot valves to control the loading and unloading of the accumulator. The fact that a portion of the accumulator can move permits a mechanical signal source to control the charging and relaxation of the charging source when the accumulator is full.

Many fail-safe devices are available for use with weighted accumulators. As the accumulator approaches the extreme loaded position, it normally sends a signal through the mechanical connection to stop the charging action. If this signal should fail, a simple positive bypass porting arrangement can become operative as movement continues. This porting arrangement diverts the fluid back to the reservoir at the rated pressure established by the accumulator weight and area ratio. This action consumes horsepower, but it prevents damage to the components. Several signal methods are available to inform the operator that the accumulator has exceeded the normal charging stroke. The signal may be a flow switch in the bypass line, another limit switch, or possibly the noise created by the flow through the bypass-valve mechanism will be sufficiently loud to alert the operator.

Some weighted accumulators are provided with integral stops to eliminate problems of overtravel. Since the usual weighted accumulator is a large piece of machinery, the energy involved precludes certain physical stop designs. It is mandatory to use bypass-type safety mechanisms or other units that are not built around forces that physically restrain the accumulator parts.

Weighted accumulators are often used on large central-power systems. Operations requiring a long holding cycle, such as molding of rubber and plastics, can take full advantage of the energy supplied by the dead weight on the column of fluid. Accumulators of the weighted type are also used for holding a steady pressure on rolling devices such as the machinery used to roll metal foils and paperboard products. Circuit design may require considerable care in avoiding possible line shocks from the large potential source of energy stored in the accumulator.

6-3 SPRING-LOADED ACCUMULATORS

Figure 6-2 shows an accumulator using the energy stored in a spring to create a constant force on the contained fluid in the adjacent ram assembly. Use of a mechanical spring or springs produces somewhat different characteristics than does the weight of a weighted-type accumulator.

The load characteristics of a spring are such that the energy storage depends on the force required to compress the spring. The free (uncompressed) length of the spring represents zero energy storage. As the spring is compressed to the maximum installed length, the minimum pressure value of the fluid in the ram assembly is established. As fluid under pressure enters the ram cylinder, causing the spring to compress, the pressure on the fluid will rise because of the increased loading required to compress the spring.

Spring-loaded accumulators may have physical stops for the ram so that pressures in excess of the load-carrying capacity of the spring will not be imposed upon the unit. The accumulator may be used in the line merely to add flow capacity at a pressure less than the maximum pressure, when high flows may be needed. An intermediate pressure is selected for accumulator discharge because the circuit may be so designed that a large volume of fluid at low pressure is needed during the start of the work cycle of the machine served. The work stroke may be at a high pressure and a very low speed. If desirable, the accumulator may store energy at a pressure less than the maximum working pressure but sufficiently higher

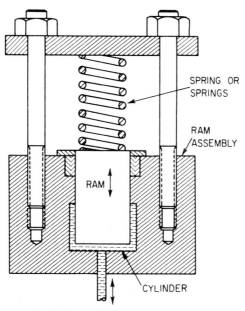

Fig. 6-2 Spring-loaded accumulator for storage of hydraulic fluid under pressure.

than the pressure required to move the machine rapidly. Then enough energy will be available to exceed breakaway friction during starting to get the machine mass in motion.

Spring-loaded accumulators can be used interchangeably with gas-charged accumulators in some installations. Physical size and weight of a spring-loaded accumulator may prevent its use in some applications. An advantage of the spring-loaded accumulator over the gas-bottle and pneumatic varieties is the elimination of possible loss of potential through gas seepage or loss of the entire gas charge. The spring, being a physical mass that resists movement in any direction but the guided one, has certain safety advantages. Transient shock waves can be absorbed by small spring-loaded accumulators. These accumulators are tee-connected to long pipelines and are strategically placed so that the waves can be most effectively trapped. If it is difficult to service the accumulators or if they must be encapsulated, the spring-loaded type provides the most dependable unit in many circuits.

An orificed check valve may permit fluid entry to a spring-loaded accumulator with a minimum of resistance. The orifice restricts movement of the fluid back into the main line when the main-line pressure has fallen to a value less than that of the fluid stored in the accumulator.

A spring-loaded, piston-type accumulator may become an integral part

Pressure Accumulators **169**

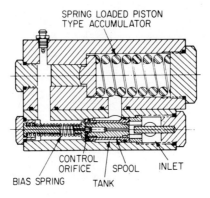

Fig. 6-3 Spring-loaded, piston-type accumulator used as part of shock-suppressor valve.

of a valve design for shock suppression. The valve shown in Fig. 6-3 contains a normally closed, two-way valve assembly and the spring-loaded accumulator. The valve consists of a balanced spool which is held normally closed by a light bias spring. Unrestricted pressure is ported to one end of the valve spool. Pressure is also connected to the opposite end

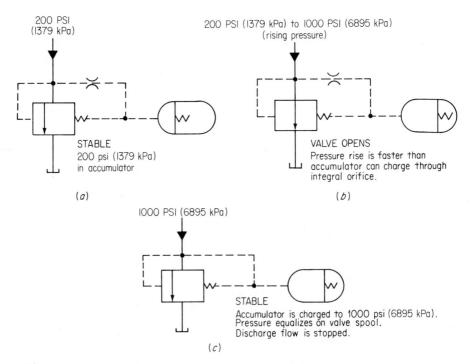

Fig. 6-4 Operation of shock suppressor with integral spring-loaded accumulator.

170 Industrial Hydraulics

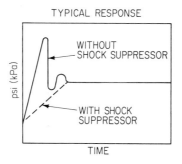

Fig. 6-5 Typical response of shock-suppression unit.

but must first pass through a control orifice. A small, spring-loaded, piston-type accumulator is also connected to this same spool end.

The size of the control orifice determines the flow rate allowed to pass into the accumulator for purposes of charging it. As long as the system pressure rise is gradual, the accumulator is able to absorb pressure equal to the system pressure, and the valve remains closed [Fig. 6-4a]. However, if the system pressure rises at a rate faster than the accumulator can charge, a pressure difference is sensed at the two spool ends, and the valve spool will move to the open position [Fig. 6-4b]. The displacement of the valve spool is free to go into the accumulator, allowing extremely fast response for valve action. The valve will stay open until the rate of pressure rise corresponds to the accumulator rate of rise; then it will close again [Fig. 6-4c]. Varying the orifice size varies the time constant, so the shock suppressor can be tuned to any given set of circumstances.

Figure 6-5 illustrates the relative response of a shock suppressor such as that illustrated in Fig. 6-3.

6-4 SPHERICAL-DIAPHRAGM-TYPE ACCUMULATORS

Figure 6-6a shows a spherical-diaphragm-type accumulator. This is one of several pneumatic- or gas-type accumulators built today. The basic design is such that the capacity is somewhat limited (usually less than 1 gal). The spherical-diaphragm-type [Fig. 6-6a] and the bag-type accumulator [Fig. 6-6b] are similar in operation.

In both the diaphragm- and bag-type accumulators there are two external connections—one for gas and one for fluid. The diaphragm or bag separates the fluid from the gas. In each type, incoming pressurized fluid compresses the gas [Fig. 6-6b] in the accumulator. When the pressure in the external circuit falls below that in the accumulator, the gas expands and forces fluid into the circuit. Since neither of these accumulators has a piston, ram, or spring, the inertia of the device is extremely small. The gas is always ready to expand to provide the desired driving force on the fluid.

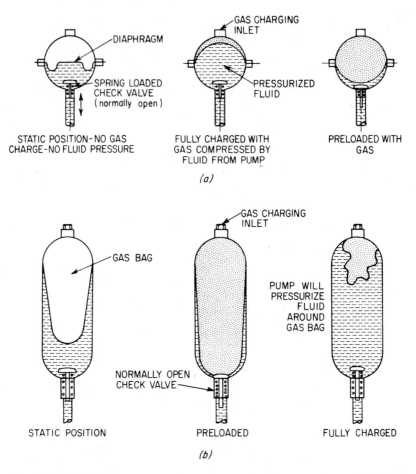

Fig. 6-6 (a) Diaphragm-type accumulator using inert gas for preload. (b) Bag-type accumulator using inert gas for preload.

By connecting one or more gas tanks to these accumulators, the effective gas volume can be increased. When additional gas tanks are used, a normally open check valve may be fitted at the gas inlet. Adding gas tanks to an accumulator changes the pressure gradients as the fluid is withdrawn. This is desirable in certain installations.

Note that both the accumulators in Fig. 6-6 have a spring-loaded, normally open check valve at the fluid-inlet opening. This check valve protects the separating element (diaphragm or bag). Without the check valve in this vital area, there is a possibility that the diaphragm or bag might be forced through the inlet connection by the gas. The check valve must be carefully maintained so that it closes fully when gas is admitted to the accumulator. This ensures that the separating element will not be forced out the inlet connection or damaged in other ways during preloading.

6-5 CYLINDRICAL GAS-CHARGED ACCUMULATORS

Figure 6-7 shows several types of gas-charged accumulators manufactured in cylindrical form. The unit in Fig. 6-7a has a floating check valve. The check poppet rides on the column of fluid. This accumulator must be installed in a vertical position. No attempt is made to separate the charging gas and the fluid. When the fluid level reaches a certain predetermined minimum height, the check poppet floats into position in the mating seat S. This prevents loss of the charging gas into the system. The pipe connection to the inlet does not need the protective devices associated with bag-type accumulators. Floating check-type accumulators are used in large systems where the accumulator capacity for fluid and gas may be in excess of 10 gal each. The primary limitation of this accumulator is the manner in which it is mounted, as it must be in a position to permit the check poppet freedom to float.

Figure 6-7b shows a cylinder with a ball (or other free-moving barrier device) that acts as a divider between the charging gas and the hydraulic fluid. Leakage is minimized because the pressures on the gas and fluid are equal. Friction between the ball and wall can be relatively high, particularly when motion is first begun. This *breakaway friction* is detrimental to the average circuit.

The cylindrical accumulator in Fig. 6-7c has a pressurized gas chamber that envelopes the fluid chamber. As in Fig. 6-7b, the gas and fluid pressures are equal. This minimizes the accumulator wall loading and the stresses on the floating barrier piston.

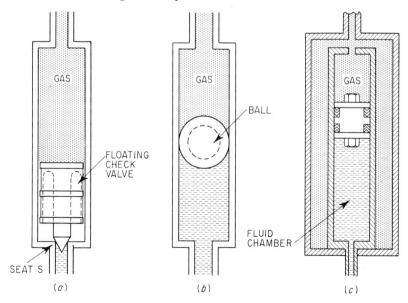

Fig. 6-7 Cylindrical accumulators. (*a*) Floating-check-valve type. (*b*) Resilient-ball type. (*c*) Floating-piston type.

Pressure Accumulators **173**

6-6 ACCUMULATOR CIRCUITS

In Chap. 10, Pressure-Control Valves, basic types of pressure-sensitive valves used in accumulator circuits are described. If an electric signal can be conveniently used, the accumulator circuit may rely on rising pressure to actuate a pressure switch. The signal transmitted from the pressure switch actuates a device that diverts the pressurized charging fluid from the accumulator. A falling pressure actuates another part of the pressure switch, providing a signal to the electrically actuated device that starts the charging operation.

Another basic type of control depends on a pressure relay system. Some designs provide a signal directly from the accumulator to a *hydraulic additive piston*. The term *additive piston* indicates that a force is added to the unloading portion of a two-way valve. This added force assists the valve function that provides the desired direction of flow back to a lesser pressure area or away from the accumulator input line. The additive piston may upset a ball or move another piston of a different area so that the pump can be diverted by relaxing the control-pressure additive in a piloted-type relief valve. The difference in the diameters of the hydraulic pistons determines the differential pressure between unloading and recharging the circuit as the accumulator pressure decreases.

The circuit in Fig. 6-8a uses an electric relay system. Note that the relief valve is shown in segments in the diagram. The main spool is in the primary block, while the control chamber is shown as another block indicating a hydraulic additive controlling the spool in conjunction with the main-spool spring. Orifice O, feeding fluid to the control chamber, is shown as it is actually located in the valve, between the input to the main spool and the spring chamber. The pilot valve is also shown as a separate entity. The connection to the solenoid-operated two-way valve is from the control chamber of the relief valve. As the solenoid is deenergized, the additive fluid is allowed to escape to the reservoir so that maximum pressure buildup will be controlled by the main-spool spring of the piloted relief valve. As the solenoid is energized, the bypass of pilot fluid to the reservoir is blocked. The relief valve then functions in the conventional manner, and the flow is diverted to the accumulator through the check valve. In the event that the contact in the pressure switch freezes, or a relay in the electric circuit fails, the pressure is limited to a safe maximum value by the piloted relief valve connected in parallel with the solenoid two-way valve. The pilot section of the relief valve is usually adjusted to a few hundred pounds per square inch higher than the pressure switch so that it will be normally inoperative but immediately available in case of emergency. The setting is usually about 20 percent higher than the accumulator relay-valve adjustment.

The pressure-switch control is accurate but prone to wear problems in the contacts and mechanical parts. If electricity cannot be used for control because of environmental reasons (such as possible fire or explosion), then

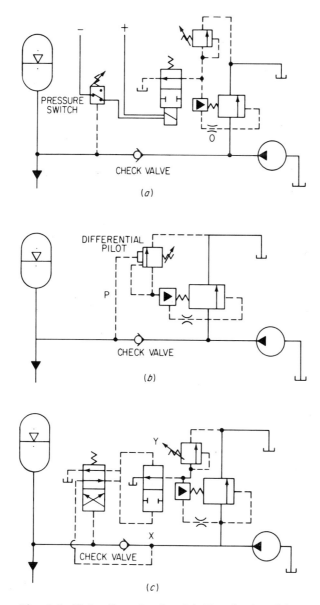

Fig. 6-8 Unloading circuits. (a) Signal relayed by pressure switch. (b) Differential piston. (c) Directional pressure-relay valve.

an all-hydraulic system performs the control function in a satisfactory manner. Pressure switches may be sensitive to line shocks that can actuate the switch prior to the desired circuit cycle time.

Figure 6-8b shows an accumulator charging circuit, using a differential pilot-piston mechanism. As in Fig. 6-8a, the relief valve is shown as being

Pressure Accumulators **175**

made up of components to clarify the circuit functions. The components of Fig. 6-8b are commercially available in envelopes, containing the main relief spool, differential pilot, and, in some models, the check valve. The pilot line P from the accumulator [Fig. 6-8b] is directed to a pilot piston that has slightly more area than the cone or ball controlling the flow from the relief-valve control chamber. The area of the additive pilot piston is often 10 to 15 percent greater than the exposed face of the spring-loaded cone used to close the passage from the control chamber to the reservoir or lower pressure area in the control section of the relief valve. A commercially available model of this valve will be discussed in Chap. 10. Hydraulic additive fluid in the control chamber of the relief valve is effective against one end of the additive pilot piston from the accumulator. The pressure in the accumulator circuit is effective against the opposite end. As the accumulator pressure rises to a value where the flow starts through the pilot-relief valve, minimizing the hydraulic additive to the main spool, the pressure in the accumulator remains static because of the check valve. With static pressure on one end of the piston pilot from the accumulator and a dropping pressure on the other end, the pilot piston becomes unbalanced and snaps the pilot relief open. The pilot-relief valve cannot close until the accumulator pressure drops a sufficient amount to permit the pilot piston to relax and restrict the passage to the tank as the cone or ball approaches the seat. As this pressure rises in the control chamber, it pushes the accumulator signal pilot piston back with a snap action. This type of control is satisfactory on relatively fast machines where a normal pressure surge in the machine assists the snap action.

The most difficult circuit to control is the type using the accumulator for standby service or merely to hold pressure on rolls, etc. This type of circuit can be charged very slowly. The circuit in Fig. 6-8a, if compatible with the other machine components, acts in a satisfactory manner. However, the circuit in Fig. 6-8b can be troubled with *sneak* operation. Sneak operation is a very slow pressure buildup or relaxation. This permits the control piston either to leak enough to make the relief valve operate as a partial unloader or to act merely as a maximum-pressure control and not as an unloading valve. Such a condition causes the unit to overheat because of incomplete unloading of the pump delivery. The snap action can be lost because the signals are extremely slow.

The circuit in Fig. 6-8c operates much like the circuit in Fig. 6-8a with the electrical components replaced by a hydraulic-pilot system. In the circuit of Fig. 6-8a, the pressure switch provides a positive signal to the solenoid valve. In the circuit of Fig. 6-8c, the pressure-sensitive, four-way pilot presents a positive signal to the pilot-operated, two-way valve, so that the signal is established prior to unloading the pump delivery back to the reservoir. The relief valve in this circuit has the piloted two-way valve in parallel with the pilot-relief valve. The pilot-relief valve must be set higher than the adjustment on the pressure-sensitive, four-way pilot valve. The pilot-relief valve here works only in an emergency. Note that the pilot pressure for the pressure-sensitive, four-way pilot can be taken ahead of the accumulator check valve so that clearance flow in the four-way pilot valve will not drain the accumulator. Further, there is always a source of pilot pressure because the check valve will provide a resistance when the load to the accumulator is light and the main-spool spring on the relief valve will always maintain a certain minimum pressure within the pump circuit. The pressure-sensitive, four-way pilot valve must be shifted over neutral before the fluid-piloted, two-way valve can shift to cause the pump either to load or to unload. The dead point of crossover is minimized because an established signal is provided by the four-way pilot valve.

Control systems, such as that illustrated in Fig. 6-8c, are usually made up of specialized components or designed as part of a packaged transmission system. The components are not of the cataloged-package variation such as those of Fig. 6-8b. To provide the utmost in dependability, the fluid-piloted, two-way valve in Fig. 6-8c may be equipped with a two-positioned detent which will provide a predetermined snap action and thus ensure freedom from *sneak* operation. When pressure is directed to end X, it will cause the relief-valve pilot fluid to establish the desired additive to the main spool so that pressure can be directed through the check valve to charge the accumulator. As the pressure increases at the pilot to the pressure-sensitive, four-way valve, there will be a proportionate movement of the directional spool. With the detent on the two-way spool, it would be necessary for the pressure-sensitive Y pilot spool to shift far enough to build up pressure on the end so that the spool can urge the detent out of the spool notch. The spool would have to clear the detent completely before it could vent the pilot additive fluid to the relief valve. At this position the piloted two-way spool would be over the high spot between the two detent notches, and the detent mechanism would tend mechanically to pull the spool into the open position.

6-7 SHOCK-WAVE ABSORPTION

An accumulator, connected directly to a fluid line and intended only to absorb shock waves, can normally be charged to some value near the expected working pressure. This means that the magnitude of the shock will be near or above the normal maximum working pressure. The shock-absorbing action of the accumulator should be such that it will not absorb needed energy during the work cycle. A high charge in the gas chamber will help prevent loss of energy at critical periods in the work cycle. The waves to be absorbed should be in the highest pressure range in most circuits.

A check valve with an integral orifice in the poppet may be used to permit a shock wave to pass virtually unrestricted into the accumulator. The energy trapped in the accumulator is returned to the circuit through the orifice in the poppet at a controlled rate. This, in effect, is similar to the shock absorber used on mobile equipment.

Shock-wave absorption and shock suppression can be considered as two different approaches to a common problem. The shock-wave-absorption principle is limited by the capabilities of the accumulator and the characteristics of the fluid-power plumbing system. This is also true of the shock-suppression devices, except that the diversion function of the suppressor of Fig. 6-3 can improve the response capabilities considerably. If the shock is of an unusually short time duration, the suppressor can be designed with specialized characteristics that will divert sufficient fluid to *clip* the shock peaks.

6-8 FLUID-STORAGE ACCUMULATORS

Figure 6-9 shows the use of an accumulator as a volume-storage unit. In this application the fluid from the fixed-delivery pump passes through the first check valve as long as the work-load pressures are less than the pressure established as a maximum by the pressure-relief valve. When the work-load pressures rise because of normal resistance in the machine cycle, the fluid can pass through the sequence valve into the accumulator. The sequence valve is adjusted so that fluid will pass through the valve at approximately 20 percent lower pressure than the setting of the maximum-pressure relief valve. When the accumulator is fully charged, the pressure will rise and fluid will then pass through the relief valve. The circuit could be arranged in a manner similar to those in Fig. 6-8 to unload the pump during the holding portion of the cycle.

When the work-load characteristics change because a new portion of the cycle calls for a large amount of fluid at a low pressure, this needed fluid can come from the accumulator and join with that being supplied by the

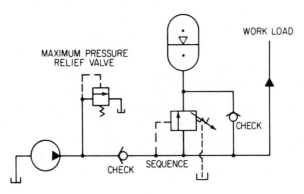

Fig. 6-9 Sequence circuit for accumulator volume-storage use.

pump. The fluid will come from the accumulator through the check valve, inserted between the work line and the accumulator, to bypass the sequence valve. When the accumulator is sufficiently discharged, or the circuit pressure starts to rise, the check valve from the accumulator will close. The pump will first satisfy the circuit needs until the circuit pressure rises to the sequence-valve adjustment value. At this setting the pump volume will be directed to the accumulator at a fixed pressure, so that clamping devices in the circuit are supplied with pressurized fluid and thus held in the correct position. As the accumulator is completely filled, or the pressure approaches the maximum relief setting, the circuit is ready for the new cycle.

By using a variable-delivery pump, the need for an auxiliary relief or safety valve can also be eliminated. This type of circuit ensures maximum performance of the pump with a minimum of energy input. The pump is always available to the work load. When the work load is not calling for volume, the pump delivery can be economically used to store the fluid in the accumulator. The rate of delivery from the accumulator back into the circuit can be easily controlled. All valving is automatic in operation.

6-9 ACCUMULATOR RELAXATION VALVES

Valving may be required to safeguard both machinery and personnel when an accumulator is incorporated in a circuit. In the circuit previously described, the accumulator might be compared to the charged capacitor of an electric circuit. If this capacitor is not discharged prior to working on the electrical device, it can be dangerous to maintenance personnel. The simple precaution of grounding the capacitor can render it harmless.

A similar situation can exist with the charged accumulator in a hydraulic system. The pump may be stopped, and the circuit may appear entirely harmless to maintenance personnel. But removing bolts from a manifold or cap may cause the fluid to exert considerable force at an inopportune moment. Resultant damage can range from nuisance leakage to actual injury of the machine components or personnel. Personnel in the vicinity may be struck by the fluid spray or by loose parts being expelled by the quick release of the stored energy.

Because of this potential danger, many machinery purchasers specify that automatic valving must be provided to discharge the accumulator when the machine is stopped for any significant length of time. Some circuits use a solenoid-actuated, two-way valve between the accumulator and the tank; this valve is opened whenever the electrical circuit is opened. When the circuit is started, the two-way valve is energized to stop the flow to the tank. As the circuit is stopped, the two-way valve is deenergized, permitting the accumulator to relax.

Because of special circuit requirements, some circuits are provided with manual valves to discharge the accumulators. Under no circumstances should a circuit be designed without some method of relaxing the accumulators for emergency service.

Accumulators are often provided with pressure gauges so that the charge is evident. Because of the automatic nature of the circuits, the gauge is normally of use only when starting a new circuit or during repairs. Unless the pressure must be known at all times because of machine requirements, gauges should be turned off. If constant monitoring is necessary, suitable *snubbing devices* must be installed to protect the delicate gauge movement. If the gauge is used only for initial installation purposes, it may be advisable to remove it before turning a machine over to production operators. If is it likely that the pressure reading will be of assistance for maintenance checking procedures, it may be desirable to leave the needle or gauge shutoff valve in the pipe terminus and remove only the gauge to protect the indicating movement.

DESIGN CONSIDERATIONS FOR PRESSURE ACCUMULATORS

While the subject is perhaps beyond the scope of this text, the terms *isothermal* and *adiabatic* may be encountered as applied to accumulator circuit usage. Generally, the terms are more likely to be encountered when dealing with compressed air systems.

An isothermal system, as the name implies, indicates an equality of temperature or indicates changes of system volume or pressure at relatively constant temperature. Because it is not practical to remove heat in an amount exactly equal to that generated by changes in volume or pressure, it is usual to recognize the adiabatic relationship in which the heat of compression is retained.

Gas-charged accumulators (usually nitrogen) are most widely used in industrial systems.

The displacement of the piston or bladder structure determines the capacity that can be withdrawn from the unit and bears a relationship to the total volume of the vessel, determined by Boyle's law (see Chapter 1). For a full understanding of the operation of accumulators, this relationship must be borne in mind, particularly the fact that every change in volume of the accumulator, such as withdrawal or addition of fluid, is accompanied by a change in pressure. The total amount of pressure fluctuation can be determined by the designer in dimensioning the relative values of piston displacement and the total volume, and is generally made from 10 to 20 percent.

Thus the total volume should be from six to eleven times the piston displacement. For instance, if the piston displacement is 100 in^3 (1639 mL), then the total available volume in the bottle should be 600 in^3 (9834 mL). When the piston is then pumped back and the vessel charged with 100 in^3 (1639 mL) of oil, the pressure will increase to six-fifths of its original value, or gain 20 percent. This, and the fact that the pistons in these bottles have a friction factor of about 15 percent total (difference between charging and discharging pressure), dictates great caution in properly selecting the pressure capacity and horsepower of pumps.

In the above case, for instance, if a minimum system pressure of 1000 psi (6895 kPa) is required, the pump must deliver 1000 (6895 kPa) × 1.15 psi (7.93 kPa) at the start and 1000 (6895 kPa) × 1.15 (7.93 kPa) × 1.20 psi (8.27 kPa) at the end of the charging operation. The pump and motor must, therefore, be capable of supplying 1380 psi (9515 kPa), so that the system will be assured of receiving the minimum of 1000 psi (6895 kPa). The piston-type accumulator is well adapted for large volumes and heavy pressures in connection with variable-displacement pumps and pressure-compensating controls. The vessels are charged with nitrogen from commercial bottles and should be equipped with suitable connections and with a fusible plug that melts in case of fire and permits the gas to escape.

For smaller capacities, accumulators of the diaphram or bladder type are suitable. The capacity-pressure relation in these accumulators is the same as in a piston-type unit, following Boyle's law.

Caution is indicated in the application of elastic-bag or bladder-type accumulators to avoid the error of using too small a unit. An oil capacity considerably in excess of the calculated demand should be provided, and a substantial amount of oil should be left in the unit after the cycle demand has been withdrawn. Pressure fluctuations should be held within reason to avoid fatigue failure of the bag. The unit should not be looked upon as a source of explosive energy but rather as a peak-demand equalizer. A pressure-compensating metering valve should be provided at the outlet so that the oil-flow velocity can be kept at a substantially constant value, established by the designer, regardless of pressure or load fluctuations.

Assuming, for instance, that the unit is to provide a demand of ΔV at a pressure of p with a pressure drop not to exceed 20 percent, then we must provide for a gas capacity of 4 ΔV plus an oil capacity of, say, 3 ΔV, or a total of 7 ΔV. The charging pressure would then be $4/7$ of the maximum operating pressure or $5/7$ of the minimum.

Experts in the design of mass-production-type machinery may provide greater safety factors. Recommendations of total capacities of 27 ΔV for bladder-type and 21 ΔV for piston-type accumulators for "average" conditions have been cited.

Since diaphram and bladder accumulators have no pistons, there is no piston-friction factor to consider; nevertheless, the efficiency is not quite 100 percent, owing to the work of expanding and compressing the diaphrams and bladders.

SUMMARY

Hydraulic accumulators are of three basic types: weighted, spring-loaded, and gas-charged.

Gas-charged accumulators are subdivided into the diaphragm, bag, piston, and floating poppet-valve types.

The laws of Charles, Boyle, and Gay-Lussac should be reviewed to understand fully the operating characteristics of accumulator devices.

An inert gas is usually used in pneumatic accumulators. All safety codes should be carefully studied prior to using accumulators or other pressure vessels in hydraulic circuits. Maximum pressures should be plainly marked and properly fused to prevent damage to personnel or machinery. Fluid under pressure has many of the characteristics of compressed gases and explosives. A broken line is a potential danger to personnel and materials, as the flow is limited only by line losses and the effective force created by the weight, kinetic energy in compressed springs, or compression of the pneumatic medium.

Accumulators may be used to absorb vibration or shock waves. Accumulators may also be used with other components to create hammerblow effects and high shock waves. The circuit in which the accumulator is used determines the functional characteristics.

A properly designed accumulator circuit can provide some of the advantages of a variable-delivery pump.

Accumulators must be discharged prior to the performance of normal maintenance work in a circuit. Failure to do so may endanger personnel and machinery.

REVIEW QUESTIONS

6-1 If a gas-charged accumulator is charged with fluid at 120°F and the fluid is blocked in the unit at 3000 psi, will the pressure be greater or less when the fluid and gas temperature drops to room temperature of 75°F?

6-2 What will the result be if hot sunlight raises the gas temperature of the accumulator of Question 6-1 to 150°F? Will the pressure be higher or lower?

6-3 What is the purpose of the outer shell in the double-shell construction used in the accumulator in Fig. 6-7c?

6-4 How could you recognize an incorrect adjustment of the maximum-pressure relief valve in Fig. 6-9? Too high? Too low?

6-5 If the sequence valve were set too high in Fig. 6-9, what type of malfunctioning could be expected? Where should gauges be located to indicate the reason for the malfunction?

6-6 Is the check valve nearest the pump esssential to operation of the circuit in Fig. 6-9? Can it be eliminated if the pump runs continuously?

6-7 Where must the valve be located in the circuit of Fig. 6-9 to discharge the accumulator when the machine is shut down?

6-8 Is there any difference in the pressure needed to start the charging of a weighted accumulator and that contained within the fully charged unit?

6-9 How does the law of Gay-Lussac affect pneumatic accumulators?

6-10 Is the charge within a spring-loaded accumulator the same as that within a weighted accumulator? How do they differ? What are the limitations of spring-loaded accumulators?

6-11 Why is an inert gas used in an accumulator? Check your local safety code. Are there any references to gases used in accumulators?

6-12 Why is a spring-loaded, piston-type accumulator used in the shock-supressor valve of Fig. 6-3?

6-13 What is the purpose of the air-bleed valve in Fig. 6-3?

6-14 If the orifice shown in Fig. 6-4 is enlarged, how will it affect the operation of the shock suppressor? What happens if it is smaller?

6-15 What is the purpose of the normally open check valve shown in Fig. 6-6?

6-16 What will happen to the pump discharge of Fig. 6-8a if the pressure-switch contacts stick closed? Suppose they fail to make contact. What will occur?

6-17 How will the circuit react if the solenoid fails in the vent valve of Fig. 6-8a?

6-18 Why is pilot pressure taken ahead of the check valve of Fig. 6-8c?

6-19 Could the circuit of Fig. 6-8c function without the pilot relief in parallel with the piloted two-way valve?

6-20 What does the hollow equilateral triangle within the symbol for an accumulator in Fig. 6-9 signify?

LABORATORY EXPERIMENTS

6-1 Precharge a pneumatic accumulator to 1500 psi. Charge the fluid portion to 3000 psi with fluid at 150°F. Insert a gauge in the line adjacent to the accumulator. Lock the fluid in the accumulator. Allow the fluid temperature to drop to ambient. Note the drop in pressure. Compare this drop to the cubic contents of the accumulator; check which of the gas laws is applicable.

6-2 Set up the circuit in Fig. 6-9. Use a long-stroke cylinder or fluid motor as a work load. Position the discharge from the relief valve so that the flow can be observed. Note the amount of fluid lost through the relief valve as the pressure rises to the point at which fluid passes through the sequence valve to recharge the accumulator. Note how an improperly set relief valve could cause both time and efficiency loss on a machine. A restriction on the fluid motor or cylinder after the pump and accumulator have discharged their load in useful work will indicate whether the load is being moved at the maximum pump capacity or whether the fluid is recharging the accumulator. A gauge in the line directly to the accumulator beyond the sequence valve will indicate where the pump delivery is being directed. Note the speed of the fluid-motor movement with a relief valve on the output as a brake with the sequence valve set at different pressure values.

6-3 Make a simple, weighted accumulator with a small cylinder and compare the fluid losses through the packing for different adjustments of gland tightness.

6-4 Build the circuit of Fig. 6-4, using a piloted, poppet-type relief valve for the two-way section. Connect the relief-valve vent connection to the accumulator. Install a shutoff valve in the line to the accumulator. Now tee this assembly to a simple circuit consisting of a constant-displacement pump, capable of pressures to at least 1000 psi, and a lever-operated two- or four-way valve. Connect the discharge of the directional-control valve to the tank and plug one port. Pressurize the line by rapidly shifting the directional-control valve. Note the pressure gauge reactions with the accumulator in the line; with the valve closed to the accumulator. Change the charge value in the accumulator and note the reaction in the circuit. Change the orifice size in the relief valve. Note the change in the response time. (*Note:* If a two-way directional-control valve is used, it will not be necessary to plug any ports.)

6-5 Procure a small air cylinder with a packed piston. Plug the blind end with a bushing and gauge. Apply hydraulic pressure at the rod end gradually until the pressure reaches approximately 100 psi. Note the distance the rod travels. Calculate the amount of compression of the free air in the blind end of the cylinder. Gradually release the hydraulic fluid. Note the drop in pressure as the cylinder extends.

6-6 Insert a Bourdon-tube pressure gauge of suitable range in a high-pressure tee. Connect a check valve so that fluid can flow to the tee and gauge. Fasten a tee ahead of the check. Insert a needle valve in a connecting line between the two tees. This needle valve will be used to relieve the fluid that will be trapped between the check valve and the gauge. Connect the input to the first tee to the circuit of Experiment 6-4 at the pump output. Close the needle valve. Repeat Experiment 6-4. Note how the maximum pressure can be retained on the gauge because of the check valve. This provides a simple and accurate way of measuring shock conditions if the gauge on the pump outlet fluctuates excessively. Pressure can be relieved to repeat the test by opening the needle valve when circuit pressure is relaxed.

6-7 Insert the gauge and valve assembly in the port of Experiment 6-5. With maximum pressure on the rod end of the cylinder, tap the rod end with a soft mallet and note the pressure surge at the gauge assembly.

6-8 Repeat Experiment 6-7 with the blind end filled with oil and note the difference in the pressure encapsulated in the gauge assembly.

6-9 Take the gauge assembly of Experiment 6-6 and check the pressure surges in available production machines. Insert the assembly in all available gauge ports. Note the difference in the reading with the snubbed gauge and with the needle valve closed. (The reading is cumulative with the needle valve closed.)

6-10 Make up an assembly such as that of Experiment 6-6 with a 100-psi gauge and connect the input to your water system. Open and close a faucet rapidly. Note the pressure surges.

7
Fluid Reservoirs

Fluid used in industrial hydraulic systems must be kept within the system and protected from contamination. The *tank* or *container* used as a *reservoir* may serve several other functions besides that of storing the fluid. Important among these other functions are dissipation of heat, trapping of foreign matter, and separation of air bubbles from the fluid.

Reservoirs can dissipate heat by radiation from the external walls or by use of integral cooling coils or more complex finned radiating devices. Trapping foreign matter may require use of filters, baffles, magnets, or *mazes*. The maze forces the fluid to travel a fixed distance to permit gravity separation. Separation of air bubbles from the fluid is accomplished by proper baffle and reservoir design that slows the flow of fluid as it returns to the reservoir. While the fluid is moving at a lower velocity the air bubbles have a greater chance of escaping to the fluid surface.

7-1 TYPES OF FLUID RESERVOIRS

For many years it was common practice to find some cavity within the machine member that could be made fluid-tight and use it for a reservoir. This machine-cavity reservoir was successful in some machinery. But in other systems, use of the machine member as a fluid reservoir caused excessive localized heating that warped precision machine castings and created sizable inaccuracies in the work being done.

Large structural pipes on certain mobile machinery make excellent fluid tanks, as they provide large heat-radiation surfaces and serve in a dual capacity as part of the fluid-power plumbing. With proper positioning and adequate baffling, they make satisfactory reservoirs, serving all the normal functions. They also add mechanical strength and functional flow paths for the power-transmitting fluid within the machine circuit.

Combination lubrication and hydraulic tanks are used in some hydraulic systems. Here the splash lubrication tank may be an integral part of a gearbox or similar device. The tank also serves as a fluid-supply source and reservoir. This arrangement is efficient if the hydraulic system is not dissipating a large amount of heat energy into the fluid medium. The lubrication pump supplying oil under pressure to bearings may also be used to pressurize the fluid hydraulically for simple gear-shifting operations or intermittent clutch operation.

A rectangular or square-shaped tank with the pump motor mounted on the top provides a compact tank-type reservoir package for many industrial applications. Modifications of the basic tank include baffles to keep foreign matter in a specific area and baffles to force the flow through the tank in such a manner that maximum heat transfer is obtained. In some designs a waterfall-like arrangement assists in freeing air trapped in the fluid. A multiplicity of baffles is effective in trapping lint and other floating impurities in certain industrial installations. The rectangular design is basically inexpensive, and the location of the pump and motor on the tank ensures a short suction line for the pump.

Figure 7-1 shows a rectangular tank with the motor and pump in place. Note that the return line enters the tank far enough above the bottom to avoid agitation of foreign matter that may have settled out from the fluid medium. Experience indicates that a large return line entering the tank must be well below the minimum fluid level. This avoids undesirable turbulence of the fluid within the reservoir and provides good dissipation of entrained air. It also tends to permit separation of solids in the fluid stream as the high-velocity flow mixes in with the larger quantity of relatively slow-moving fluid in the reservoir.

The return line is located on the opposite side of the baffle from the submerged self-indicating suction filter. The suction line, connected to the subplate-mounted pump, is fabricated from flexible metal hose. This provides flexibility for installation and maintenance. The metal hose can be rigidly attached to the submerged self-indicating suction filter. A rigid pipe may be used as the suction line if a filter is not installed in this area. Where cold start-up conditions are encountered, such as on mobile equipment, a suction filter may create too much resistance to flow. Under these circumstances, filtration may be designed into another portion of the circuit. A suction filter installed in the reservoir without an indicator or easy-changing facility can be forgotten, with resulting loss of efficiency or damage to the pump as the filter element becomes saturated with foreign matter.

Replacement oil added to the reservoir can introduce contaminants. The combination tank-fill-cap assembly shown in Fig. 7-2a provides a relatively fine screen, mechanical protection, and a closure for the opening. The closure cap also serves as an air breather. It is equipped with an air

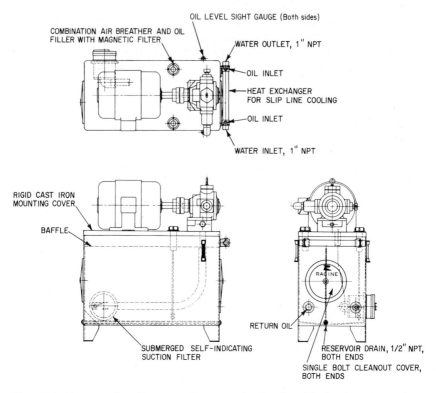

Fig. 7-1 Rectangular flat-topped reservoir fitted with basic accessories. (*Racine Hydraulics, Inc.*)

filter unit to permit movement of air into and out of the tank without excessive fluid contamination. The tank-fill-cap unit shown in Fig. 7-2*b* also includes a magnet assembly to collect ferrous materials within the fluid stream. It can be easily removed for inspection.

The single-bolt cleanout cover on each end of the tank [Fig. 7-1] provides adequate access to the reservoir for cleaning purposes. The access covers are located so that both sides of the baffle can be reached. Larger reservoirs or special-purpose tanks may be fitted with additional cleanout doors. Introduction of single-bolt cleanout doors minimizes sealing requirements. Fluid must, however, be below the level of the opening prior to cover removal. Because of this, filters or strainers located within the tank on the suction line may be difficult to maintain and are likely to be neglected. The feet formed for the tank to rest upon hold the tank far enough above the floor to provide drain space, free movement of air for maximum heat radiation, and space for cleaning the floor beneath the tank. Preventing an accumulation of debris under the tank can effectively assist in keeping the fluid medium cooler.

A rigid, cast-iron mounting cover or heavy metal plate is recommended for the tank cover because adequate strength must be provided in the plate upon which the motor and pump are mounted so there will be no misalignment. Note in Fig. 7-1 how the motor and pump are mounted upon a one-piece, cast-iron plate which serves as the reservoir cover. The cast-iron-cover material assists in preventing amplification of noise by the resonant drumlike fluid container.

Fluid level and reservoir temperature can materially affect the operation of a hydraulic circuit. A sight gauge, such as that shown in Fig. 7-3, can indicate the fluid level within the reservoir. The addition of a thermometer, as shown, assists in providing essential temperature information. Because of the need for maximum accessibility, some purchasers specify that a fill-cap assembly and sight-gauge be provided on both sides of standard reservoir units.

By mounting the electric motor or gasoline-drive engine and pump in a vertical plane, integral with the top cover of the reservoir, it is possible to make a very compact package. A typical compact power package is shown in Fig. 7-4. The suction filter may be replaced by an external return-line filter if a suitable access hole is not readily available.

The compact package design offers complete mechanical protection for the pump and flexible-coupling drive. Concentricity and correct position

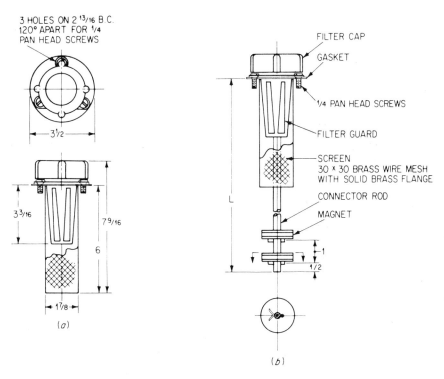

Fig. 7-2 Reservoir filler assembly. (*a*) Basic unit with screen, guard, and filter cap. (*b*) Basic unit with addition of magnet assembly to attract ferrous materials.

Fluid Reservoirs **189**

Fig. 7-3 Fluid sight gauge and thermometer. *(Racine Hydraulics, Inc.)*

of the pump and motor assembly are assured. Manifold structures to interconnect valving can be easily fastened to the rigid steel cover.

Reservoirs may be *pressurized*, usually to values less than 100 psi, to improve the suction characteristics of the pump or to eliminate contamination from the surrounding atmosphere. The source of pressurization may be compressed air in a suitable volume to meet the differential requirements produced by the rod areas of the cylinders, or the pressurization may be with an inert gas to eliminate contamination by moisture. Pressurized-tank systems often use an auxiliary scavenger system to collect the fluid from atmospheric drains and pump this fluid back into the pressurized area.

Hydraulic reservoirs should be sealed and provided with a suitable air-breather device if they are not pressurized. This breather will remove contaminants as air enters and leaves the tank during filling and emptying of the cylinders and rams. In the pressurized reservoir it is necessary that the unit be sealed. When atmospheric air is to enter the tank, it is imperative that the breather be of adequate capacity to permit air to enter as fast as the components are moving and requiring fluid from the reservoir.

A modification of the pressurized reservoir includes a filtered low-

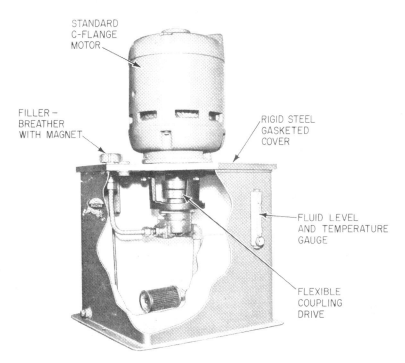

Fig. 7-4 Vertical power-package assembly. (*Racine Hydraulics, Inc.*)

pressure air supply directed into the reservoir of hydraulic systems operating in highly contaminated areas (such as in foundries, near sea water, or in locations of high humidity). Air constantly escapes from the breather so that the contaminants are blown away from the hydraulic system. This low-pressure system is essential around certain types of polishing machines, where lint would normally plug the breather mechanisms within a short period of time. Simple blowers, much like household hand vacuums, take air piped from a nearby noncontaminated area through filters and keep the reservoir under a few ounces of pressure at all times.

Power packages, designed for portability, may require some modification to make them compatible with the expected working conditions. The power unit shown in Fig. 7-5 is fabricated in two sections. A relatively large, automotive-type radiator is permanently fastened to the reservoir. The hydraulic pump shown in Fig. 7-6 is manifolded to a bracket to which the interconnecting lines are attached. Forced air circulation through the radiator is provided by a fan attached to a through-pump shaft as shown by the dotted lines of Fig. 7-6. Isolation of the pump and connecting lines removes stresses that can be expected from rough service for which the unit is designed. Quick connectors permit separation of the two parts of the unit for mobility in difficult areas.

Fluid Reservoirs

Fig. 7-5 Portable power package. *(Racine Hydraulics, Inc.)*

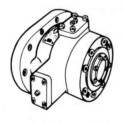

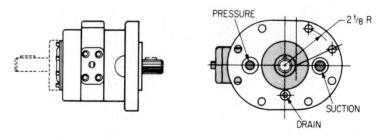

Fig. 7-6 Face-mounted pump used on portable power unit. *(Racine Hydraulics, Inc.)*

The unit is designed to power portable construction and maintenance tools, such as power spades, tampers, chain saws, drills, impact wrenches, shears, and pipe-fabricating devices. Note the lever at the upper part of the unit just above the pressure gauge. This is a two-way valve that is opened as the engine is cranked to relieve pressure during the start-up procedure. The pump is of the variable-delivery-vane type so that the pump delivery equals the supply need for each of the specific tools as they are attached. Compensated flow-regulating valves are usually employed to operate the tool at the most efficient speed. The pump automatically compensates for the minimum horsepower usage in conformity with the pattern established by the flow-control valve.

The return hoses from the tools are connected to the manifold block at the top of the radiator. All return fluid is circulated through the radiator and cooled prior to entering the relatively small reservoir.

Power units of relatively small size may provide the functions of direction, pressure, and flow control within the basic package. The unit shown schematically in Fig. 7-7 employs a constant-displacement pump to pressurize the fluid. The drive motor is reversed to change the direction of flow from the pump. Suction check valves automatically provide the correct suction characteristics regardless of rotation of the pump unit.

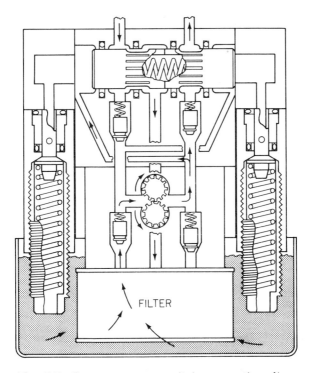

Fig. 7-7 Compact power unit incorporating direction-, pressure-, and flow-control functions.

Fluid delivery as shown directs pressurized liquid to the right-hand outlet port. Fluid is also directed to the left end of a spool-type, directional-control valve. This causes the spool to shift and create a path to the tank for return fluid entering the left port. By reversing the pump drive motor, the flow is directed out the left port and the spool is shifted to the right, providing a return flow path for fluid entering the right port. As the pump motor is stopped, the check valves close. The spool is urged to a stop position by the bias spring, so that the fluid is locked in the working ports. Seals are provided between the dividing lands in the valve mechanism, so that the unit is essentially locked. The circuit is shown symbolically in Fig. 7-8a. An external relief valve or pressure switch can be installed at point 1. An internal relief valve or pressure switch can be connected at point 2. The relief valve serves as a safety device for thermal protection or an operational malfunction. Usual control is by pressure or limit-switch signal.

Two pilot operated check valves can be employed to create the locking function with the reversing pump, shown in Fig. 7-8b. Small series-wound electric motors are used to drive the pump if a high delivery rate is wanted at low pressures and a decreasing flow as the pressure is increased. Prior to stalling out the motor, a pressure switch stops the unit. Pressures to 5000 psi are commercially available in compact units of this type. Plastic-faced seals with resilient backing plus antiextrusion backup washers provide the needed seal to contain the high pressure and maintain the leak-free valving function.

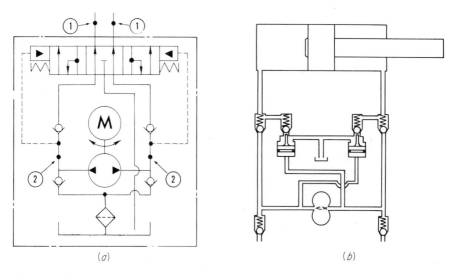

Fig. 7-8 (a) Symbolic representation of power-package circuitry. (b) Schematic circuit of reversing pump employing pilot check valves.

7-2 SCAVENGER TANKS

Use of pressurized tanks or reservoirs located on top of machinery may pose problems with drain lines. To collect the drain fluid and get it back into pressurized or elevated reservoirs. tanks are installed at a low point in the installation. These tanks are provided with a low-pressure pump to transfer the drain fluid back to the desired area. Some installations pass this fluid through filtering mechanisms to ensure cleanliness of the fluid picked up by the scavenging lines. These scavenger reservoirs may be fitted with float-switch control or certain types of centrifugal pumps that can run continuously. A jet pump may be used in this type of service.

7-3 SETTLING TANKS

Where contamination is difficult to control, it is common practice to include a separate *settling tank* in the system. This tank is popular in large hydraulic systems that use a common reservoir system to provide fluid to several different pumping stations. The settling tank is usually of sufficient size to permit passage of the fluid through it at a very low velocity so that foreign matter in the fluid will settle out before the fluid reaches the high-velocity portion of the circuit.

Figure 7-9 shows a typical settling tank. Fluid enters at the left, and its velocity immediately decreases. As the fluid moves slowly toward the outlet, sediment is deposited in the conical settling area. The sediment is drained from this area through two gate valves. Valve B, which is a low-pressure gate valve, is normally closed; valve A, of similar construction, is normally open. At an established time the settling tank is drained of sediment and foreign matter by closing valve A and opening valve B. This permits the cavity between the two valves to drain into the sludge catch basin, after which the sludge can be discarded. By closing valve B and opening valve A, the unit is put back into service until time for the next cleaning.

A properly designed settling tank can be as effective in removing contaminants from a fluid-power system as the most intricate systems employing filtering devices. The settling tank is effective in removing water and condensation from oil fluid systems. A settling tank may be a part of a more intricate fluid-conditioning system. Return fluid from the system may enter the settling tank so that large contaminants can be removed prior to passing through other filtering media. There is a large radiation area in a settling tank which can also help cool the fluid.

Sight gauges may be installed in the section of pipe between valves A and B so the water level in the bottom of the tank can be visually checked as often as necessary. Cleanout valve B may have to be removed periodically in highly contaminated systems to permit removal of solids that precipitate out of the fluid and form a cake in the area above valve B. This

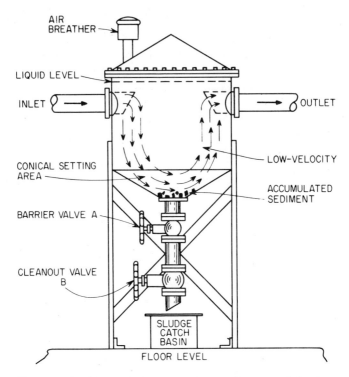

Fig. 7-9 Settling tank for removing moisture and foreign matter from hydraulic fluids.

has been found necessary in systems where phenolic resin molding compounds are handled.

7-4 RESERVOIR PIPING

Pipes in the reservoir are given maximum protection from damage or contamination. Piping to circuit valves that can be physically located at the reservoir can be minimized with effective use of manifolds and other completely trouble-free connecting devices.

Directional-, flow-, and pressure-control valves are all available with manifold mounts so that piping can terminate in a separate plate. The valve is then mounted on this plate. The plate may be cross-drilled to minimize plumbing by having several valves on one plate, and it also may have port connections on either the side or back or on both side and back. Thus, it may be desirable to have all pressure and pilot piping within the tank and the cylinder lines connected to the sides of the plates on the exterior of the reservoir.

Figure 7-10 illustrates how valves are connected within the reservoir with the cylinder connections conveniently located for piping. Figure 7-11

Fig. 7-10 Power package with major valve piping within the reservoir. *(Racine Hydraulics, Inc.)*

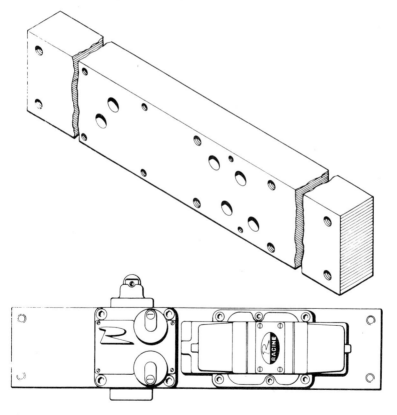

Fig. 7-11 Manifold for mounting hydraulic-control valves. *(Racine Hydraulics, Inc.)*

Fluid Reservoirs **197**

illustrates a manifold block used to mount and interconnect hydraulic control valves. This unit is designed to be installed on a machine base adjacent to a moving member to control rapid traverse and feed cycle.

7-5 MOUNTING OF PUMPS

A cast base, such as that shown in Fig. 7-1, may be provided to mount the motor and pump. This may be separate from the reservoir. The reservoir may be a part of a machine which does not lend itself to serve as a pump-and-motor mounting surface. A single reservoir may supply several pump-and-motor sets so that the several units may be remotely located. By mounting the pump on a separate base, the mechanical strength requirements of the reservoir are minimized.

To ensure accurate alignment and eliminate mechanical vibration, it may be desirable to mount the pump to the end bell of an electric motor [Fig. 7-12] or use an adapter bracket [Figs. 7-4 and 7-5].

Pump pulley or gear drives may require auxiliary outboard bearings to provide sufficient strength to avoid deflection of the pump shaft and subsequent bearing and seal wear. The capability of the pump shaft to tolerate an overhung load should be carefully analyzed prior to installing a pulley or gear drive.

Several types of couplings are available to provide a degree of flexibility between the prime mover (electric motor, gasoline engine, power takeoff, etc.) and the pump drive shaft. A spline drive is quite common for pumps employed in mobile equipment circuits, where the drive is a power takeoff from the engine. Direct engine drive and electric-motor adaptations generally use a coupling specifically designed to permit easy separation of the shafts and protection from slight misalignment. Figure 7-13a shows one-half of a drive coupling attached to the electric-motor shaft. Figure 7-13b shows the three parts of a flexible insert-type coupling. If the pump and motor should be misaligned, the replaceable resilient insert will be destroyed before the pump bearings. Usually, no lubrication is required for this type of coupling.

Since the coupling halves are true, alignment can be accomplished by setting a straight edge across the halves [Fig. 7-14]. This check must be performed in both the vertical and the horizontal planes. This type of coupling usually has an open center in the insert, which allows variations in the distance between the end of the pump and motor shaft.

The coupling of Fig. 7-15a employs a solid center section. A flexible metal insert is employed in the coupling of Fig. 7-15b to couple the two hubs. This insert is so designed as to absorb any twisting motion between the pump and motor. Other designs include a chain to couple the two halves, drive pins with resilient bushings, and vertical-face keys in two planes, known as an Oldham coupling. Tang drives are employed in small units.

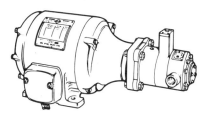

Fig. 7-12 End-bell mount on electric motor.

Fig. 7-13 (*a*) Coupling half attached to electric motor. (*b*) Resilient insert-type coupling.

Fluid Reservoirs

Fig. 7-14 Checking flexible coupling alignment.

(a)

(b)

Fig. 7-15 (a) Solid insert-type coupling. (b) Flexible spring-type coupling.

200 *Industrial Hydraulics*

The angularity of the drive should be carefully adjusted so that chain and pulley drives are at right angles and flexible couplings are in direct alignment. The amount of misalignment that can be tolerated is usually specified by both the pump and coupling manufacturers. Deviation from their recommendations may shorten equipment life.

7-6 OVERHEAD RESERVOIRS FOR GRAVITY-FILL PRESSES

Press applications may require a large supply of fluid at a high level in the machine structure. For example, with high-speed movement of large rams, the flow of the fluid into the ram may approach several hundred gallons per minute, or more. This may require location of the reservoir such that the ram will be partially submerged in the fluid. A two-way valve is located in the ram area so that the valve can alternately open and close the path provided between the reservoir and the ram cavity. These rapid traverse movements usually occur with a minimum of pressure drop but a large flow of fluid.

In some designs the ram drops by gravity, and atmospheric pressure forces the fluid to enter the cavity created by the ram movement. This flow of the fluid can be assisted if the reservoir fluid level is above the ram and two-way valve. The fluid then flows into the cavity with minimum assistance from atmospheric pressure. This is referred to as a *prefill function*. The ram must be filled rapidly and the two-way valve closed so that the ram can be pressurized to the right amount to provide the work function. This type of reservoir requires careful design to avoid undesired turbulence from the high velocities involved in the rapid movement of large quantities of fluid.

The two-way poppet valve in Fig. 4-8 is of similar design to the high-flow devices employed in overhead-tank systems. The overhead supply facilitates rapid movement of the fluid, particularly where the design of the machine dictates a power ram assembly in the upper portion or crown of the press. The hydraulic circuit for a press with an overhead reservoir can be quite similar to that shown in Fig. 4-9. The major difference may be the added scavenger-tank assembly often needed to drain certain valving or packing assemblies. An additional advantage of an overhead reservoir is the saving of productive floor space in manufacturing areas.

7-7 SIZING OF RESERVOIRS

Many factors are involved in deciding what the size and general configuration of a hydraulic reservoir shall be. The first consideration in sizing a reservoir is the amount of fluid necessary to fill all the components and ensure a place for the fluid in the event that it all drains back into the tank.

Second, a sufficient fluid capacity is necessary to ensure that there will be enough fluid available to the pump suction line at all times, even though differential cylinder areas and ram displacements consume fluid momentarily during the machine cycle. Enough fluid must be provided to prevent a vortex forming around the suction line when a minimum fluid level during the machine cycle may permit air to enter the pump or ram.

7-8 RESERVOIR CONSTRUCTION

The container for the fluid medium in the hydraulic system can directly affect the performance and life span of the system and its components. Reservoirs should be designed to give adequate protection to the fluid, to dissipate heat, to remove or trap foreign matter, and to store enough fluid to ensure continuous performance despite slight losses during operation.

The amount of heat that can be transferred by radiation from the reservoir walls and plumbing may be small because of the changes in the temperature of the surrounding atmosphere. Heat-transfer conditions must be based on known facts. Experience can be of assistance in computing the horsepower that will be dissipated in the form of heat. Heat exchangers can be incorporated in a circuit with little difficulty; so the design of the reservoir structure may not involve heat transfer as a basic consideration, leaving this function to heat exchangers in the system.

Available space may limit the amount of fluid that can be accommodated. If space is a problem, it may be necessary to add extra baffling to prevent the high-velocity flow from the return lines from creating suction area turbulence that may result in air entering the pump intake. Generally, low flat tanks are not desirable for hydraulic service because of the difficulty of eliminating entrained air. Also, a relatively high-velocity flow path can exist in tanks of this type.

Noise pollution is a significant factor in fluid-power transmission systems. *Fluid pollution makes noise pollution.* The two major noise-producing pollutants in hydraulic fluids are air and water. Air forms bubbles within the fluid, and the bubbles make noise when they collapse. Water forms vapor bubbles in the fluid whenever temperatures are high enough and pressures low enough—and vapor bubbles are even more violent than air bubbles.

Dissolved air makes no noise. But you can rarely depend on the air to stay dissolved. Most cavitation in pumps operating with conventional hydraulic fluids is caused by dissolved air that forms bubbles at the low pressure side of the pump.

There are other fluid pollutants that can also cause noise, usually because they increase pump wear. Even fluid viscosities that are too low can cause noise by increasing wear. These pollutants are controlled—and unwanted water reduced—by servicing filters regularly, using the right fluid, and maintaining the appropriate reservoir level.

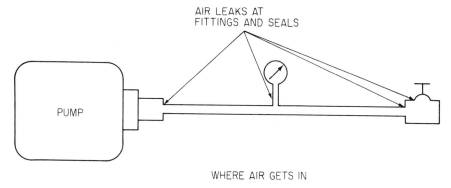

Fig. 7-16 Air enters fluid-power systems in several areas.

Air control, on the other hand, is more complex. Figure 7-16 shows where air gets in. Basic principles are to avoid air-pressurized reservoirs where possible and air leaks, and to design reservoirs that dissipate air. Figure 7-17 shows other areas where air can easily get into the system.

We recognize that sometimes air-pressurized reservoirs are used to avoid inlet vacuums. Unfortunately, this raises the saturation pressure of the fluid, so high-pressure drops through the inlet line still cause outgassing or movement of the air in the fluid. Outgassing is produced by the drop from reservoir pressure, not by low absolute pressure.

Other sources of air contamination are air leaks, which can show up anywhere the system pressure drops below atmospheric pressure. Air leaks are commonly found around cylinder-rod packing, fittings in pump-inlet lines, and pump-shaft seals. These areas can leak even when they are oil-tight. One quick check for such leaks is to flood the suspected leak point with hydraulic fluid. If noise is reduced, the noise probably was caused by an air leak.

Despite all efforts to keep air out of hydraulic fluid, a certain amount of air is almost inevitable. The best place to remove entrained air bubbles is the reservoir. Reservoirs should be large enough and have sufficient baffling to provide time for buffles to separate out. A volume equivalent to 2 min of maximum system flow is generally considered adequate. Lower-

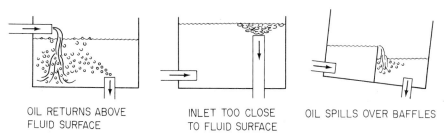

Fig. 7-17 Conductor positioning can cause air entry into the fluid.

Fluid Reservoirs

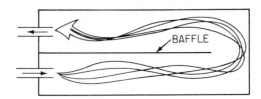

Fig. 7-18 Extended flow by use of linear baffle.

capacity reservoirs can be used, but baffling systems must be carefully designed to maximize dwell time. Figure 7-18 shows another baffle design to maximize dwell time within the reservoir.

When adequate size of baffling cannot be provided, sloping screens, as shown in Fig. 7-19, can be used to speed up bubble separation. Experiments have shown that a 60 mesh screen installed at 30° to the horizontal is quite effective in removing up to 90 percent of the air bubbles.

In addition, tank lines should be terminated as far from the pump inlet as possible and their discharge directed to produce the longest flow path to the tank outlet. Neither return nor outlet lines should be located close to the fluid surface of the reservoir.

From past experience several practical rules have been developed for hydraulic-system reservoirs. These rules do not cover all installations, but they are helpful guides:

1 The tank should be as high as its narrowest dimension.

2 Minimum fluid capacity should be three times the delivery of the pump plus the displacement of the piston rods or rams for machine tool hydraulic systems. Reservoirs for mobile equipment systems may be much smaller. Because of the smaller size, it is important to carefully position baffles, intake and return lines, and—in many installations—a slightly pressurized reservoir. Special valving may be used to permit free input of air in a mobile-type reservoir as pistons are extended. As

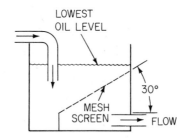

Fig. 7-19 Mesh screw assists in removing air.

204 Industrial Hydraulics

the pistons retract, a resistance valve exhausts air at a predetermined pressure level and maintains the desired reservoir pressurization.

3 Fluid-level gauges should show the minimum and maximum permissible levels. Fluid-level gauges must be protected from mechanical damage.
4 Return-to-tank lines and certain drain lines from pumps should be below the fluid level to prevent undue aeration of the fluid.
5 Suction lines should terminate 1½ pipe diameters above the bottom of the tank. Suction lines should be carefully isolated from return-to-tank flows.
6 Gravity drains from seal cavities on directional-control and pressure-control valves are normally connected above the fluid level in the reservoir.

In planning a reservoir, good practice dictates the use of sufficient connections in the tank to eliminate the need for pipeline headers. This ensures that surge pressures from one drain will not adversely affect the others that are connected to the common header. Some drain lines offer visual indication of the valve operating cycle and can be used to indicate when a valve is malfunctioning.

A narrow, flat, metal box in a vertical plane, with a clear plastic front that acts as a junction for the drain lines, will provide visual inspection facilities. The drain connections enter through connectors in the top of the box. From these inlet connections, the fluid drops freely to a funnel-shaped receptacle connected directly to the tank. This receptacle is of sufficient capacity to prevent a rise of the fluid level within it during the maximum drain flow. Visual indication supplies valuable information, thus minimizing maintenance time. Excessive drain fluid can indicate both wear of and damage to seals.

If cooling coils are located in a reservoir, the fluid level in the reservoir must never fall below the level of the top of the coil containing cooling water. Condensate will form on coils that are not immersed. A water line should *never* be positioned directly against the fluid reservoir, as this can also cause collection of condensate in a localized spot.

Paints, lacquers, and other surface finishes must be compatible with the fluid contained in the reservoir. Certain fire-resistant and other synthetic fluids will remove some finishes. Before using any paint, sealer, etc., it should be tested and proved compatible with the fluid. If there is any doubt, the usual practice is to leave the metal bare or construct the tank from stainless materials.

Auxiliary devices, such as drain plugs with integral magnets or bar magnets hanging within the fluid stream, have proved effective in catching and holding metallic particles that have washed back to the reservoir.

SUMMARY OF DESIGN CONSIDERATIONS FOR FLUID-POWER RESERVOIRS

General-purpose reservoirs are usually designed to contain three times the flow of the pump in gallons per minute. Thus a 5-gpm pump would be provided with a reservoir containing approximately 15 gallons of fluid plus sufficient capacity to hold all of the fluid that can return from the cylinders if all of the fluid is returned to the reservoir. The temperature within machine-tool reservoirs is generally limited to 120F (50C). Mobile equipment hydraulic circuits are generally limited to 170F (75C). Mobile reservoirs are usually of smaller capacity in gallons than the delivery of the pump in gpm.

Pressurized reservoirs are often used with mobile hydraulic systems. Filtration of mobile hydraulic reservoirs is generally of the return-line type.

SUMMARY

Fluid-power reservoirs function as containers for the fluid medium and as localized areas that can assist in temperature conditioning and maintenance of fluid cleanliness.

Use of a cavity within the machine for a reservoir may create undesirable localized heating conditions. Protection against contamination may require pressurization of the reservoir. Judicious use of settling tanks can improve the fluid conditioning function. Piping within the reservoir area can be minimized by use of manifolding and mounting of components on the reservoir walls.

Pumps and motors must be installed so that vibration is minimum and alignment is assured. Reservoir size must be adequate to contain the fluid needed during functional operation and the fluid returned to the tank when all components are drained to it. Incompatible surface finishes can contaminate components as well as the fluid medium.

REVIEW QUESTIONS

7-1 What is the purpose of a permanent magnet mounted in a drain plug or hanging in the reservoir?

7-2 Why is barrier valve A normally open in the settling tank in Fig. 7-9?

7-3 Why would a reservoir be pressurized?

7-4 What is a reservoir baffle?

7-5 What is a recommended distance between the suction line to a pump and the bottom of the reservoir?

7-6 How are air bubbles extracted from the fluid in a hydraulic circuit?

7-7 Why is an air breather necessary on a hydraulic reservoir?
7-8 What is a scavenger tank?
7-9 What is a prefill function?
7-10 What problems may be encountered when flat and shallow tanks are used as hydraulic reservoirs?
7-11 Why is it often desirable to eliminate combined drain functions?
7-12 What does an excessive flow from a drain line indicate?
7-13 Why must cooling coils always be below the fluid level in a reservoir?
7-14 Why is it undesirable to run water lines against reservoirs?
7-15 What is the purpose of the lever-operated, two-way valve located in the line between the pump outlet and reservoir in Fig. 7-5?
7-16 How is the flow reversed in the circuit of Fig. 7-8a?
7-17 What is the purpose of the manifold block shown in Fig. 7-11?
7-18 What happens to a motor coupling with a resilient insert if misalignment is excessive?
7-19 How can a motor coupling be aligned?
7-20 Why is a flexible, metallic line used as a suction conductor in Fig. 7-1?

LABORATORY EXPERIMENTS

7-1 Position the suction line of a small pump-motor combination 1½ pipe diameters from the bottom of a shallow baking pan. Return the fluid to the opposite corner. Note the vortex formed with various amounts of fluid in the pan around the suction line.
7-2 Hang a permanent magnet within a power unit on an operating machine, in such a position that it does not touch any metal. Make certain it is within the flow path of the fluid. Examine the magnet after one week and after one month.
7-3 Take a sample of fluid from any reservoir within the flow path while the machine is in operation. Allow it to settle and examine the foreign matter that precipitates from the fluid. Pass the sample through filter paper. Examine the residue under a microscope. Pass a permanent magnet across the residue. Note the amount of magnetic impurities. Calculate the percentage of magnetic and nonmagnetic impurities.
7-4 Connect a return to the tank header for all drains on a working hydraulic system as explained in Sec. 7-8. Watch the amount of fluid returning from each drain through the plastic window. Change the working pressures on the machine and note the change in drain flow. Hold a thermometer under each drain (after removing the plastic window) and compare the temperature of each flow.
7-5 Check several flexible couplings on working machines as shown in Fig. 7-14. Using feeler gauges, tabulate the amount of misalignment

on those inspected. Check maintenance records of these machines to determine if pump failures have occurred.

7-6 Note the temperature within a reservoir on a production machine at 30-min intervals for a full working day and plot this on graph paper. Repeat with several other machines. Note variance in patterns from morning start-up to the end of the day.

7-7 Reroute a major return line to a reservoir to a position above the fluid level. Operate the machine for a short period of time. Note the foaming of the oil. Record the time required to start pump cavitation (noisy pump). Allow fluid to stand. Record time necessary for foam to dissipate.

7-8 Immerse similar pieces of painted metal in several different types of hydraulic fluid. Include petroleum oil, phosphate-ester-type fluids, and water emulsions. Note the time for the surface finish to deteriorate.

7-9 Repeat Experiment 7-8 with different types of resilient seals. Note the degree of compatibility. Include plain tap water as one of the fluids in this test.

7-10 Disassemble an air breather from a hydraulic reservoir selected at random. Note the amount of contamination. Insert an air line at the fill pipe. Check for air leaks around the tank. Dried-out gaskets can permit entry of foreign matter if air can escape as indicated by this test. Dirty air filters will increase the flow through other potential air-entry sources.

8
Filtration of Hydraulic Fluids

Industrial hydraulic systems operate best when the fluid is free of all impurities and other foreign matter. Hydraulic fluids, however, can tolerate some foreign matter. The amount and kind of foreign matter that can be tolerated depend on the type of hydraulic system and the components used in it.

Filters are used to remove the smallest particles of foreign matter from the hydraulic fluid. Large particles are removed by *strainers*. The amount of contamination that can be tolerated by a hydraulic system determines if filters, strainers, or other fluid-cleaning devices are needed. Proper reservoir design and use of simple components may eliminate the need for filters and strainers if the fluid is clean when added to the system. Automatic controls use somewhat more complex pilot systems having fine orifices. The more complex the controls, the finer the degree of fluid cleaning required.

In this chapter you will learn about the many different types of filters and strainers used in hydraulic systems. You will also learn how these devices are applied to different types of industrial hydraulic circuits.

8-1 TYPES OF FLUID FILTERS AND CLEANING DEVICES

The simplest cleaning and renovating systems use baffles and settling tanks. The fluid is passed through a tank such as that in Fig. 7-9. The high velocity of the fluid in the pipeline is suddenly terminated in the tank, where slow movement of the fluid allows the impurities to settle mechanically. Baffles in the tank can block the passage of floating impurities. The settling tank and baffle mechanisms are highly effective and are the least expensive method of conditioning hydraulic fluids if space and quantities of fluid are not at a premium.

If large quantities of fluid and considerable space are not available, the next step in the cleaning and renewing process may consist of straining the fluid through mechanical filters or cleaners. These mechanical filters can be made of simple, heavy wire mesh enclosed in a suitable housing so that all the fluid must pass through the screen. In other filter designs, intricate multiple passes with safety bypass connections may be used. Or a series of stamped disks that can be moved to dislodge foreign matter and cause it to collect in a sediment pocket at the bottom of the filter housing may be used. Also, ribbons of paper can be mechanically wound to provide a filtering element. The effectiveness of any filter is measured by the degree of filtration it produces.

The term *degree of filtration* is increasingly important in industrial hydraulic applications as closer-fitting apparatus finds wider usage. The precisely mated parts in servomechanisms and similar controls require an increasingly finer degree of cleanliness of the fluid. *Degree of filtration means that a specific filter element, for example, 25 microns in size, will, when new and clean, stop 98 percent of the particles measuring 25 microns or more while operating under its designed flow conditions.*

Micron (μ) is the measuring unit used in expressing the degree of filtration in sizes less than 0.001 in. A micron equals one-millionth of a meter, or 0.00004 in. A convenient way to remember micron size is to think of 25 μ as equal to about one-thousandth (0.001) of an inch.

Commonly used mesh sizes for filter elements are 60 mesh (238 μ), 100 mesh (149 μ), and 200 mesh (74 μ). For a comparative value, consider that the normal lower limit of visibility to the naked eye is about 40 μ. White blood cells are about 25 μ, and red blood cells are about 8 μ.

8-2 METAL-ELEMENT FILTERS

Sintered particles of powdered metal formed into the cartridge of a filter will provide nominal 98 percent removal ratings of 2 to 60 μ. An estimated absolute rating for 100 percent removal is 4 to 60 μ.

Granular (porous) media have adjacent grains, and these particles join to form a porous, rigid structure. The grains can be joined with a compatible adhesive or by sintering at temperatures slightly below the melting point of the particular material. Sintering may be done with or without the simultaneous application of pressure. The degree of filtration depends on the compression and smallness of the grain chosen. Dirt particles are trapped in the interstices.

Media made of bonded granular materials are normally thicker in cross section than woven-wire media for the same fineness of filtration. Pore sizes are more mixed in granular media than in ordinary unsintered square-weave mesh, but special interlocking weaves (dutch twill, for example) overcome this disadvantage.

Smaller contaminant particles can be filtered out by granular media than by woven types, but there is wider variation in particle size stopped and a greater tendency toward contaminant migration. Extreme fineness is possible; a leached-glass porous medium has a removal rating better than 0.01 μ—it is not, however, an industrial-type filter. Granular media are most suitable where solids concentration is very low.

Porous *depth-type filters* can be cleaned to a limited extent by *backwashing*. Progressive fouling will occur after many backwash cycles, but the element can be restored to its original state with a suitable cleaning solution.

Edge-type filter media are manufactured in disk type, ribbon type, and wound-wire type. The disk type is rated at 98 percent for removal of particles from 40 to 125 μ, the ribbon type for particles from 25 to 500 μ, and the wire type for particles from 75 to 700 μ. The 100 percent rating for ribbon types is 50 to 500 μ and for the wire type 125 to 700 μ.

Edge-type filter media may be either metal or nonmetal. The fluid filters between edges of flat disks or wires of circular or other cross section. Spacing determines the degree of filtration. In the disk type, spacing is with alternate disks that are smaller in diameter or with a central ring structure that has radial projections. The wire type is spaced by winding the wire around a mandrel or by using projections on the wire itself.

Long particles can get through an edge filter if one dimension (diameter or thickness) of the particle is smaller than the edge opening. Round or cubical particles are stopped. If downstream devices have small orifices or small-clearance moving parts, just a few fibers passing downstream can readily bridge the orifice or clearance, restricting flow or movement.

Mesh- and cloth-type media, either sintered or unsintered, are rated at 98 percent for particles 2 μ and over. At 100 percent rating the micron size rises to 12 μ and over.

Woven media are cloths of wire or yarn (natural or synthetic). Both are produced on looms and use loom terminology: *warp* is the wire or yarn that runs the length of the cloth; *shoot* (sometimes called *woof*) runs perpendicular to and is woven into the warp by means of a shuttle. Many weaves are available. The most common are square (plain), dutch, twill, and twilled dutch. Plain and twilled weaves present straight-through openings to fluid flow; twilled-dutch and dutch weaves present a somewhat tortuous path, hence finer filtration.

Wire cloths may be sintered or unsintered. Sintered wire cloth has warp and shoot wires sintered together where they cross, fixing the pore size permanently. In addition, sintering allows integral bonding of coarser meshes to finer facings to produce a stronger cloth.

Wound-wire mesh is a form of cloth made by helically cross-winding a fine flattened wire on a mandrel. Wires are then sintered where they cross, and the finished mesh can be pulled from a mandrel to be slit into sheet stock or kept as a cylinder.

8-3 NONMETAL-ELEMENT FILTERS

Natural and synthetic fibers are of four categories: felted, nonfelted, woven cloth, and wound yarn. Felted elements provide a filtering action that involves both depth and surface. The 98 percent rating shows particle removal in sizes from 5 to 100 μ. Felted media are masses of fibers matted together to produce a porous structure. Good felts are carefully designed to have interlocking fibers. They may or may not be resin-bonded. The degree of filtration obtained, regardless of the fiber type, depends on fiber diameter and length. Smaller fibers give finer filtration. Paper filters are similar in structure to felted media.

Felted media, being of the depth type, are thrown away after use and are limited to relatively low pressure differentials. Media migration will occur under some circumstances, particularly if fibers are loosely interlocked.

Nonfelted fibers depend entirely on depth filtering action. The micron rating at 98 percent is 10 and over. Unfelted fibers are initially loose, but are packed into a woven or knitted bag. Cotton or other cellulose fibers are the types usually chosen, and their major application is in the filtering of oil. Application is limited by temperature and pressure, and media migration is likely to occur. Filtering ability is improved with fillers such as diatomaceous earth. Fillers are not recommended in most hydraulic systems as the filler may remove additives from the hydraulic fluid.

Wound yarn provides a depth filtering action that can filter particles from 1 to 100 μ on a 98 percent rating and 10 to 150 at a 100 percent rating. Wound-yarn media are commonly made in cylindrical form, helically wound around a support core that is part of the filter. The weave gives a diamond-shape pattern externally, without pores in the strict sense of the term. Filtration here is the depth type and takes place through (and in) tight channels formed by the nap (furry surfaces) of the yarn.

The approximate filtering capabilities of other nonmetallic filtering media are shown in the following table.

Approximate Filtering Capabilities of Various Nonmetallic Media

Type of filter	Rating, μ 98%	Rating, μ 100%	Type of filter	Rating, μ 98%	Rating, μ 100%
Sintered plastic	3–15	13–20	Bonded stone	5–100	
Fired porcelain	0.2–25	0.2–25	Ribbon	40	
Fitted glass	0.1–125	0.1–125	Disk	0.5–10	5–10
Bonded carbon	10–70		Membrane	0.1–12	0.1–12

Membrane media are essentially thin porous sheets produced by the evaporation of solvents from a solution of cellulose acetate or cellulose nitrate. The resulting structure is monolithic (a single piece). Because it is not a fused or bonded mass of fibers or granules, it is not subject to media migration.

Composites of various media improve overall characteristics for some applications. Without going into details of manufacturers' special products, here are several useful combinations: (1) edge-type media with metal powder (the powder is bonded to itself and to the edge-type medium by sintering), (2) sintered stainless-steel mesh with metal powder, and (3) glass fiber with membrane. In the first two the coarser medium is downstream, the finer is upstream. The two media are bonded together to prevent bypass leakage around one or the other. In a glass fiber and membrane combination, the fiber is upstream and is not bonded to the membrane.

There are also magnetic composite filters which combine mechanical separation of nonmagnetic solids with magnetic removal of iron particles. Two common constructions are (1) Alnico magnets with ordinary mesh and (2) mesh with magnetic wires. Extremely fine filtration is possible. Efficiency of the magnetic portion is inversely proportional to the velocity of the fluid past the magnet. One drawback is that as the filter cake increases in thickness during onstream time, particles are more likely to break loose.

8-4 FILTER RATINGS FOR FLOW

The flow pattern through filter media is usually of the *laminar type*, and the pressure drop is directly proportional to viscosity and velocity. By comparison, flow through the housing is usually *turbulent*, and the pressure drop is proportional to the square of the velocity—normal viscosities have no effect.

Flow rates depend on open area, total thickness of the filter medium, and pore size. It is possible to have two wire cloths of the same size with substantially the same removal rating but with widely different effective open areas—giving widely differing pressure drops for the same flow rate. For example, each of two cloths might be rated at 43 μ. The first could be calendered 200 × 200 mesh with 0.0026-in-diameter wires; the second could be a 325 × 325 mesh with 0.0014-in-diameter wires. The first has about 40 000 holes per square inch and 11.4 percent open area; the second has about 105 000 holes per square inch and 30 percent open area. It is obvious that the second with more holes and open area has the lower pressure drop and the greater dirt capacity at a given rate of flow.

Most filters can be designed to support almost any pressure drop encountered in service, but much depends on the size of the element and the design of the support structure. Absolute pressure has very little effect as long as the differential pressure across the filter element is low. If the differential pressure is high, compressible media might deform or even flake or break, thus releasing contaminants; in extreme cases the media might be compressed enough to restrict flow. Small filter sizes are the least affected, because filtering area is small compared with other dimensions and flexing is less likely to occur. However, even in large sizes, support structures can be built to help media withstand high differentials.

In practice, filter media can usually be categorized as high pressure or low pressure—though almost any medium can be adapted for almost any differential pressure. Fiberglas, glass fibers, paper, felts, random-packed materials in the fiber family, and membranes are usually considered 75- to 125-psi media. Wire and sintered-metal media are normally considered applicable to 10 000-psi differentials. Between these 125- and 10 000-psi categories are found media such as edge-type, ceramic, stone, wound yarn, and sintered plastics.

8-5 FILTER TEMPERATURE RATINGS

The temperature limit of a filter is usually a direct function of the lowest-temperature material in the fiber assembly. Glass fibers can serve at temperatures over 1000°F, but if resin binders are used, the filter might have lower limits. Temperatures in the usual hydraulic circuit rarely go above 150°F; so temperature is usually not a critical item.

8-6 FILTER CIRCUITS

In selecting strainers and filters for use in hydraulic equipment, there are a number of factors that must be considered—long life, amount of filtration, particle size, type of fluid, etc. In the past, many troubles attributed to poor-quality components should have been charged to foreign matter in the fluid. When filters clog, some systems may be damaged by undetected cavitation of the pump. A study of typical filtering practices can assist in determining causes of malfunction in hydraulic systems.

Figure 8-1 shows a block diagram of a typical hydraulic circuit. The circuit contains a reservoir, pump, relief valve, pressure filter, load, and the return to the reservoir. Note that the pump-inlet strainer is omitted. This circuit may malfunction at some time because the filtration occurs

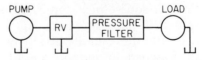

Fig. 8-1 Pressure filter in pump discharge line.

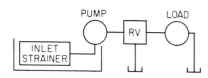

Fig. 8-2 Inlet strainer in pump suction line.

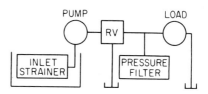

Fig. 8-3 Inlet strainer and pressure filter.

after the relief valve, which is a relatively sensitive device. Even though the fluid is cleaned at some point in the circuit, this is not sufficient or satisfactory for many applications. The relief valve may malfunction because of small particles entering the orifices in the pilot mechanism. Where the circuit pressure characteristics are not critical and the filter is protecting a sensitive control device, this circuit might be satisfactory.

Figure 8-2 shows a similar block diagram of a circuit with the filtration furnished by an inlet strainer. The strainer cleans the fluid before the fluid enters the circuit. There is danger of the filter becoming clogged and damaging the pump by resulting cavitation.

Figure 8-3 shows a strainer at the pump inlet and a filter in the line. This arrangement may provide needed protection for the pump and the relief valve at a degree of filtration different from that needed for precise components within the directional-valve network.

The primary circuit used in precision missile-testing systems is shown in Fig. 8-4. Note that the suction line to the variable-delivery pump is taken at the base of the tank vortex. The tank is pressurized with compressed air. A heat exchanger is used to maintain a stable temperature of the fluid. A flowmeter is provided to measure fluid movement in the external circuit. This meter is not affected by fluid passed through the main relief valve.

Figure 8-5 shows the same circuit with several filter stations added. Note that the supply air is passed through an air drier after passing through the reducing valve. The dry air is then filtered through a 2-μ nominal, 10-μ absolute unit. Many devices similar to this unit will use about 2-psi dry nitrogen to prevent condensation and introduction of contaminated air through leaks.

One source of contamination frequently overlooked is the case drain from the hydraulic pump. This can be a source of considerable contamination because it cannot readily be cleaned when the system is flushed. Furthermore, the pump is a prolific dirt generator, continually adding dirt to the pump case. For this reason, a 2-μ nominal, 10-μ absolute filter is included in the pump-case drain line between the pump and reservoir.

The line filters in this system have silver-soldered stainless-steel wire-cloth elements. Two filters are used in series (cascade) at each filtration point. The first filter is rated at 10 μ nominal, 25 μ absolute. The second is rated 2 μ nominal, 10 μ absolute. Cascade filtration has been found more

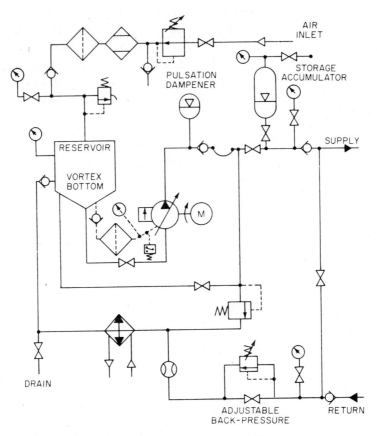

Fig. 8-4 Filtration employed in a primary missile-testing hydraulic circuit.

satisfactory from the standpoint of filtration effectiveness and the time between element replacements. The filter units in this system will withstand 4500 psi without danger of rupture.

The 10-μ absolute filter elements used in the primary circuit give the finest degree of filtration that is practical for the required flow rates and pressures. However, as discussed in Sec. 8-3, there is a chemically inert membrane filter material available with absolute ratings less than 1 μ. While its dirt-retention capacity and structural strength are poor, it can be used at low flow rates and low pressures. A bypass filter of this type rated at 3 μ absolute has been added to the system as shown in Fig. 8-5. A 10-μ absolute filter has been cascaded ahead of it to lengthen the service life. The purpose of this filter is twofold: (1) to filter all hydraulic fluid introduced into the system, (2) to clean the fluid in the system to a greater degree than possible with the primary filters. This ensures that the contamination level of the circuit is constantly being upgraded while the auxiliary bypass mechanism is in use.

8-7 CIRCUIT REFINEMENTS

Filters alone will not ensure the degree of cleanliness needed in the circuit in Fig. 8-5. Dirt is constantly being generated or being introduced into the system, and since the filters are not 100 percent effective, precautions must be taken to hold this introduction or generation to a minimum. To accomplish this, practices which were formerly acceptable on less complex industrial hydraulic systems have now been discarded and improvements developed. As a result of experience and gradual evolution, the following refinements have been incorporated into many circuits.

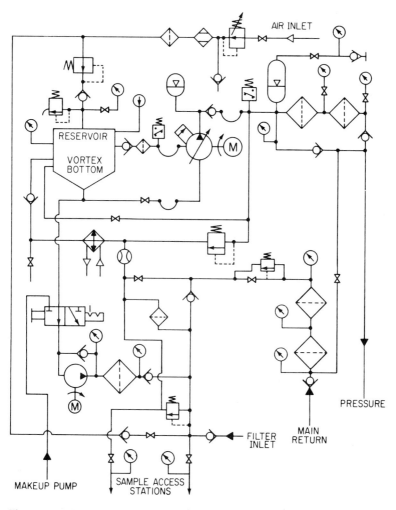

Fig. 8-5 Filtration employed in a primary hydraulic circuit fitted with main-line filters and auxiliary conditioning devices.

Filtration of Hydraulic Fluids **217**

1 Stainless-steel plumbing is widely used because it produces the least contamination. Carbon-steel tubing would be a second choice. Black pipe provided directly from the mill or possibly from storage is a prolific source of system contaminants over a long period of time.

2 Threaded parts are eliminated wherever possible, as the threading operation creates particles which cannot be completely removed by cleaning. The brittle apex of a thread can break loose long after the system has been put into service.

3 Stainless-steel reservoirs are used because they cause a minimum of contamination. Contrary to current acceptable industrial hydraulic reservoir design, the tank in Fig. 8-5 is designed to prevent dirt from settling within it. The tank in this circuit is not considered a repository for foreign matter. Pipe connections and contour are designed to impart a swirling motion to the fluid, and the tank bottom is sloped with the outlet pipe at the lowest point. All inside seams are welded and ground smooth to eliminate dirt pockets. In intricate hydraulic systems such as that in Fig. 8-5, accumulations of dirt in pockets may lie unnoticed for an extended period and then become dislodged and discharged into the main fluid stream. Even if this occurs upstream of the pressure-line filter, the dirt is still a problem since the contamination level downstream of the filter also increases.

4 Instead of using a cluster of fittings where a number of lines connect to a common point, drilled manifold blocks are used to minimize dirt traps.

5 Cleanliness of all component parts is strictly maintained.

6 Heat exchangers are used to keep the fluid at a fixed temperature so that orifice flows are constant and the flow through the filters is at a suitable rate.

7 In extreme situations, specially lined hoses are used. Materials such as Teflon will eliminate the sloughing off of particles from the hose.

8 All terminal connection points are always protected with individual caps when removed for servicing. Linty rags or dirty corks should never be used to cover or plug open ends.

8-8 FILTER HOUSING

The housing in which the filter medium is contained will depend on specific needs. When the filter is in a pump suction line, the prime consideration may be protection from crushing because atmospheric pressure is attempting to push the fluid through the filter medium to satisfy the vacuum being created by the pump.

Note in Fig. 8-6 the perforated metal used to support the wire mesh as

Fig. 8-6 Sectional view of perforated metal backup for wire mesh.

used in an indicating-type suction filter. This suction filter is shown as it is installed in the reservoir of Fig. 7-1. The pleats in the mesh increase the filtering capacity without sacrificing mechanical strength. Figure 7-4 shows an installation of a nonindicating suction filter submerged in the reservoir.

Figure 8-7 shows a pleated filtering element in a throwaway-type can that screws onto the head assembly to which the pipes are connected. A

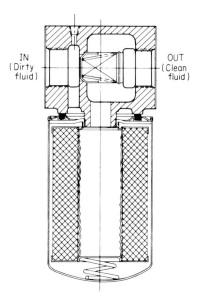

Fig. 8-7 Spin-on-type filter with throwaway-type cartridge.

Filtration of Hydraulic Fluids **219**

unit of this type can be installed external to the main reservoir in either a suction or a return line. This provides for a quick change of the filter cartridge with minimum downtime. Usual pressure rating is a maximum of approximately 100 psi. The spring-loaded poppet in the head assembly may have a much lower spring rating if it is used for suction service. This helps to prevent pump cavitation, resulting from cold, stiff oil or from a saturated filter cartridge. The simplicity of the throwaway design makes it practical to replace filter cartridges on a service-time or operating-hour basis, rather than by some type of mechanical or pressure-drop indicator reference.

Figure 8-8 shows a pleated-type filter unit in a housing having a relief valve similar to that used in the filter assembly of Fig. 8-7. Single-bolt design of the filter housing of Fig. 8-8 permits rapid change of the filter

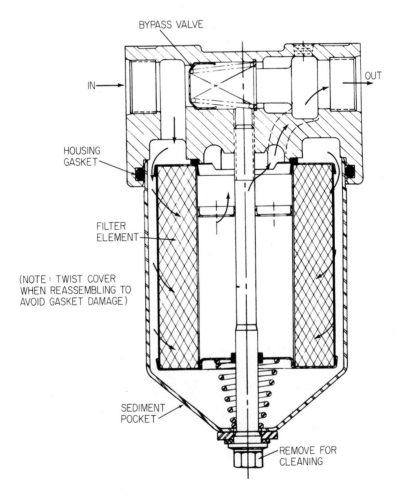

Fig. 8-8 Single-bolt, cartridge-type filter with relief valve.

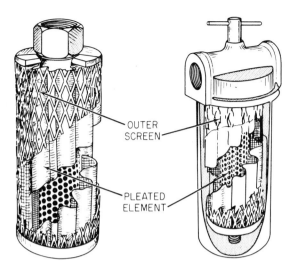

Fig. 8-9 Pleated-type filters for use in a hydraulic reservoir or separate housings.

cartridge. The outer shell slides into an encapsulated O ring in the head assembly. This permits higher pressure within the housing than is usually possible with the spin-on design. The center spring holds the filter element in position.

Note the gauge port at the *in* port of the filter assembly of Fig. 8-7. Two ports can be provided in the filter head assembly (one of the two is shown in dotted lines in Fig. 8-8) so that the pressure drop across the filter cartridge can be read on sensitive gauges to indicate the degree of cartridge saturation.

The filter in Fig. 8-9 has an outer screen for mechanical protection of the pleated element, whether it is used in the reservoir or in a separate housing.

A ribbon-type element using edge filtration is shown in Fig. 8-10. Figure 8-11 compares the disk-, ribbon-, and pleated-type elements. The disk-type element in Fig. 8-12 can be cleaned while it is in operation by revolving the intermediate disks that are keyed to a central shaft. The thickness of the space between the disks determines the degree of filtration.

8-9 INDICATORS FOR FILTERS

Figure 8-13a shows a cross section of a surface-type filter fitted with an indicator to show the condition of the element by registering the pressure drop through the unit. The filter cartridge floats within the housing. An internal spring forces the cartridge against seat *A*. As the cartridge

Filtration of Hydraulic Fluids

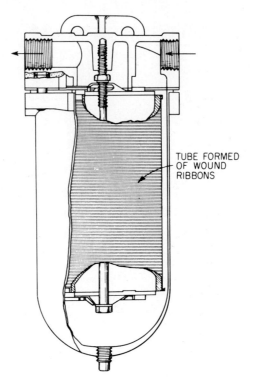

Fig. 8-10 Ribbon-type filter unit.

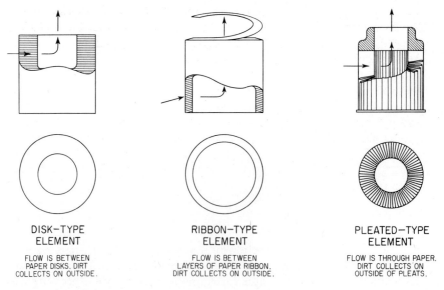

Fig. 8-11 Comparison of disk, ribbon, and pleated elements.

222 *Industrial Hydraulics*

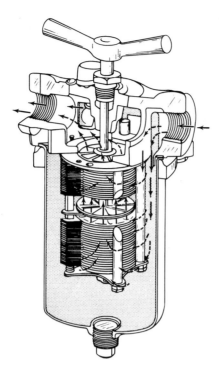

Fig. 8-12 Disk-type element that can be cleaned while in use.

becomes contaminated, a greater pressure drop is required to pass fluid through it. This greater pressure drop urges the floating filter element to the left against the internal spring. The movement of the cartridge can be viewed through an inspection window, and the need for cleaning can be indicated by an arrow-shaped device mechanically connected to the cartridge, as shown in Fig. 8-13*b*.

Where the filter is not readily accessible for inspection, the indicating element can be replaced by a cam which will actuate a limit switch at the desired point in the contamination cycle, thus indicating need for attention. The manner in which the limit switch is mounted is shown in Fig. 8-13*c*. Filters of this type with the moving cartridge automatically bypass the filter screen if the unit is not cleaned. At no time will the pump be deprived of fluid.

8-10 PRESSURE DROP

A review of the term *pressure drop* will be helpful at this time. The term means *pressure loss*. For example, if a fluid stream passing through a screen has a pressure of 10 psi before entering the screen and a pressure of

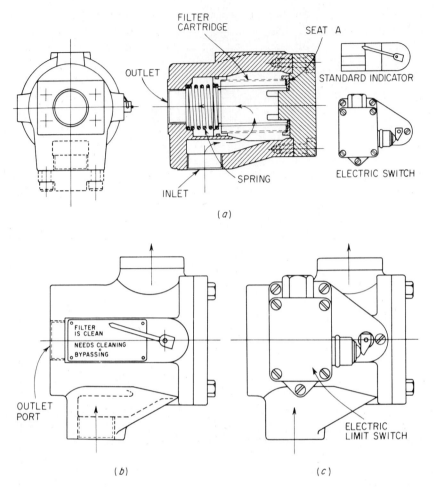

Fig. 8-13 (*a*) Pressure filter with manual or electrical indicator. (*b*) Intake filter with manual indicator. (*c*) Intake filter with electrical indicator.

9 psi after passing through the screen, the resultant 1-psi difference (10 − 9) or loss of pressure is the *pressure drop*—or *pressure differential* as it is sometimes called. As another example—a fluid stream enters a pipe at 10 psi and leaves the other end at 9 psi. A pressure differential or pressure drop of 1 psi exists along the length of the pipe.

Pressure drop is present everywhere a fluid stream flows and is especially pronounced when the fluid must pass through a restriction where it gains velocity while it loses pressure. Such gain in velocity is clearly demonstrated with a garden hose. As the nozzle is gradually closed, the outlet is restricted. The fluid velocity through the nozzle increases, causing the water to travel farther in a smaller stream. Pressure drop during flow of the water through the nozzle creates the high velocity. The pres-

224 Industrial Hydraulics

sure converted to velocity of the water stream is then dissipated in the work load moving the water through the nozzle and then through the air.

To illustrate how pressure drop works in a filter, consider a wad of glazier's putty that is gently pushed against a window screen. The putty sticks to the screen but does not go through. If more pressure is applied on the putty, it will be forced through the screen. Considerable pressure is necessary to force the putty through the screen against the resistance to flow, even though the only pressure after the screen is atmospheric. The pressure drop, therefore, is very large. A concentrated stream of fluid could exert pressure on the putty in the same way. In doing this, some of the energy of the stream is used up and the fluid pressure would decrease in direct proportion to the amount of force required to clear the path.

Pressure drop in a filter is a direct measure of the amount of force required for the fluid to penetrate the mesh or other filter material. Anything obstructing the flow of a stream of fluid through the mesh will be subjected to whatever pressure is necessary, up to the point where the obstruction will be removed or pushed through or the fluid is stopped. In a filter the obstruction has no place to go except through the mesh. So, if the obstruction is dirt, the primary purpose of the filter is nullified if the dirt can go through the filter.

The amount of force necessary to nullify a filter depends on several factors: fineness of the filter, size of the particles, strength or toughness of the particles, and rigidity of the filter element. Experience indicates that the kind of dirt commonly found in hydraulic and coolant systems will be forced through a filter if the pressure drop exceeds 12 or 15 psi.

There are other limiting factors. If the filter is to be used on the suction side of a pump, not more than 5-psi total pressure drop in the suction line can be tolerated by the pump. This allowance includes the drop through the filter and the drop through the intake pipe. Some pumps cannot stand more than 3-psi total pressure drop in the suction system. The pressure drop of a suction filter must be low enough, when clean, to allow for some extra capacity through the intake pipe and also some capacity to tolerate the foreign matter as it accumulates. If a filter has a port-to-port drop of 1 psi and the intake pipe has a drop of 0.5 psi, then the filter has a tolerance ratio of 2:1 if a maximum drop of 3 psi could be tolerated (3:1.5). This ratio would be considered adequate in most commercial applications.

The rate of contamination deposit and the amount of dirt in the system may dictate a higher ratio to lengthen the time between cleanings. In such cases the tolerance ratio is best determined by actual experience. The tolerance ratio can be increased by starting with a lower initial (when the filter is clean) pressure drop, a larger filter, or a coarser screen. Never exceed the upper ratio limits. A review of manufacturers' recommendations for suction characteristics of the pump in a circuit plus the recommendations for filter size and grade can add life and effectiveness to a hydraulic system.

8-11 FILTER OPERATION AND MAINTENANCE

Bypass or proportional filters are easily neglected if a plan for systematic maintenance is not adopted. Only a portion of the fluid passes through these filters, usually somewhat haphazardly, as there is no method of ensuring complete cleaning of the fluid other than continuous flow. This flow can be expected to involve all the fluid in the system eventually.

Indicating devices, such as gauges, flow switches, and pressure switches, should be provided for visual check of the fluid filter. If these devices are not used, it is necessary to set up a program based on specific time intervals for checking the filter.

Here is a check list that will provide a basis for a satisfactory maintenance program:

1. Provide mechanisms to indicate when the filter is saturated and needs cleaning.

2. Become acquainted with the filter and clearly identify it on the machine. Make certain that all personnel understand that a filter must be serviced at some predictable or definite time.

3. Make provisions for changing cartridges or servicing the filter without undue loss of fluid and with a minimum of machine downtime.

4. Keep a record and see if the time between cleanings is satisfactory. Possibly a larger filter or one with more capacity will save maintenance time and costs over a relatively short period of time.

5. Evaluate the effectiveness of the filter. Filters of different capabilities may eliminate machine or component malfunctioning.

8-12 FLUID STORAGE AND HANDLING

The handling and storage of industrial hydraulic fluid are as important as the filtering devices used as an integral part of the hydraulic system. A clean-fluid program starts in the storage area. The following points are especially important:

1. Drums of fluid should be stored on their sides to prevent water and other foreign material from collecting on top and leaking past the bung seals into the drum.

2. Clean the top of the drum before opening it. The pump or faucet to be used should not be laid on a dirty surface, and parts to be inserted into the new drum should be wiped off carefully before use.

3. Use only clean containers in handling or transferring fluid from drums to machines. Containers used for one fluid should not contain fluid of any other type or viscosity. Galvanized containers can affect many of the

industrial hydraulic fluids containing additives—these additives might react with the zinc to form soaps. Use of galvanized containers should therefore be avoided.

4 Do not draw fluid out of storage until it can be placed in a hydraulic reservoir or other carefully protected area.

5 Keep the fluid clean at all times.

Keeping the fluid clean at the reservoir depends on several factors. Perhaps the most important is a well-protected yet accessible filler hole. The filler pipe cover should be designed so that it can easily be cleaned and removed without getting dirt into the opening. A portable fluid-transfer unit such as that in Fig. 8-14a protects the fluid from common contaminants around a factory and ensures a supply of clean fluid to the machine. Figure 8-14b shows a filter commonly used in installations of this type. When fluid is being removed from a machine for repair purposes, the filter will prevent contaminants from being transferred from the reservoir into the clean storage barrels. The unit in Fig. 8-14a can also be used to clean the fluid in a reservoir while the machine is operating.

Fluid returned to storage can often be processed through centrifuges to remove water and some other contaminants. Absorbent filters using finely divided materials such as activated clay or the natural clay called fuller's earth in the form of a bed or renewable cartridge are not generally recommended for hydraulic fluids. These materials have a tendency to remove desirable additives from the hydraulic fluid.

8-13 STARTING NEW MACHINERY

A filter like that in Fig. 8-14a is convenient for removing foreign matter that may settle in the tank of a new machine. Some machine manufacturers connect the suction hose from the pump shown in Fig. 8-14a to the drain connection in the hydraulic reservoir where all the foreign matter will collect. As the machine is cycled, the scavenger cleaning unit picks up all the debris and deposits it in the filter. Because of the filters on the machine, the scavenger action of the portable transfer unit, and the magnetic devices in the reservoir, a satisfactorily clean supply of fluid is ensured after the initial break-in period. If the machine, after installation, is refilled with clean fluid, it can be started with a minimum of difficulty. Long, satisfactory service can be anticipated.

SUMMARY

Filters are used to maintain hydraulic fluid in clean, contaminant-free condition. Filters are needed to counteract the generation of foreign matter

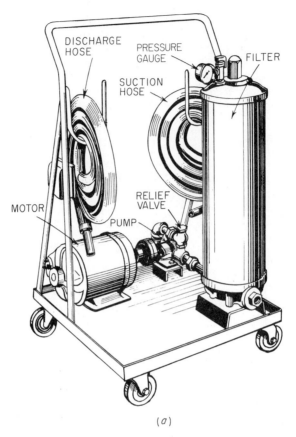

Fig. 8-14 (*a*) Portable fluid filter and transfer unit.
(*b*) Filter for the unit in *a*.

in the circuit. Foreign matter is produced by normal wear of machine parts and by contamination from external sources.

Filter types vary from coarse-screen devices sometimes called strainers to microscopic orifices in leached-glass devices. The filter medium may be either metallic or nonmetallic. The degree of contaminant removal is determined by the type of filter.

The allowable pressure drop that can be tolerated through the filter will determine many of the physical aspects of the installation.

Scheduled maintenance programs are essential when filters are used in an industrial hydraulic system.

Fluid-storage and fluid-handling techniques can directly affect the ultimate need for filtration in a hydraulic circuit. Filtration may be most important when starting up new systems.

Filtering devices may be supplemented with permanent magnets located within the reservoir or as a part of the filter mechanism. Specific

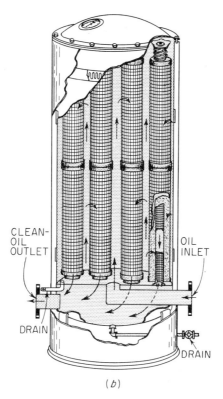

Fig. 8-14 *(Cont.)*

cleaning schedules may be necessary to avoid migration of collected foreign matter that may lodge on the surface when magnets are employed to collect ferrous materials within a system.

Static seals, gaskets, and filter media must be compatible when the unit is used with certain fire-resistant hydraulic fluids. The filtering characteristics of fire-resistant hydraulic fluid may be different from those for petroleum-base fluids. The manufacturer's recommendations should be carefully followed.

REVIEW QUESTIONS

8-1 Why is a filler such as diatomaceous or fuller's earth not recommended for use in filters for hydraulic fluid?

8-2 What type of malfunctioning can be expected if an input filter to a pump is plugged by foreign matter?

8-3 Why is the reservoir in Fig. 8-4 constructed with the vortex at the bottom?

Filtration of Hydraulic Fluids **229**

8-4 What parts of the filter should be checked for compatibility if fire-resistant hydraulic fluids are to be used in a fluid-power system?

8-5 What is the purpose of the spin-on feature of the filter shown in Fig. 8-7?

8-6 Why would nitrogen be used to pressurize a hydraulic-fluid tank?

8-7 Why is the relief valve provided in the housing of the filter?

8-8 What is the advantage of the pleated mesh illustrated in Fig. 8-9?

8-9 What is a settling tank?

8-10 What is the difference between a nominal micron rating and an absolute micron rating?

8-11 How big is a micron?

8-12 Can porous-type granular-medium filter elements be cleaned and reused?

8-13 Will an edge-type filter or a granular-medium-type filter be more effective for long, thin contaminant particles?

8-14 What is sintering? Why is wire mesh sintered?

8-15 Why are filters cascaded?

8-16 Is pressure drop present in all hydraulic systems? Explain.

8-17 Why should fluid drums be stored on their sides?

8-18 How can water be separated from hydraulic fluid?

8-19 Does a bypass-type filter ensure 100 percent cleaning of all contained fluid?

8-20 Why are two test stations used at a filter?

LABORATORY EXPERIMENTS

8-1 Obtain a pressure-type filter housing and several elements with different orifice sizes. Note the pressure drop when passing equal amounts of fluid through the different-size orifices.

8-2 Block varying portions of the element in Experiment 8-1 with tape or suitable material and note the effect on pressure drop. Introduce foreign matter into the filter and note the change in pressure drop.

8-3 Install precision vacuum gauges on several intake lines to pumps on various machines, with and without intake filtration. Note the differences in readings.

8-4 Operate a piloted relief valve for several hours with unfiltered fluid. Remove the valve and insert a 50-μ or better filter in the line to the relief valve. Note the difference in performance.

8-5 Energize a spring-offset pilot valve of small capacity with a sliding spool and leave the pressure on at a relatively high level. For example, use 3000 psi and leave the valve energized for about one hour. Then release the pressure and observe the time needed for the spring to return the valve spool to the original position.

8-6 Install a 10-μ or better filter directly ahead of the valve in Experiment 8-5. Shift the valve several times for a reasonable flushing action. Repeat Experiment 8-5 and observe the time necessary for the spring to return the spool.

8-7 Immerse various filters under hydraulic fluid in an open container. Connect the input of the filter to a source of compressed air and a manometer to measure pressure. Note the difference in pressure necessary to cause a bubble or stream of bubbles to rise with different filter materials.

8-8 Remove the fluid samples from various machines with different types of filtration. Check the degree of contamination in each. Repeat this experiment shortly after new filter elements have been added or the existing ones cleaned.

8-9 Remove a sample quantity of oil from a new drum. Pass the oil through filter paper. Check the amount of contamination with a microscope.

8-10 Procure a spin-on filter assembly such as that illustrated in Fig. 8-7. Tap the center passage so that a large setscrew or pipe plug can be inserted. Reassemble the unit. Gradually apply pressurized oil to the input with a sensitive gauge teed in at the entry. Note the pressure level as the flow commences. Note the relation of pressure to flow. Compare the results with the nominal pressure rating of the valve within the filter assembly.

8-11 Repeat the test of Experiment 8-10 with the plug removed and a clean filter element in the normal assembled position. Note the pressure needed to start flow. Increase the flow to determine the actual flow through the filter prior to actuating the bypass mechanism.

8-12 Pipe a second filter identical to that of Experiment 8-11 in parallel. Note the pressure drop through the combined filters with a manometer or sensitive gauge.

9
Fluid-Temperature Control

Every hydraulic system operates most efficiently when the fluid temperature is held within a certain band. There is an ideal temperature at which the resistance to fluid flow is minimum while the fluid still retains its lubricating and sealing characteristics. All hydraulic fluids—oil, water, synthetics, etc.—are designed to have these desirable characteristics of minimum flow resistance with suitable lubricating and sealing properties when the fluid temperature is maintained correctly. Operating at a temperature below that of the correct band results in slower movement of the fluid through orifices and other restrictions in the piping system. This slower movement requires that the pump develop a higher discharge pressure to force the fluid through the piping. Incorrect time responses of equipment result from the slower fluid movement.

Temperatures higher than the desirable level can reduce the lubricating characteristics of the hydraulic fluid. Some fluids may *break down*, forming sludge, varnish, or other undesirable contaminants. This can lead to plugged orifices, piping, or valves or to other troubles. When the fluid viscosity decreases at higher temperatures, as usually happens, the lubricating film formed by the fluid may be destroyed. This allows metal-to-metal contact in various parts of the equipment. Heat is generated by this contact, and the fluid temperature rises. If the fluid reaches the breakdown temperature, the contaminants mentioned above may form.

To control the temperature of hydraulic fluids, many circuits use a *heat exchanger*. This heat exchanger may either heat or cool the hydraulic fluid, depending on the requirements of the circuit. Various types of heat exchangers used in hydraulic systems are discussed in this chapter.

9-1 TYPES OF HEAT EXCHANGERS

To *heat* the hydraulic fluid the following types of heat exchangers are used: (1) shell and tube, (2) plate type, (3) immersion-type electric, (4) immersed steam coils, and (5) surface heaters using electricity or steam. To *cool* the hydraulic fluid these heat exchangers are used: (1) finned radiating, (2) honeycomb, (3) shell and tube, (4) plate type, and (5) immersion coils.

9-2 SHELL-AND-TUBE HEAT EXCHANGERS

This type [Fig. 9-1] consists of a cylindrical metal *shell* containing a bundle of straight metal *tubes*.

The hydraulic fluid usually flows through the shell around the tubes while the heating or cooling medium flows through the tubes. *Baffles* in the shell [Fig. 9-1] cause the hydraulic fluid to follow a circuitous path. In following this path the fluid contacts all the tubes, ensuring maximum transfer of heat between the heating or cooling medium and the hydraulic fluid.

Shell-and-tube heat exchangers can withstand high pressures. This heat exchanger, sometimes termed a *liquid type,* can use steam or hot water as the heating medium. For cooling, refrigerated liquids or water can be used. By proper design of piping connections and valves, heating or cooling mediums can be alternately directed through the heat exchanger. With this arrangement, the heat exchanger can effectively control the temperature of hydraulic fluids through extremely wide temperature bands. The need for heating a hydraulic fluid in devices operating at low ambient temperatures is as important as providing cooling at high ambient temperatures.

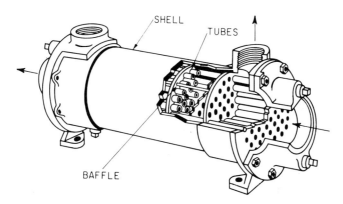

Fig. 9-1 Shell-and-tube type of heat exchanger.

Fluid-Temperature Control

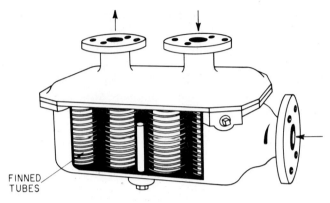

Fig. 9-2 Plate-type cooler.

9-3 PLATE-TYPE HEAT EXCHANGERS

This type [Fig. 9-2] contains finned tubes (or plates) through which the hydraulic fluid flows. The cooling or heating medium flows around the outside of the tubes and fins. Since the fins are tightly connected to the tubes, the surface area for the transfer of heat is large. This large area makes the plate-type heat exchanger a useful device in hydraulic circuits. The operating pressure of the exchanger is, however, less than the shell-and-tube type because the structure is not as strong.

9-4 IMMERSION-TYPE ELECTRIC HEATERS

This type [Fig. 9-3] consists of resistance wire encased in a suitable protective shell. The heater is positioned in the hydraulic-fluid reservoir to provide direct heat transfer to the fluid. Caution must be taken to provide some movement of fluid around the heater so that fluid adjacent to the heating element is not overheated. Fluid deterioration can be caused by overheating. In some hydraulic systems, auxiliary equipment may be used to agitate or move the fluid near the electric immersion heater. Local safety codes must be carefully reviewed before using electric immersion heaters with flammable hydraulic fluids.

Electric immersion heaters for fluid-power applications may have their wattage limited to a value that will not cause deterioration of the fluid.

Fig. 9-3 Electric immersion heat exchanger.

Ratings of the heater should be carefully compared with the recommendations furnished by the fluid supplier to ensure compatibility.

9-5 IMMERSED STEAM OR HOT-WATER COILS

These heat exchangers [Fig. 9-4] consist of a coil of pipe or tubing around which the hydraulic fluid is circulated. Positive movement of the fluid is necessary to avoid overheating. As with the electric immersion heater, steam or hot-water coils are usually located in the fluid reservoir.

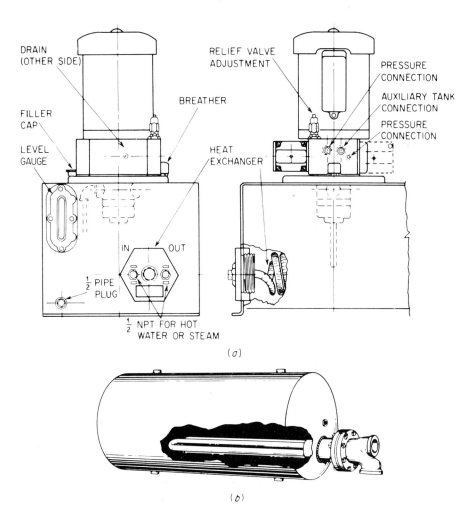

Fig. 9-4 (a) Immersed steam or hot-water heating coils in power unit. (b) In-storage tank.

Fluid-Temperature Control 235

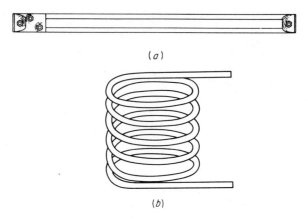

Fig. 9-5 (*a*) Electric surface heater. (*b*) Steam jacket.

9-6 SURFACE HEATERS

Two different types of surface heaters are popular: (1) electric [Fig. 9-5*a*] and (2) steam jacket [Fig. 9-5*b*]. In either type, heat is applied to the external surface of the fluid reservoir or to components in the system. These heat exchangers are effective if the hydraulic fluid is circulated so that fluid at a lower temperature is constantly reaching the heating surface. Proper baffling of the flow of fluid through the reservoir will ensure effective heat transfer.

9-7 FINNED RADIATING CONDUCTORS

The conductors connected to the large directional-control valve in Fig. 9-6 are the finned radiating type. These conductors, located in the air stream, have fins to increase the heat-transfer area. The fins conduct heat to the air stream more effectively, giving a larger heat-transfer capacity.

9-8 HONEYCOMB-TYPE RADIATOR

This type [Fig. 9-7*a*] resembles the radiator used to cool automobile and truck engines. In mobile equipment the hydraulic-fluid radiator may be placed near the water-cooling radiator. Air is forced through the fluid radiator by a belt-driven fan. This flow of air increases the heat-transfer rate. In stationary equipment the fan for the radiator may be driven by a separate electric motor to provide adequate heat transfer. Where water or other cooling media are not available, the fan-cooled radiator can produce good results. Honeycomb-type radiators must be protected against excessive pressure because of the lightweight construction needed to provide the best heat-transfer rate.

Figure 9-7b illustrates a cooling unit encircling the motor drive coupling. A squirrel-cage-type fan unit is attached to one-half of the coupling. Air is forced through the heat-transfer section by the integral squirrel-cage blower. The cooler is designed to be mounted on the electric-motor shaft in the coupling area between the pump and motor. It may, however, also be used with other engine drives where a coupling of the type shown is employed. The unit may also be constructed of ⅜-in finned annular tubing. The tubing can be formed into a wraparound crescent shape and become a multipass exchanger. Either the pump-drain or return-line oil can be directed through the inside of the tubes before going back to the tank, cooling the oil. The cooler also serves as a coupling guard. Fans on this type of cooler are dynamically balanced. About ½ hp is required to drive the fan of a unit such as that shown in Fig. 9-7b.

Curve A [Fig. 9-7c] shows pressure drop vs. flow rate for applying the cooler to either the pump-case drain or system return line. Maximum recommended backpressure when the cooler is applied to the case drain of a vane pump is 7 psi, because of backpressure exerted on the pump shaft seal.

Curve B shows the heat dissipation characteristics of the cooler at 1800 rpm; when a 1200-rpm drive is used, two-thirds of plotted data will apply.

Some hydraulic circuits having light-duty cycles and favorable operating conditions may operate satisfactory with simple radiation of heat from

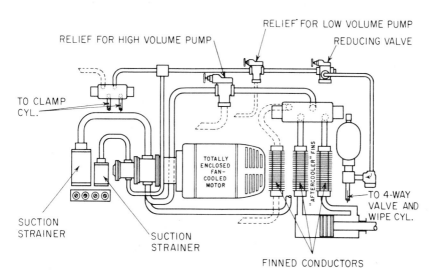

Fig. 9-6 Air-cooling arrangement used with a totally enclosed, fan-cooled motor.

Fluid-Temperature Control

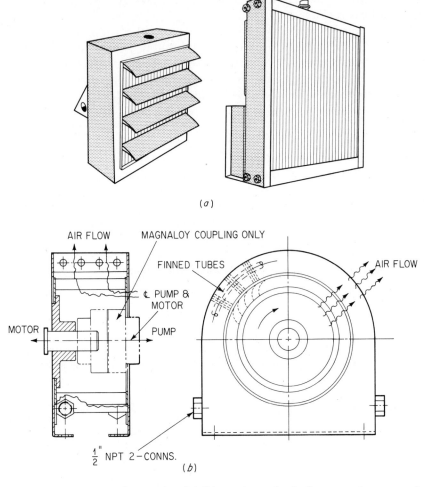

Fig. 9-7 Radiator cooling units. (*a*) Direct-through air-flow-type honeycomb radiator cooling unit. (*b*) Radial cooling with squirrel-cage blower. (*c*) Characteristic curves of coupling guard-type coolers.

the circuit components. Other circuits, however, require additional heat-transfer surface under almost every operating condition.

9-9 RESERVOIR IMMERSION COILS

Cooling coils can be immersed in the reservoir fluid or cast into the reservoir side walls. Water is pumped through the coil to cool the fluid. Either arrangement usually provides excellent heat transfer. Large quantities of heat can be absorbed by the water and conveyed to the discharge area. It is essential that any tubes in the fluid tank be fully immersed at all

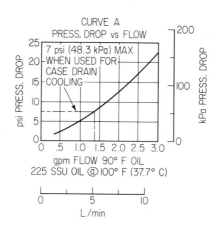

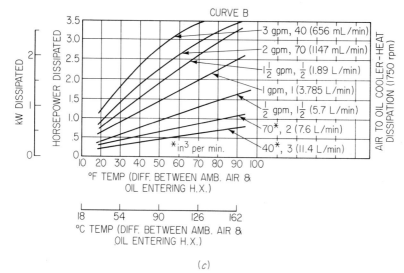

Fig. 9-7 *(Cont.)*

times. If the tubes are not immersed, water vapor in the air may condense on the tube surface. This moisture may mix with the hydraulic fluid and cause operating troubles.

9-10 AUTOMATIC CONTROL OF FLUID TEMPERATURES

Figure 9-8 shows a common method of controlling the temperature of the hydraulic fluid when cooling water is available. Note that a control bulb is inserted in the fluid sump. It is important that this control bulb be installed in a position and location where the fluid flow past the unit will reflect a true picture of the fluid temperature. If the bulb is located in a static fluid area, the control will be much less effective. The temperature may be controlled by a thermal switch that will actuate a solenoid-operated water valve rather than by the self-contained temperature-control valve shown in Fig. 9-8. In some installations it is desirable to install the fluid inlet at the bottom and the output at the top of the heat exchanger to ensure a constant supply of fluid in the exchanger. This will prevent any possibility of condensation should the fluid level partially uncover the tubes carrying the cooling water. A check valve in the bottom fluid supply line can prevent draining of the fluid, which would uncover tubes and cause condensation of moisture in this area.

Figure 9-9 shows a plastic extrusion press circuit that uses a fluid motor to drive the extrusion screw. Note that the exhaust fluid from both the fluid motor and the clamp mechanism is directed through the heat exchanger with pressure limits established by a spring-loaded check valve.

Pilot pressure for both directional-control valves is taken from a point ahead of the 50-psi check valve. The directional-control valve within the motor circuit takes pilot pressure externally from the 2-gpm pump circuit. The pilot pressure is connected internally for the clamp directional-control valve.

Pressures in excess of 20 psi are diverted through the spring-loaded check valve to prevent overloading of the heat exchanger. The amount of water passing through the heat exchanger is controlled in a manner similar to that shown in Fig. 9-8.

Figure 9-6 shows a circuit using air cooling as the major method of temperature control. For maximum heat dissipation, the finned tubes would be installed in the air intake of the fan-cooled motor. If there were less need for heat transfer, the finned tubes might be installed in the air discharge from the fan-cooled motor. The pump body is in the discharge slipstream from the motor fan so that it will be cooled by this air flow. The fins on the cylinder lines to the large valve will conduct heat as the fluid flows in each direction. The temperature of hydraulic fluid passing through these lines having relatively large heat-transfer areas will be materially reduced. A finned line directs the fluid back to the tank. This flow will be in only one direction and will usually be at a high rate. Since the valve is of the open-center type, there is a nearly constant flow through this tank conductor. The tubing eliminates many heat-generating devices in the fluid conductors by minimizing the unions, elbows, and orifice effect of certain types of fittings.

9-11 REDUCING FLUID-TEMPERATURE RISE

The most economical heat-transfer system will carry away the heat gained in functional operation through the tank, conductors, and components. Assuming that the hydraulic system has been carefully engineered and that a certain amount of heat will be necessary for proper functioning of the equipment, some installation procedures can aid in heat dissipation.

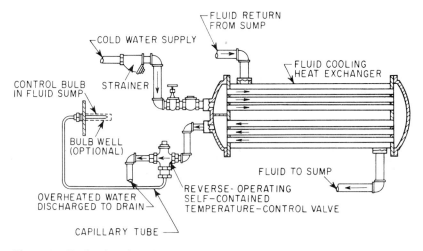

Fig. 9-8 Cooler fitted with an automatic temperature control.

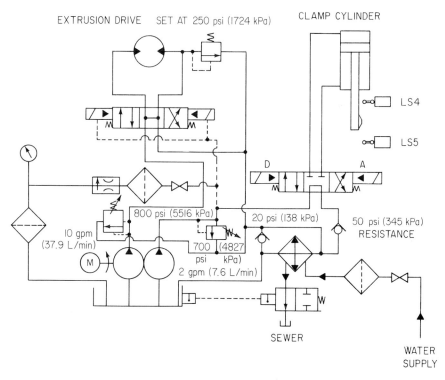

Fig. 9-9 Plastic extrusion press circuit using a heat exchanger with water as the cooling agent.

Fluid-Temperature Control **241**

The following list of suggestions is useful in making maximum use of heat dissipation by radiation:

1. Provide a light sheet-metal hood over the electric motors so that the air passing through the motor also passes by the pump to pick up heat radiated from the pump.
2. Keep the tank and plumbing away from steam lines, cleaning lines using hot water, cleaning tanks, and other sources of heat.
3. Insulate the platens of steam presses, etc., from the ram or other hydraulic parts to minimize objectionable heat transfer.
4. Do not operate at fluid pressures higher than necessary. Keep fluid pressures as low as is consistent with good machine functioning.
5. Keep line sizes large enough to ensure a minimum of frictional losses and subsequent heating.
6. Baffle the reservoir so that fluid will be forced to flow along the outside walls of the tank.
7. Keep the bottom of the tank off the floor so that air can flow around all tank surfaces.
8. Do not pile objects against the tank. They can stop air flow to the tank surfaces.
9. Remove from the top of the tank all debris that may be acting as a natural insulator.
10. Consider light-metal lines for return flows to increase radiation capabilities.

DESIGN CONSIDERATIONS FOR FLUID-TEMPERATURE CONTROL

Heat in a hydraulic system is, of course, caused by power losses. While an elaborate analysis of power losses of a given system can be made, it will generally be found sufficient to rely on empirical data. The heat that is generated in a hydraulic system can, in general, be assumed to be the equivalent of about 20 percent of the connected horsepower (nameplate rating of the driving motor). This figure may vary considerably below or above this amount, depending on the efficiency of the pump and circuit, the existence of heat-generating devices, etc. The pumping unit, hydraulic cylinders, reservoirs, and piping may serve to dissipate most or all of this amount, depending on conditions. For medium-pressure installations with moderate service, oil tanks will generally dissipate the heat generated in the hydraulic system. A heat-transfer coefficient of $1\frac{1}{2}$ to 2 Btu/ (F)(ft^2)(hr) (8.52 to 11.34 W/m^{2}K) can be used to calculate heat dissipation.

Water cooling must be resorted to in some installations. Whether water cooling should be used on any particular installation is determined by requirements, type of service, and custom. Generally, medium-pressure circuits employing rotary-type pumps at pressures of 1000 psi (6895 kPa) and less do not require water cooling. Hydraulically operated machine tools are rarely equipped with cooling systems, although recently some heavy-duty metalworking machines have appeared employing this means of oil-temperature control. Oil hydraulic presses utilizing high pressures are generally equipped with water coolers. Hydraulic transmissions operating continuously under heavy load are sometimes water cooled.

Heat dissipation by natural radiation can be assisted by installation of water-cooling coils in the oil tank. This is a low-cost, but not very efficient, means and is economical only for small installations. Copper coils can be used for this purpose. Coils of this type will dissipate 10 to 15 Btu/per hr per °F per ft^2 (56.8 to 85.2 W/m^{2}K) of surface.

The following general information will assist the reader in the selection of a suitable oil-cooling unit. The heat load can be determined as a percentage of connected horsepower or, if desired, after a detailed analysis of the power losses. It can be expressed in Btu per minute (kJ/min) or its horsepower equivalent (watts). A maximum inlet oil temperature of 125 F (51.7 C) should be provided for, and cooling-water temperature according to local conditions (75 F average) (23.9 C). An oil flow of about 25 gallons (94.6 L) per 1000 Btu (1055 kJ) should be provided. This will result in a temperature drop of the oil of $1000/(25 \times 7.35 \times 0.45) = 12$ F (6.67 C), so that the oil leaves the cooler at about 113 F (45 C).

Water-to-oil rates for plate-type coolers can be chosen from ½:1 up to 2:1, balancing water economy against cooler costs, and ¼:1 to 1:1 for shell-and-tube-type coolers.

This will result in a minimum water consumption of about 50 lb (22.7 kg) per 1000 Btu (1055 kJ) to a maximum of 400 lb (181.8 kg) per 1000 Btu (1055 kJ), or a water-temperature rise from 2½ to 20F (1.39 to 11.1 C).

In the designing of a cooling system, adequate protection for the cooler against surge pressures and shocks must be provided. Normal operating pressure should not exceed 10 to 15 psi (68.9 to 103.4 kP). Considerably higher pressures may be developed at lower oil temperature or higher viscosity. Circulating pumps, if separately driven, should not be under-motored for this reason, and care must be taken not to exceed the pressure ratings of the coolers, especially when the oil flow is subject to rapid variation. Protection of the cooler by surge relief valves is advisable.

Most axial- and radial-piston pumps are equipped with auxiliary pumps that can be utilized to circulate oil through the cooler.

Here is an example combining a few rules of thumb: Allowing 20 percent of connected horsepower as heat load, approximately 1 gpm (3.79 L/min) of oil circulation per heatload horsepower would lead to 1 gpm (3.79 L) for each 5 connected horsepower (3.73 kW). For instance, a pump operated by a 125-hp motor (93.2 kW) would have a 25-gpm (94.6 L/m) auxiliary pump, etc. This capacity will also suffice for efficient operation of auxiliary hydraulic controls, servomotors, etc.

Oil cooling can be made automatic by the use of any of the number of thermostatically controlled valves on the market.

The scarcity and ever-increasing cost of water make the use of waterless or water-saving oil coolers attractive. Air-to-oil unit coolers of the finned-tube type with forced-air circulation have been developed for this purpose. They are suitable for applications where oil temperatures are limited to a minimum of 130 (54.4 C) entering the cooler. The evaporative-type cooler is suitable for a dry climate. An evaporative cooler is made up of a coil section through which oil is circulated; over this coil section water is sprayed, while air is drawn up from the bottom, counterflow to the water, and blown out the top. Part of the water in passing through the air stream evaporates, thereby cooling the remainder of the water. The water temperature will approach the wet-bulb temperature; therefore, the oil will approach the wet-bulb temperature. The water consumption of this type of cooler is about 1 lb (0.45 kg) for 1000 Btu (1055 kJ), contrasted to 100 lb (45.5 kg) for 1000 Btu (1055 kJ) average for a shell-and-tube-type heat exchanger.

SUMMARY

When a heat exchanger is found to be a necessary part of the fluid-power system, several basic factors will assist in determining the type to be used. If water is not readily available, it may be necessary to use an air-transfer type of heat exchanger with a suitable radiator and fan. Specific localized problems will dictate other considerations. In the event that the amount of space available and the heat generation are not compatible with either air- or water-cooled heat exchangers, it may be necessary to resort to mechanical refrigeration equipment.

Inadequate temperature control will cause sluggish operation when the fluid medium is at too low a temperature. Cavitation of the pump and slow response of hydraulic signal sources can be expected. Temperatures above the normal level will increase slippage or clearance flow in pumps, valves, and piston rings. Elevated temperatures can reduce the lubricating value of the fluid medium and cause expansion of closely fitted parts to a point where binding and scuffing can occur.

High shocks can often be traced to excessively high temperatures. The damping effect of fixed orifices becomes less effective as the temperature increases. Another ill effect of high temperature can be traced directly to rapid deterioration of sealing mechanisms. Most packing and seals for commercial hydraulic equipment are designed for a temperature not in excess of 170°F and usually for much lower temperatures. An extreme temperature can be one of the first indications of circuit malfunction.

Localized heating will often pinpoint the area in which repairs or replacements are needed.

The direct effect of temperature on the effectiveness of fluid-power devices indicates the importance of both the design and maintenance of heat-exchange equipment. For this reason many manufacturers provide temperature-indicating devices within the hydraulic circuit.

REVIEW QUESTIONS

9-1 What effect do baffles within reservoirs have on heat transfer?
9-2 Why would the fluid enter a cooler from the low point and be exhausted from a high point?
9-3 What types of circuits would most likely require honeycomb heat-transfer equipment?
9-4 What is the reason for using the check valve on the tank line in the circuit in Fig. 9-9?
9-5 What effect will cold fluid have upon a simple orifice?
9-6 Why might an electric immersion heater require special precautions when installed in a fluid-power system?
9-7 Can a heat exchanger be used for both heating and cooling?
9-8 Is the position of a temperature-sensing device within the reservoir of importance? Explain.
9-9 What happens to fluid that is heated to an excessive temperature?
9-10 Can a hydraulic circuit be designed so that it generates no heat?
9-11 Does the heat exchanger of Fig. 9-9 provide both heating and cooling? What indicates the capability?
9-12 How is a symbol modified to show that air is the cooling medium?

LABORATORY EXPERIMENTS

9-1 Install a 5-gpm pump on a 15-gal capacity reservoir. Pass the fluid through a relief valve set at 500 psi until the temperature in the tank becomes stabilized. Note the temperature. Repeat at 750 and 1000 psi. Note the rate of temperature change and the stabilization points.
9-2 Repeat Experiment 9-1 with an open-center valve that will relax the pressure to 10 psi for one-half the time. Note the new stabilization temperature of the fluid in the reservoir.
9-3 Install the suction and discharge pipes within 6 in of each other in Experiments 9-1 and 9-2. Note the change in stabilization temperature.
9-4 Carefully isolate the intake and discharge lines with baffling to make the fluid travel the longest distance against the sides of the

reservoir and note the difference in temperature at which the tank is stabilized.

9-5 Note the local temperature at the relief valve and at the pump in Experiment 9-1. Install a temperature-measuring device in a tee immediately below the discharge of the relief valve and compare the temperature here with the average temperature in the other parts of the reservoir.

9-6 Install a vacuum gauge in the suction line of a production machine. Record the vacuum at machine start-up. Repeat at the highest operating temperature when the fluid temperature has stabilized. Note the effect of temperature on the ability of atmospheric pressure to push fluid into the pump.

9-7 Repeat Experiment 9-6 with a mobile machine that is operating outdoors.

9-8 Install an immersion heater, such as that shown in Fig. 9-3, in an open tank. Measure the temperature gradients within the tank without circulation. Circulate the oil through the tank. Note the change in temperature gradients.

9-9 Repeat Experiment 9-8 with a surface heating element applied to one wall of the tank.

9-10 Pass 5 gpm of No. 10 motor oil through a fixed orifice of 0.125-in diameter. Record the pressure required at 70, 90, 120, and 150°F. Repeat the experiment with No. 30 motor oil. Note the difference in pressure. Calculate the effect this can have on machine tool operation.

10
Pressure-Control Valves

Pressure-control valves are used in hydraulic circuits to maintain desired pressure levels in various parts of the circuits. A pressure-control valve maintains the desired pressure level by (1) diverting higher-pressure fluid to a lower-pressure area, thereby limiting the pressure in the higher-pressure area, or (2) restricting flow into another area. Valves that divert fluid can be *safety, relief, counterbalance, sequence,* and *unloading* types. Valves that restrict flow into another area can be of the *reducing* type.

A pressure-control valve may also be defined as either a *normally closed* or *normally open* two-way valve. Relief, sequence, unloading, and counterbalance valves are normally closed, two-way valves that are partially or fully opened while performing their design function. A reducing valve is a normally open valve that restricts and finally blocks fluid flow into a secondary area. With either type of operation, the valve can be said to create automatically an *orifice* to provide the desired pressure control. An orifice is not always created when the valve is *piloted* from an external source. One valve of this type is the unloading valve—it is not self-operating; it depends on a signal from an external source. Relief, reducing, counterbalance, and sequence valves can be fully automatic in operation, with the operating signal taken from within the envelope. In this chapter we shall study the different types of pressure-control valves and learn how they are used in various hydraulic circuits.

10-1 TYPES OF PRESSURE-CONTROL VALVES

Eight popular devices for pressure-control service are:

Safety valve Usually a poppet-type two-way valve intended to release fluid to a secondary area when the fluid pressure approaches the set opening pressure of the valve. This type of valve protects piping and equipment from excessive pressure.

Relief valve Valve which limits the maximum pressure that can be applied in that portion of the circuit to which it is connected.

Counterbalance valve Valve which maintains resistance against flow in one direction but permits free flow in the other direction.

Sequence valve Valve which directs flow to more than one portion of a fluid circuit, in sequence.

Unloading valve Valve which allows pressure to build up to an adjustable setting, then bypasses the flow as long as a remote source maintains the preset pressure on the pilot port.

Pressure-reducing valve Valve which maintains a reduced pressure at its outlet regardless of the higher inlet pressure.

Hydraulic fuse Device equipped with a frangible disk which establishes the maximum pressure in a hydraulic circuit by rupturing at a preset pressure value.

Pressure switch Switch operated by fluid pressure and responsive to a rise or fall in fluid pressure.

10-2 SPRING-LOADED SAFETY VALVES

The simplest and most inexpensive type of safety valve is spring loaded. Circuits that can utilize the characteristics of spring-loaded valves can be adequately protected at minimum cost.

Figure 10-1a shows a typical spring-loaded safety valve. The body is designed to manifold as a sandwich in a mobile-type *valve stack*. The valve assembly can be housed in any convenient position because of the simplicity of construction. Fluid from the circuit to be protected enters the hole under the ball and acts against that portion of the surface of the ball serving to close the passage from the pressurized area to the tank conductor. The bias spring, holding the ball against the seat, is precompressed to a specific setting. The ball will not lift off its seat until the fluid pressure reaches the value for which the bias spring is set. Should the fluid pressure exceed the spring setting, the ball will lift off its seat and allow fluid to pass through the outlet to the tank port.

The cap-shaped guide bearing against the ball is carefully designed to create a predetermined *pressure gradient*, or controlled release of pressure as the flow is directed to the tank. This assists in preventing erratic motion of the ball and spring assembly. The ball creates a tight seal until the

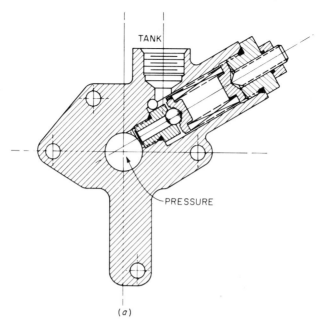

Fig. 10-1 (*a*) Adjustable safety valve to sandwich into directional valve stack.

desired pressure is attained. The guide cap assembly guides and supports the spring as well as serving as a damping agent. As the valve functions, it creates an orifice of the proper size to divert sufficient fluid to the tank to prevent further pressure rise. The time required to actuate a valve such as that shown in Fig. 10-1*a* is so small that overpressures, or pressures beyond the set value, are normally insignificant. The relatively heavy spring assists in reclosing the orifice very rapidly when the pressure surge or peak has passed. This type of valve is very fast and provides the safety function as indicated.

Increasing the spring tension by turning the adjusting mechanism [Fig. 10-1*a*] will raise the fluid pressure needed to force fluid through the orifice. An increase in the amount of fluid to be returned to the tank, without a change in spring tension, will raise the pressure needed to force fluid through the orifice. A change in the viscosity or density of the fluid will also change the pressure required to force fluid through the orifice. Thus, a heavy, viscous fluid will require a higher pressure to force it through a certain size orifice than will a thin, watery fluid.

Spring-loaded safety valves are usually small in size and limited in pressure range for a given spring design. Operating pressures for this type of valve are generally limited to 3000 psi or less.

Safety valves are built to be as dependable as possible because failure of a safety valve could lead to damage of expensive equipment. Some of the

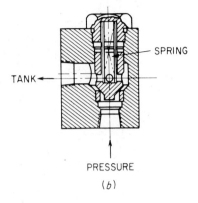

Fig. 10-1 (*b*) Nonadjustable safety valve.

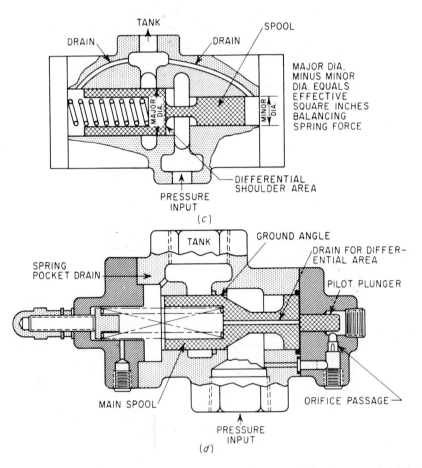

Fig. 10-1 (*c*) Relief valve having a spool with two different areas. (*d*) Pilot piston actuates the spool of this relief valve.

250 *Industrial Hydraulics*

refinements used in relief valves to reduce valve noise and increase valve stability are omitted in safety valves.

Safety valves are not normally considered as functional parts expected to operate during each cycle of the machine. A safety valve is usually set to open at a pressure about 25 percent higher than the maximum pressure expected in a circuit. Spring-loaded safety valves vary somewhat in their opening pressure for a given spring setting. If a more exact opening pressure is desired, the spring is sometimes replaced by a deadweight mechanism. The weighted unit must be mounted in a fixed position. Spring-loaded valves can be mounted in almost any position.

10-3 SPRING-LOADED RELIEF VALVES

The spring-loaded valves [Fig. 10-1a and b] can be used to limit the maximum flow pressure applied to a portion of a hydraulic circuit. An angle valve [Fig. 10-1b] is popular for replacing an elbow in the circuit, thereby reducing costs. When used as relief valves, both these valves operate in the same way as described for safety valves.

Where it is desired to reduce the piston-face area while maintaining a large passageway for fluid, a relief valve like that in Fig. 10-1c can be used. The *net area* (major area minus minor area) between major and minor diameters is the effective area against which the pressure acts in this relief valve. Both the large and small ends of the spool must be drained to the tank.

A small-diameter piston [Fig. 10-1d] provides similar results. The pilot piston actuates the main spool of this relief valve. Fluid quantities handled by the valve are larger than the quantities handled by the other three valves shown.

Smooth operation with a minimum of noisy *chatter* and uncontrolled *oscillation* (movement of spool back and forth in the valve body) depends directly on the relationship of operating components—particularly piston-face area and spring setting. To control these variables, a restrictive passage is used to dampen the acceleration and deceleration of the spool or poppet. The drain lines in the valve in Fig. 10-1c can each be fitted with an orifice to serve this same purpose. These drain lines, too, are in a relatively clean area away from high-velocity fluid. The orifice passage in the valve in Fig. 10-1d connects directly to the input pressure line. It is vulnerable to dirt or foreign matter in the fluid. This is a hazard when using a valve of this type.

Note that the spring cavities of the valves in Fig. 10-1c and d have a passage to the tank port. Flow resistance caused by dirt in this passage will be reflected into the spring cavity. This pressure acts against the pressure on the piston face, holding the valve more tightly on its seat. As a result, the input pressure must be higher before the valve will open. Such a condition is undesirable.

10-4 COMPOUND RELIEF VALVES

In the study of ISO hydraulic symbols in Chap. 2 it was stated that simplified symbols are widely used. Because of this, pressure-relief valves used in common hydraulic circuits are rarely shown complete with all auxiliary devices and connections. Instead, the simplified symbol shows only the basic relief valve, pressure input, tank connection, valve spring, and the offset arrow indicating that the valve is normally closed. A slash arrow as shown on the bias spring of the pilot-relief valve in Fig. 10-2a may be added to the bias-spring symbol of the simplified valve as shown in Fig. 10-2b if the valve is adjustable, particularly if this information is significant to circuit operation. Figure 10-2a shows the *complete* symbol for a compound relief valve. All adjacent controls are shown, along with the main relief element. The envelope surrounding all the elements may have five connections. These are (1) pressure input, (2) tank connection, (3) remote-control-station connections, (4) test station, and (5) external drain for the pilot relief (not shown) that is provided only on special order.

The input pressure and tank connection provide the major flow through the valve. Only enough fluid need flow to the test-station and remote-control connections for the respective functions. The test station is generally used for a gauge connection to check fluid pressure. This does not require a flow of fluid. The remote-control connection passes the quantity of fluid coming through the fixed internal orifice at the rate established by the spring in the main relief element. An external drain from the pilot-relief valve, if fitted, will not pass more fluid than passes through the fixed internal orifice.

Figure 10-3 shows a cutaway view of a compound relief valve. Note that the main spool is held by the spring in a position that blocks the passage

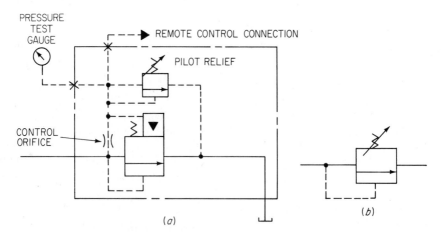

Fig. 10-2 (*a*) Complete symbol for a compound relief valve. (*b*) Simplified symbol for a compound relief valve.

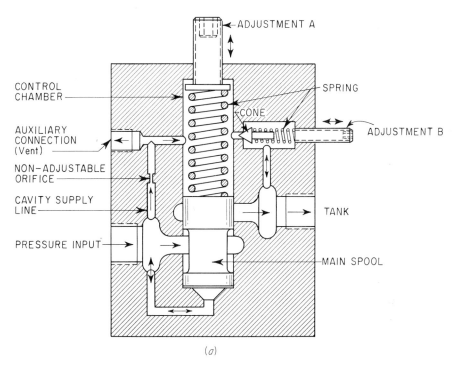

Fig. 10-3 (a) Sliding-spool-type compound relief valve.

from the pressure input port to the tank port, just as the symbol in Fig. 10-2 shows. Input pressure is directed to the bottom of the spool below the spring cavity without restriction. The supply line to the spring cavity is restricted by an orifice in the line. The area of each end of the main spool is the same. In certain poppet designs, the areas may not be exactly equal. One end may have a larger area to ensure certain functional actions. In operation, if fluid cannot escape through adjustment B, a balance is provided by the equal areas at each end of the spool. The spring then maintains the spool in the position where it blocks the path through the valve from pressure input to tank. When the pressure in the spring chamber above the main spool rises to a point where it can unseat the cone at adjustment B, a portion of the fluid will be passed to the tank port. When the volume of this flow exceeds that passing through the orifice into the main spring chamber, the pressure that is effective on the upper spool area may be less than that on the lower area so that balance is no longer maintained. As the pressure continues to rise at the lower end of the valve spool and the flow continues to increase through adjustment B while the degree of unbalance of the main spool becomes more pronounced, the pressure will force the main spool against the spring. This creates a path from the pressure input to the tank much like that created in a direct spring-operated relief valve.

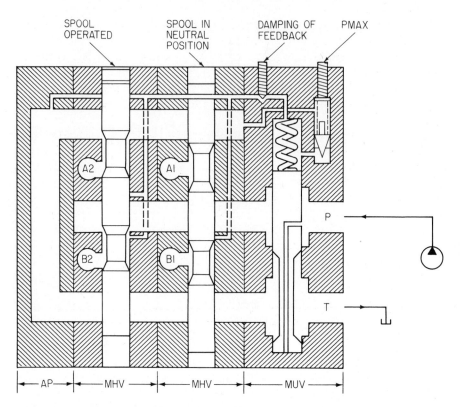

Fig. 10-3 (*b*) Pilot-type relief valve integrated into multiple control assembly.

Adjustment *A* would provide a specific maximum relieving pressure if adjustment *B* were completely relaxed. When adjustment *B* is in use, providing an additive pressure to the main-spool spring, the minimum relieving pressure will be fixed by adjustment *A*. The relieving pressure can never be less than that established by adjustment *A* in this valve. Pressure in the circuit could be less if there were relaxation through some other path. Resistance to pilot-fluid flow created by adjustment *B* may be considered as a hydraulic additive to the value of adjustment spring *A*. In many valves, such as that shown in Fig. 10-3*b*, the main-spool spring is not adjustable. The adjustment is limited to that made by the mechanism at *B* or through a parallel pilot valve piped to the auxiliary connection.

The pocket containing the main-spool spring is called the *control chamber*. It will be well to remember this term, as it is widely used in industrial hydraulics.

Note the auxiliary vent connection in the upper left side of the valve in Fig. 10-3*a*. This port (which is connected to the control chamber) permits the escape of fluid directly to the tank without restriction. Thus, there can be no hydraulic additive pressure to the main-spool spring.

If a small relief valve [Fig. 10-1*a*] is placed in the circuit with a connection to the vent port, the maximum relieving pressure will be established by this additive at a remote point.

Figure 10-3b shows how a relief valve can be integrated into a control assembly to maintain a uniform pressure drop across a directional-control valve. Port $A2$ is connected to one end of a hydraulic cylinder or one port of a hydraulic-fluid motor. Port $B2$ is connected to the other end of the cylinder or to the other fluid motor port.

Fluid entering the common pressure conductor, as shown, is directed to port $B2$ and exhaust fluid is directed to the tank porting through channel $A2$. This is a common flow-directing condition within a hydraulic circuit.

Pressurized pilot fluid to assist the main relief-valve bias spring is sensed at port $B2$. Note the small pilot passages at either side of the input pressure channel. The left four-way directional-control spool has been shifted so that the upper pilot line is blocked and the sensing line at port $B2$ is in communication with the control chamber of the relief valve.

The rate of pilot flow is controllable by the screw in this conductor, which is indicated as a damping device. This limits unwanted oscillation of the relief-valve spool.

The bias spring in the relief-valve control chamber determines the magnitude of the pressure in the input flow line. The sensing line at port

Fig. 10-3 (c) Relief valve becomes an integral part of control assembly.

B2 reflects the resistance to flow encountered by the fluid as it is directed to the cylinder or motor to transmit power. This resistance to flow is reflected as a pressure additive to the bias spring in the relief-valve control chamber. Thus a balance is created so that the net pressure across the directional-control valve spool is that created by the bias spring in the control chamber of the relief valve.

Should the reflected load at port B2 be greater than the setting of the pilot valve labeled P_{max}, which is located adjacent to and communicating with the relief-valve control chamber, then a maximum pressure level will be established. This will prevent operation of the system at unsafe pressure levels. The discharge of the small pilot-relief valve communicates with the return to tank passage.

A passage from the relief-valve control chamber is established when both directional-control-valve spools are placed in the neutral position. This passageway communicates and terminates in the return to tank conductor. The sensing lines from A1, B1, A2, and B2 are blocked in this neutral position. Pressure in the supply line is limited to that created by the relief-valve bias spring. Thus the pump will be unloaded or is in idle standby conditions, creating minimum energy loss.

Sensing lines are immediately available when either directional-valve spool is actuated and the pilot bypass to tank is simultaneously blocked.

Many directional-control valves can be integrated in a circuit, as shown in Fig. 10-3c.

Control of the directional-valve spool can be by lever, as shown in Fig. 10-3c, or by other means, such as electric solenoid, hydraulic pilot pressure, or mechanical means.

Additive pressure flow to the control chamber of the relief valve of Fig. 10-3a is continuously available through the nonadjustable orifice communicating with the pressure input. Additive pressure flow to the control chamber of the relief valve of Fig. 10-3b is only available when the sensing line from one of the cylinder ports is uncovered and the series bypass line is blocked during normal functional operation.

The relief valve of Fig. 10-3b helps to conserve energy. It also ensures smooth control of the load because of the uniform pressure drop across the directional-control-valve spool regardless of working load.

Directional-control valves not employing this or a similar system of piloting can be quite difficult to actuate with erratic operation of the work load. This occurs because the relief valve will function at maximum pressure as controlled by the pilot valve communicating with the control chamber [Fig. 10-3a] if the full pump flow cannot pass through to the load. Thus any metering or restriction at the directional-control valve to control the speed of the actuated load occurs at pressures far in excess of the load values.

The relief valve of Fig. 10-3b serves three basic functions: (1) When all directional spools are in neutral position, the relief valve serves as a minimum pressure level control or pump unloading valve. (2) As a directional-control spool is actuated, pressure from the control chamber communicating to the return to tank port is blocked as the sensing line to a cylinder port is uncovered, so that the relief valve becomes a compensating device to direct all pump fluid not being employed to do work back to the return to tank line at a pressure level created by the bias spring plus the pressure level sensed at the load. Pressure ahead of the directional-control-valve spool equals bias-spring value plus load-sensing pressure. Pressure beyond the directional spool is only load value. Thus the pressure across the directional-control-valve spool is always at the value of the relief-valve bias spring, which ensures a smooth, accurate flow-control function of the flow-directing member. (3) Upon encountering a load or malfunction beyond the value of the P_{max} setting, the relief valve will act as a safety valve limiting pressure to this predetermined value.

The auxiliary connection (vent line) from the control chamber of a piloted relief valve may be connected to a small relief valve such as the different types shown in Fig. 10-4. The remote valve can adjust to any pressure less than that set at adjustment B [Fig. 10-3a], or P_{max} [Fig. 10-3b].

The relief valve of Fig. 10-3a or b could be modified so that the discharge from the pilot-valve assembly is directed to an external connection rather than the return to tank line as shown. This external connection can then be directed to the pressure input of a pilot valve [Fig. 10-4] so that the spring-loaded cone assembly is in series with the internal member. The external valve would function to create an additive pressure to the integral pilot valve. Thus the integral pilot valve would then establish the minimum pressure controlled by the relief valve, and higher values would be determined by the adjustment of the remote-control valve.

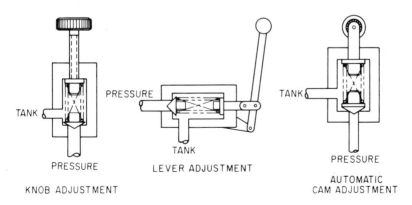

Fig. 10-4 Three adjustments commonly used in relief valves—knob, lever, and cam.

Pressure-Control Valves 257

10-5 ELECTRICALLY OPERATED RELIEF VALVES

A solenoid-operated, two-way valve may be designed as an integral part of a compound relief valve. The vent from the relief-valve control chamber is directed to the inlet of the two-way valve. An alternate design directs the control fluid to an external two-way valve through associated conductors. If the two-way valve is normally open, the relief pressure will be at the minimum established by the main-spool bias spring until the solenoid is energized. If the two-way valve is normally closed, the pressure will remain at the maximum established by the control pilot relief until the solenoid is energized. Both designs direct the discharge fluid from the two-way valve back to the reservoir with minimum backpressure. All backpressure will affect the pressure setting of the main valve. The solenoid-operated, two-way valve serves to *vent* the control fluid to the tank. It is a digital type of control, providing an ON and OFF function. There is no practical degree of modulation incorporated in the design.

An electrically operated relief valve, designed to provide modulation or variable remote pressure control, employs a direct-current solenoid to actuate a pilot-control assembly.

An 80-V dc solenoid with a resistance of 280 Ω is employed in the solenoid-operated relief valve shown in Fig. 10-5a. When used with a 115-V, 60-Hz ac power supply through suitable controls, the maximum pressure setting is ensured.

Pressures less than maximum are attained by varying the current to the dc solenoid coil. The body of the sliding-spool-type two-way valve, shown in Fig. 10-5a, is drilled to conduct fluid from the input port to the end of the spool opposite the bias spring. This fluid is also directed through an external tube to a fixed-flow, pressure-compensated, flow-control valve, shown at 3. Fluid can pass through orifice 6 concentrically located in the main spool and then on to the control chamber adjacent to pilot 2. The poppet in pilot 2 is urged against the adjacent seat by a spring. The spring is compressed a controlled amount by a hydraulic piston. The pressure applied to the piston is determined by the poppet assembly in valve 4. The poppet assembly of 4 is biased by dc solenoid 5. Reducing valve 3 provides a continuous, uniform, small flow to valve 4, regardless of input or output pressure in the major flow path. If the current applied to solenoid 5 is reduced, the bias force holding the poppet against the seat in valve 4 is reduced. This reduces the force on the bias spring of valve 2 which, in turn, reduces the pressure assist to the bias spring of the main valve 1, and the major pressure is reduced accordingly.

This type of control can be applied to most piloted pressure-control valves, such as sequence and reducing valves, as can the relief function described.

Functions shown in the schematic circuit of Fig. 10-5b are numbered as in Fig. 10-5a so that the actions can be followed. Drain and return fluid from valve 4 is directed through an external tube to join with that from

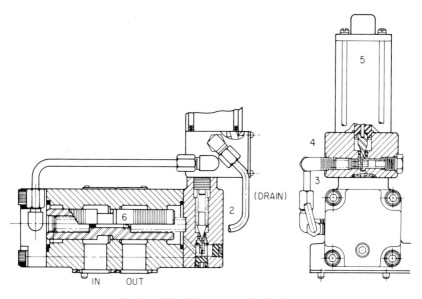

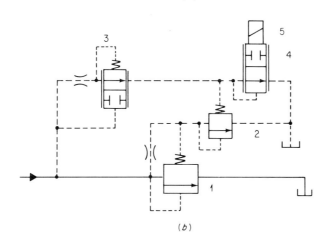

Fig. 10-5 (a) Electrohydraulic pressure-control valve. (b) Schematic of electrohydraulic modulating-type remote relief-valve control.

valve section 2; it proceeds to a drain port in the valve interface, where it passes through a connecting subplate to a conductor which directs the return fluid to the reservoir.

Within the schematic circuit of Fig. 10-5b, note the parallel lines incorporated in valve symbols 3 and 4. These lines indicate the fact that the valve-control element modulates or moves to an infinite number of positions, unlike the ON-OFF function of the relief valve vented directly by a

Pressure-Control Valves **259**

solenoid-operated, two-way valve, which has been previously described. The symbol for the ON-OFF valve would not employ the parallel lines at the side of the symbol where the ports terminate.

The use of a piston to create the desired load on the bias spring of valve 2 adds a control over the degree of sensitivity of adjustment. A large-diameter piston requires less pressure to provide a given load to the spring. A small piston will respond more rapidly with the same quantity of fluid. Generally, gradual adjustment is desirable. Because of this, the piston is as large as existing space will justify.

The controls recommended for this type of valve are normally of the current-control type. Voltage-control circuits, directly applied to the valve coils, are quite large in control-component size and can be expected to be heat-generating in nature. Typical repeatability for a valve of this type at pressures of 1000 to 3000 psi is plus or minus 20 psi. Because of the control mechanism involved, a minimum pressure may be 100 psi for a unit designed to operate up to 3500 psi maximum.

The chart in Fig. 10-6a shows the relation of the input electrical signal to the resulting pressure. The module of Fig. 10-6b provides the needed control. Pins 1 and 2 of this module are connected to the 115-V, 60-Hz power supply. This power is rectified in a silicon-controlled rectifier (SCR) circuit as a full-wave rectification. The dc power is applied to the amplifier and also, through a regulator and trim control R, to pin 8.

Note that the potentiometer has been connected to pins 8, 6, and 3, providing a variable control for power to be applied through $R20$, $R2$, and the summing junction to the amplifier. By varying the potentiometer from low resistance between pins 8 and 6 to a higher resistance, we decrease the power signal to the amplifier and therefore decrease the amplifier output to the valve coil.

The coil is connected to the pins 4 and 5 of the module. Between pins 3 (common ground for the module circuit) and 7, we have shown a capacitor. This capacitor provides a time delay proportional to its fixed size and the fixed resistance size of the resistance $R2$. This delay is known as a *resistance capacitance (RC) circuit time delay*. The time delay works on either an increase or a decrease. The time delay obtained is approximately 10 milliseconds per microfared (10 ms/μF) of capacitance.

The ground for the valve coil passes through a current-sensing resistor $R3$. Note that this coil pole is also applied to the negative pole of the summing junction. This provides temperature-resistance compensation for the coil.

Figure 10-6c shows the physical characteristics of a control assembly used to provide the signal for the valve shown in Fig. 10-5. Figure 10-6d through f shows the symbolic representation for relief, reducing, and sequence valve control.

The main spool of the relief valve in Fig. 10-3 is a sliding element in a single-diameter bore. This design is most common in sequence-type

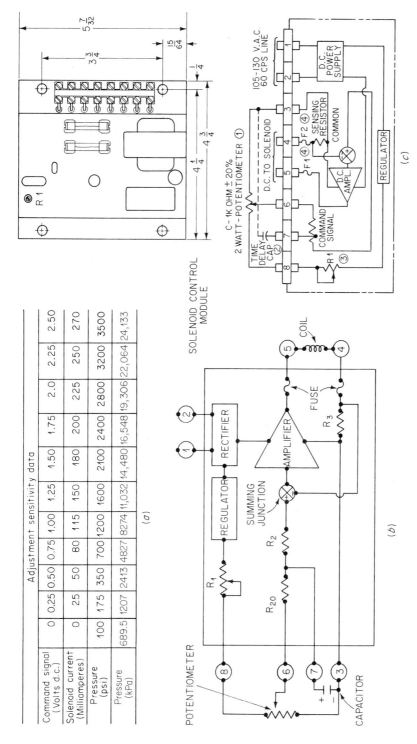

Fig. 10-6 Electrically modulated pressure-control valves. (*a*) Sensitivity data. (*b*) Schematic control circuit. (*c*) Typical control module.

Pressure-Control Valves   **261**

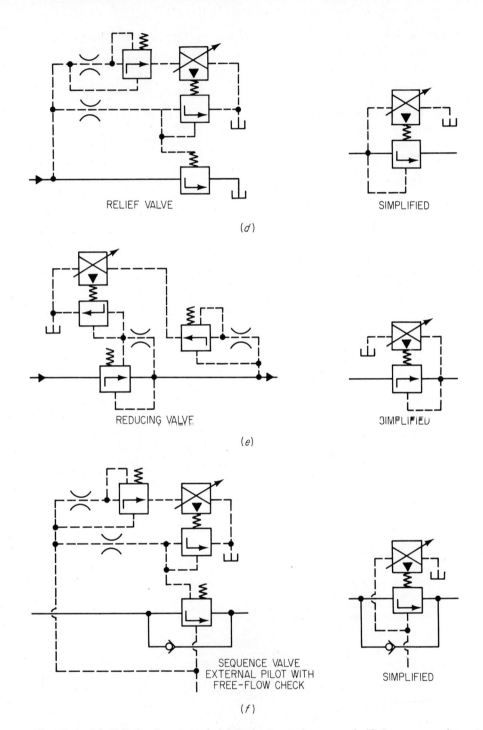

Fig. 10-6 (d) Relief-valve control. (e) Reducing-valve control. (f) Sequence-valve control.

262 Industrial Hydraulics

valves. Relief valves usually have a poppet-type seat for maximum speed of response when a rising pressure indicates the need for relief.

Note the appendage at the bottom of the relief spool in Fig. 10-7. This serves to assist in moving the spool to the closed position after the vent connection is closed. The fluid impinging on this button will assist in closing the poppet structure while high flows are passing through the opening.

Rapid operation and extreme dependability are required of the valve shown in Fig. 10-7. To avoid accidental adjustment by unqualified personnel, the pilot mechanism adjustment is recessed under a protective cap. There is a greater area holding the poppet closed in this unit than that available to unseat the poppet. Therefore, the pressure in the control chamber can be expected to be somewhat less than that in the main working lines. The ratio of the large-diameter skirt to the seat area determines the pressure values.

To obtain balancing values similar to those obtained with the single-diameter spool and still maintain the poppet construction, a balancing extension above the main-spool spring may be used. In this design the fluid passing through the auxiliary pilot control can be directed back to the reservoir through the center of the spool or through an external drain. If the flows are large, it may be desirable to use the external drain. Pressures encountered in forcing the drain fluid through the pipe to the tank will reflect back to the valve and act as an additive to the springs. If this flow is of any magnitude, it can directly affect the stability of the valve. Where the return to the tank port of the main relief valve is to be pressurized, it is essential that the pilot fluid be returned to the tank through a separate line.

Three ways of adjusting relief valves are shown in Fig. 10-4. The *knob adjustment* is the most common. The knob is used to turn a threaded

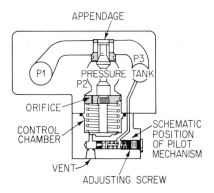

Fig. 10-7 Poppet-type compound relief valve with protected adjusting screw.

Pressure-Control Valves **263**

plunger into or out of the valve body to increase or decrease the spring tension, thereby increasing or decreasing the relief-pressure setting of the valve. In the *lever adjustment,* the spring tension is set by the position of the lever. The lever can be set either manually or automatically. *Cam adjustment* uses a circular- or spherical-shaped cam to maintain the desired tension in the valve spring. The cam can be set either manually or mechanically.

10-6 PRESSURE-REDUCING VALVES

Every pressure-reducing valve has at least two ports—an inlet port and an outlet port.

High-pressure fluid enters the inlet port and passes directly through the valve to the outlet port until resistance at the outlet port approaches the set value of the pilot or the bias mechanism holding the passage through the valve in the open position. As pressure rises at the outlet port because of resistance to flow, the control spool moves and restricts flow in a pattern commensurate with the rate of pressure rise in the outlet port.

The control spool can completely close the path from inlet to outlet. As pressure in the secondary or outlet port diminishes, the bias mechanism will urge the spool to an open position so that pressurized fluid can enter and raise the pressure to the set value. In this manner the valve maintains a desired reduced pressure at its outlet, regardless of the higher inlet pressure potential.

Figure 10-8 shows a typical pilot-actuated and pilot-controlled pressure-reducing valve. High-pressure fluid enters the upper port and is directed through the passage formed by the spool and valve body to the low-pressure outlet connection. It is within this part of the valve structure that the major control function is provided. The valve spool is normally urged to a position where a free passageway is provided through the valve by a nonadjustable bias spring. The pilot signal is taken from the discharge or outlet of the valve (low-pressure area). It is the limiting of the passage resulting from a signal emanating from the low-pressure area that determines the level of the secondary outlet pressure.

The nonadjustable spring urging the main spool to the open position is assisted by pressurized fluid in much the same manner as are piloted relief valves. Resistance to flow at the low-pressure outlet, caused by a load on the circuit at that point (such as the piston rod and connected machine member contacting the work load), will increase the pressure value at the low-pressure outlet. This pressure is reflected through orifice A to the control chamber and also to the bottom of the spool through the passage within the casting to permit passage through a filter prior to entering passage A. As the spool has equal areas on both ends, the nonadjustable spring will hold the main spool in the open position as long as pressures are equal at both ends. As the low pressure increases at the

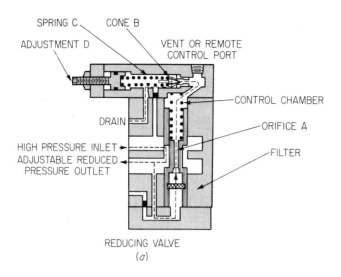

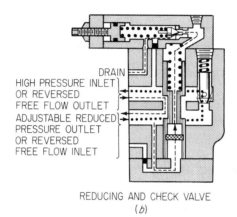

Fig. 10-8 Compound pressure-reducing valve. (*a*) For normal flow in one direction only or with limited return flow. (*b*) Equipped with integral check for free return flow through the valve.

low-pressure outlet, pressure also increases in the control chamber because of the flow through orifice *A* that equalizes pressure on each end of the main spool. When the pressures in the control chamber and low-pressure outlet exceed the setting of spring *C* and cone *B*, a path is established to the tank. When the flow is equal to the flow through orifice *A*, a maximum pressure assist is established. As the flow past cone *B* exceeds the flow through orifice *A*, the main spool will be unbalanced and will be urged against the nonadjustable spring. This will restrict the flow to the low-pressure outlet, establishing the low-pressure value. The resulting orifice that is created by the main spool and body will permit a

Pressure-Control Valves **265**

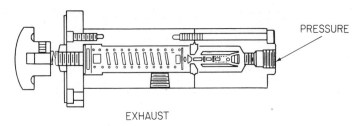

Fig. 10-9 High-pressure, remote-control relief valve.

flow of fluid equal to that going through orifice A at the value established by the nonadjustable spring in the control chamber.

A compact, high-pressure, remote-control relief valve [Fig. 10-9] can be connected to the drain connection in series with the integral pilot-relief valve. The adjustment at D will determine a minimum pressure additive. The remote valve can be adjusted for higher pressures.

A precision control pilot-relief valve such as that shown in Fig. 10-9 with a guided poppet can provide accurate pressure-level adjustment at a remote location (usually 15 ft or less). It is usually connected to the remote-control (or vent) connection. This will parallel the integral pilot relief. The pilot relief, set to the lowest value, will control the pressure assist to the main-spool spring and the low pressure at the outlet of the valve.

The size of orifice A will determine the rate at which the main spool will return to an open position after having closed during functional operation. External forces tending to increase the pressure on the low-pressure outlet port will be dissipated through orifice A and the integral pilot-relief valve. High shock waves encountered in the low-pressure lines because of external loads may require additional relief capacity (by the use of a separate relief valve in the low-pressure line) or enlarged flow capacity through orifice A. The flow through orifice A must be less than the maximum flow capabilities through the auxiliary pilot-relief control that is used to establish maximum pressure assistance.

10-7 COUNTERBALANCE, SEQUENCE, AND UNLOADING VALVES

These similar valves differ mainly in their application and in some basic construction features. However, they may be considered as a family of valves.

Counterbalance valves serve an extremely useful function in hydraulic circuits associated with mobile equipment. Many machine tools can be equipped with counterweights for vertical or angular slides so that counterbalance valves are not subject to extreme service usage. Because of this auxiliary-type service, valves such as shown in Fig. 10-1d or pilot-operated equivalents can be used in machine tool circuits.

Features of particular importance in mobile applications include thermal protection, tight seal, and extreme dependability. Because of these requirements, counterbalance valves used primarily for mobile service are often referred to as *holding valves*.

Figure 10-10 illustrates what may be termed a *counterbalance-type holding valve*. A check-valve poppet is included to permit a free return flow through the valve. Fluid returning from the cylinder is checked by the poppet in one passage and contained between the relief poppet and attached piston in the other passage. The attached piston is smaller in diameter than the seat where the relief poppet rests. The bias spring urges the relief poppet against this seat. Because of this difference in diameter, there is an area which opposes the bias spring as pressure is applied. A safety relief is thus provided to release pressures that may rise because of thermal expansion when the unit is static and the cylinder is locked in position. Generally, the flow needed to provide relief is barely a seepage. This thermal relief area is usually about one-fifth the area of the exposed lower face of the attached piston. Because of the poppet seat and the dynamic seal of the attached piston, there is little loss of fluid from the cylinder in normal operation and it is considered *dead-tight*. The 5:1 ratio between the area of the piston and the seal diameter of the relief poppet means that a bias-spring adjustment to provide a safety-relief pressure of 2500 psi would still permit the attached piston to open the poppet when 500 psi is directed to the pilot-assist port. Any pressure developed because of a load on the controlled cylinder will reduce the amount of pilot-assist pressure required to unseat the relief or counterbalance poppet.

Because of the safety implications of the holding function, the valve may be designed in cartridge form so that it can be inserted directly into a cylinder-head casting. This completely eliminates potential line breakage

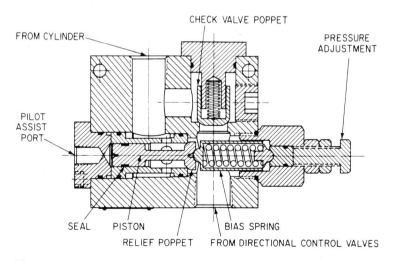

Fig. 10-10 Counterbalance-type holding valve.

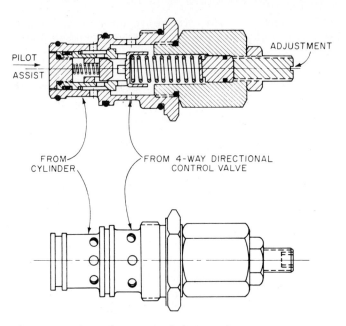

Fig. 10-11 Cartridge counterbalance valve.

between the cylinder and the holding valve. Note the check valve concentrically located in the holding-valve assembly of Fig. 10-11.

Figure 10-12 shows how these valves are applied to a utility service truck. In position 1 the unit is static. Both holding valves are relaxed, checks are closed, and the relief piston is closed. Fluid is positively locked in the cylinders. As pressure is applied to raise the basket, flow passes through the check valve to the blind end of both cylinders. The release is piloted open within the valve on the rod end to permit return of exhaust fluid. This condition continues as long as power is applied to the cylinder to create movement from left to right.

The relief piston is usually relatively wide-open during the dynamic lift function. As the boom traverses over center, the pilot-assist pressure becomes less as the load tends to *run away*. At this time the relief piston starts to close, so that positive pressure must be applied at the pilot-assist port to create an opening through the holding valve to permit return of oil in the lowering function. The load may continue as shown, so that various motion combinations can be effected with a positive lock on the unit not functioning. Use of the holding valve ensures movement at the desired rate determined by the operator without loss of control as the load changes.

An external pilot pressure can be applied to the pilot plunger of the valve in Fig. 10-1d to divert fluid through the valve. This is quite satisfactory for unloading or diversion functions. Because of the sliding spool it is not considered a dead-tight seal.

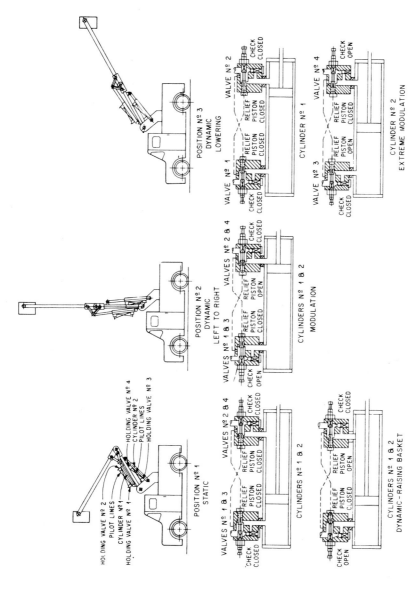

Fig. 10-12 Holding valve applied to utility truck hydraulic circuit.

Pressure-Control Valves **269**

Relief, safety, and unloading valves [Fig. 10-1d] rarely contain integral check valves, as they are not normally installed in lines where a return flow is needed.

One feature these valves have in common is that they are all normally closed, two-way valves. The point at which the actuating signal originates may be different. Also, the manner in which they are drained may be different.

The press circuit in Fig. 10-13 shows the application of these valves. Counterbalance valve C receives its signal from the main inlet line. This signal is normally taken from a point within the counterbalance-valve structure. If desirable, this signal could come from the opposite cylinder line so that a pressure would be applied prior to opening the valve and allowing the ram to descend. Valve C is shown externally drained. It can be internally drained if there is to be no resistance on the cylinder line when flow is diverted to the tank. If there is a question as to the amount of backpressure that may be encountered, it is safest to provide an external drain.

Valve D in Fig. 10-13 is a sequence valve. The signal is taken from ahead

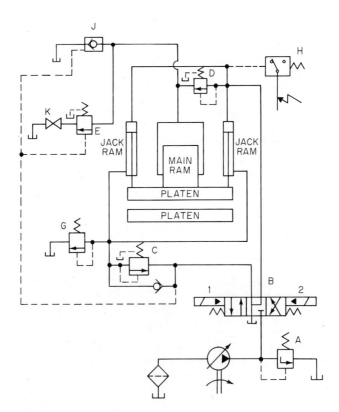

Fig. 10-13 Hydraulic press circuit.

270 Industrial Hydraulics

of the valve and is picked up within the valve envelope. The drain in this valve must be independently connected to the tank because there will be pressure exerted on the discharge of the valve. An integral check valve may or may not be used in a sequence valve applied as indicated in this circuit. If valve E were not included in the circuit, the amount of fluid passing back through the check valve incorporated in valve D as an alternate circuit design might be used to assist in decompression of the press. Then the check valve would pass a certain part of the flow back through the four-way valve to the tank. Generally, the sequence valve does not contain the check valve, and a valve such as E is used for the decompressing function.

Valve E is used in the circuit as an unloading valve. It will be externally piloted and externally drained if there is any resistance in the return to the tank port. If needle valve K were on the upstream side of valve E, the latter could be internally drained. With valve K on the downstream side as shown, the spring pocket must be externally drained so that backpressure does not provide an additive pressure to the spring. An unloading valve is never internally piloted directly from the upstream source. Subsequent circuits will show how the unloading valve can be used both to bypass pump delivery back to the tank when a signal source is available and also to act as a pilot-operated, two-way valve, as in the circuit in Fig. 10-13. The needle valve, used in conjunction with the unloading valve, establishes the amount of flow necessary to decompress the press in the proper time interval. The unloading valve receives the signal and starts and stops the flow of fluid. Special modifications can be made to the unloading valve to incorporate both functions. Pilot check valve J might be considered an unloading valve during this portion of the cycle.

The press circuit in Fig. 10-13 also shows some of the other pressure-control valves discussed in this chapter. A safety valve G is used at the rod end of the jack rams to prevent accidental intensification of pressure. The variable-delivery pump has a pressure compensator to short-stroke the pump. This pump provides the pressurized fluid for the circuit. Full flow volume is available for rapid press movement. The pressure compensator short-strokes the pump (holding the pressure) when a desired maximum pressure is attained.

Valve A is a safety valve; it functions only if the pump compensating mechanism fails. This valve would be adjusted to a pressure valve about 20 percent higher than that established by the pump limiting mechanism.

Valve B is the directional-control valve. This valve is shown as a solenoid-controlled, hydraulically actuated unit. When side 1 is actuated, it directs fluid to the head end of the small pilot cylinders or jack rams. Valve D will prevent fluid flow into the main ram until a certain minimum pressure is reached. As the fluid entering the jack rams causes the platen to move downward, the main ram acts as a pump and the vacuum thus created in the main cylinder is nullified by atmospheric pressure, causing fluid to flow through valve J. In many installations the reservoir is above the main ram, and valve J is in the bottom of the tank so that gravity also assists the fluid flow. When the platens meet and resistance causes a rise in fluid pressure, the flow from the pump is diverted through valve D at its set value, and the full pressure will be effective on both jack rams and the main ram. Full tonnage will then be provided between the platens.

Next, the compensator on the pump establishes the maximum pressure, and pressure switch H creates a signal at a pressure value somewhat less than the maximum value created by the pump control. The signal from pressure switch H will start a timer that times out at a predetermined period. This deenergizes side 1 of valve B and activates side 2. Fluid is directed by side 2 to the rod side of the jack rams through the check valve around valve C. This check valve may be built integral with valve C.

The pressure opens valve E, a relatively small two-way valve that allows all stresses in the press to be relieved and the pressure to drop to a low value. After this decompression the pilot pressure opens pilot check valve J and permits fluid to exhaust from the main ram. At this time the platen can rise and open the press, ready for unloading and reloading. When the platen reaches the upper level of travel, it contacts a limit switch, electric eye, or some other sensing device that causes the solenoid on the directional-control valve to deenergize. Deenergizing the solenoid causes the platen to stop its upward travel.

The press may not produce at its maximum rate because of the time necessary to move the ram. If the ram needs to be raised only halfway up to clear the workpiece, the moving time can be halved. The press can then function at a more economical rate. When changing dies or clearing jam-ups, the automatic controls may have to be blocked in order to raise the ram to its full travel. When the press opens the predetermined amount and the directional control has been returned to the neutral position, the pump will be short-stroked, holding pressure on other devices. The flow from the pump may also be used for other functional operations such as stock feeding or ejecting.

At this time the student is not expected to be completely familiar with all the details of this press circuit. The entire circuit is shown to illustrate the relative positions of the safety valve and other pressure-control valves. As described in Chap. 2, Hydraulic Symbols, the center block of the descriptive symbol of valve B shows that the cylinder lines to the main ram and the jack rams are connected together in the neutral position of the valve. The pressure port is blocked in neutral. Lines in the block to the left indicate that as side 1 directs the pressure source to the main ram and the head end of the jack rams, the fluid from the rod end of the jack rams is directed to the reservoir through counterbalance valve C. When side 1 is indicated as being actuated, it is necessary to visualize the cylinder lines, tank line, and pressure line as being connected to the arrows placed in the box adjacent to the actuating mechanism on side 1.

The box at the extreme left in the symbol for valve B indicates the type of actuation. The box containing a diagonal line plus a solid equilateral triangle indicates that pilot pressure is provided from an internal source and that the pilot mechanism is solenoid actuated. The main valve spool is then shifted into position against the spring that normally urges the main spool to neutral position. With the actuation of side 1, the block in the symbol adjacent to this side shows the path of the fluid through the valve at this time. If side 2 is actuated and side 1 is relaxed, the flow through the valve is as shown by the box adjacent to the actuator on side 2. This shows that fluid is directed to the rod end of the jack rams, to the pilot lines that control decompression valve E, and to the pilot mechanism that opens the pilot check valve, permitting a large flow of fluid to the reservoir from the main ram. Pilot check valve J is generally of large enough flow capacity so that little energy is needed to move the fluid to the tank and out of the main ram.

If a valve spool in B is used to block the ports in neutral, the weight of the main ram and platen can cause the machine to close too fast and damage the machine parts when valve B is shifted. Further, if the spool in B were to develop a slight leak, there might be undesirable movement of the machine members. For this reason it is usual to relax the fluid to both the top and bottom of the main cylinder and hold the ram in position with the counterbalance valve, as shown by the circuit position of valve C.

If for any reason valve C failed to open, fluid would be locked within the rod side of the jack rams. The resistance caused by this locked fluid would cause the supply fluid to open a path through valve D, and full force would be exerted on the main ram. With an intensification ratio of 100:1 and an input pressure of 3000 psi, a final pressure of 300 000 psi might be developed. The rod end of the jack ram would burst at some value much less than this. To prevent such damage to the rod end of the jack ram, a safety valve G is provided at this point. As shown in Fig. 10-1a, these valves, when used at pressures above the normally expected working levels, are virtually leakproof. There is no loss of fluid unless a malfunction occurs. Valve G is often a simple ball-and-spring-type safety valve that squeals or whistles loudly when flow passes through it. Noise in the operation of a safety valve can be a definite advantage because the noise indicates to the operator that there has been a malfunction.

Figure 10-14a shows how a relief valve can be connected from one cylinder line to another to prevent pumping action of a fluid motor from creating excessively high pressures. If this condition can exist in both flow directions, it may be necessary to provide the relief action in both directions. Figure 10-14b shows a valve with two differential-area relief valves across the cylinder lines. The two check valves in Fig. 10-14a provide fluid in the event that fluid loss through the fluid-motor drain makes additional fluid necessary to prevent cavitation. Certain motors must be maintained in a pressurized state, and this circuit is not always usable with them. The two relief and check valves may be in a common envelope similar to the assembly shown in Fig. 10-14b. A single envelope with all these valves included is often referred to as a *relief-and-replenishing unit*.

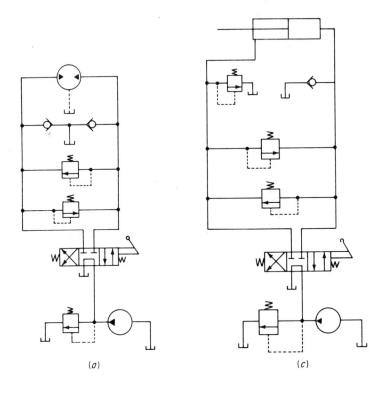

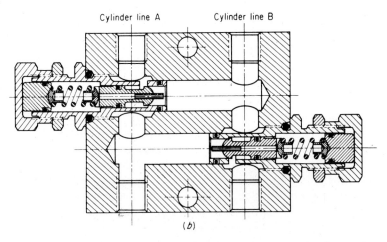

Fig. 10-14 (*a*) Fluid-motor control circuit. (*b*) Dual relief-valve assembly. (*c*) Circuit to compensate for differential cylinder.

Figure 10-14c shows a circuit used on heavy machinery where a swing cylinder is actuated. As the directional valve reaches neutral, the cylinder ports are blocked. One relief valve on the rod end is directed to the opposite cylinder line, and flow from the second relief valve is diverted to the tank. The relief valve directed to the tank is set at a higher pressure value than the unit directing the fluid to the other cylinder line. In the circuit of Fig. 10-14c, the relief valves are acting as a deceleration control, much as in the fluid-motor circuit of Fig. 10-14a. In Fig. 10-14c, however, there is the added need to provide for the different piston areas created because the piston rod is at only one end of the cylinder. The relief valve directing fluid to the tank takes care of the area differential, while the check valve in the head-end line delivers fluid from the tank to the cylinder to avoid cavitation and subsequent bounciness. When the piston moves to the right, there is more fluid leaving the head end than can be used in the rod end of the cylinder. Here again, the excess fluid can pass through the relief valves to the tank. As in the previous circuit, all these components are built into one envelope structure to minimize plumbing requirements.

Unloading valves associated with fluid-power accumulators and those used to unload pumps in a multiple-pump circuit can be similar. As mentioned in Sec. 6-6, a differential pilot can be provided for an unloading valve to create a specific difference between the unloading and reset pressure. A circuit containing the valve shown in Fig. 10-15 is illustrated in Fig. 6-8b. This unit resembles a conventional pilot-type relief valve. An additional pilot piston is introduced into the pilot-head assembly. Note the flat nose on the poppet within the pilot valve. The pilot piston abuts this flat spot. Pilot pressure in the relief-valve control chamber tends to balance this auxiliary piston until the pressure rises to a value where the cone is unseated and flow to the tank commences past this closure. This creates a drop in pressure at the face of the auxiliary piston. The opposite end of this piston is connected to a positive pressure source (accumulator or other pump circuit) so that it follows the cone and forcefully holds the poppet off the seat until the pressure recedes in this auxiliary pilot circuit. As the pressure drops and the cone again approaches the seat, there is an effective pressure increase on the face of the auxiliary piston so that the valve usually closes with a *snap action*. As the pilot assembly closes, the line pressure in the primary circuit is ready to increase because the bypass to the tank ceases.

Figure 10-16 shows how relief and unloading valves, such as those shown in Fig. 10-15a, can be incorporated in a common envelope with two pumps so that a variety of circuit requirements can be provided with a minimum of plumbing. These units can have solenoid control sections manifolded into the same package. In the circuit shown, the unloading valve can be internally drained, as there is no restriction on the tank line. Two pumps are used in the circuit. One is a large-volume pump for rapid

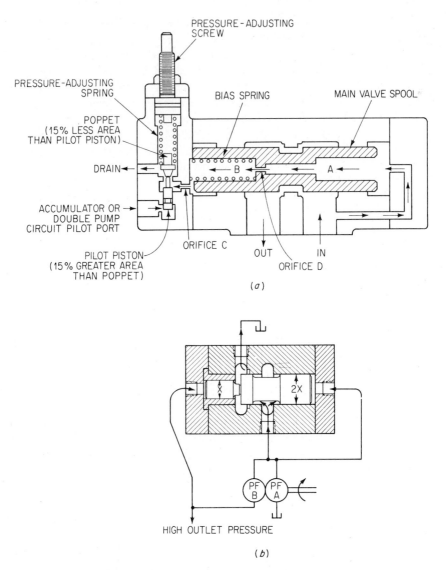

Fig. 10-15 (*a*) Differential unloading valve. (*b*) Pressure-dividing valve.

movement of a cylinder. A smaller-volume pump maintains the higher pressures needed during a feed or holding cycle. Use of the two pumps provides an inexpensive answer to this type of circuit with easy-to-maintain equipment.

Valves used to control the charging of accumulator circuits incorporate many of the features of piloted relief valves, unloading valves, and sequence valves. A knowledge of the basic functions of these valves makes it possible to develop suitable circuits to provide needed operations.

Figure 10-15b shows a pressure-dividing valve used to equalize the pressure to two pumps piped in series. The delivery of pump A [Fig. 10-15b] is greater than that of B to compensate for leakage. A pilot line is taken from the area between the two pumps and directed to a piston on one end of the two-way valve spool. At the other end, the piston has one-half the area of the pilot end. The line to this smaller piston is taken from the output of the second or B pump. If the pressure between A and B starts to rise to more than one-half the value of the discharge, it will cause the two-way valve spool to move an amount sufficient to start a flow to the tank. This flow, in turn, will drop the pressure. The bypass feature ensures equal pressure from each section of the pump. The input pressure to pump B will always be equal to one-half the discharge pressure. The strength of the pump housings is suitable for the pressures they are expected to contain.

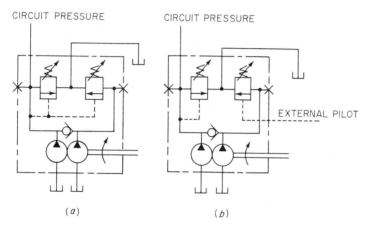

Fig. 10-16 Dual pump controls. High-low pressure circuits with a common pressure outlet. (a) Internally piloted (automatic-control) circuit. Deliveries of both pumps are discharged into the circuit until the pressure approaches the setting of the unloading valve. Large pump then discharges to tank at minimum pressure while small pump continues to furnish oil to the circuit until the pressure setting of the integral relief valve is reached. If the small-volume, high-pressure pump is of the pressure-compensated, variable-displacement type, the flow will decrease to that necessary to make up for slippage. The relief valve then serves as a safety valve set at approximately 25 percent more than the pressure-compensation control within the high-pressure pump. (b) Externally piloted (remote-control) circuit. Function is similar to a, except that the unloading valve is controlled by an external pilot and unloading pressure can thus be remotely controlled, permitting operation at pressures above the setting of the unloading valve with both pumps as long as the external pilot pressure source is programmed accordingly.

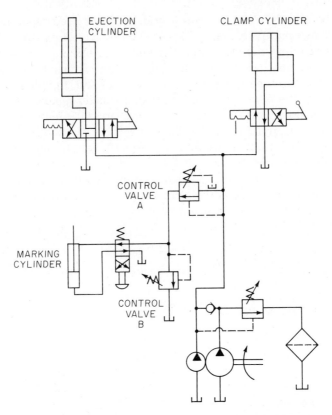

Fig. 10-17 Cascade circuit used in a manufacturing process.

Where fluid pressures vary in a circuit, it may be desirable to use pressure-control valves in series rather than reducing the pressure in a conventional manner. In Fig. 10-17, the functions in the machine cycle include a high-pressure, short-stroke clamp cylinder and an ejection cylinder. The ejection cylinder employs a 2:1 ratio of rod to piston-head area. The four-way control valve shifted to the right maintains the cylinder in the retracted position. Shifting the four-way control valve to the neutral detent position as shown causes the rod to extend at maximum speed with delivery of both pumps effective in displacing the rod area of the cylinder. Fluid from the rod end joins with that delivered by the pumps.

When resistance is encountered, the high-volume pump is unloaded and pressure rises to the value established by control valve A. If the part does not eject at this force value, the four-way valve can be shifted to the position shown in the block at the left. In this position, forces are doubled to eject the most difficult parts prior to rapid return of the cylinder as the four-way valve is shifted to the opposite extreme where the detent holds the flow-directing mechanism in its static position.

There is also a marking cylinder that actuates a delicate mechanism. The workpiece is clamped fast by the clamp cylinder. To do this the delivery of two pumps is combined in the high-low circuit. As the workpiece is firmly clamped, the fluid pressure rises and the high-volume pump is unloaded by the pilot line from the small-volume pump. Valve A is set at a pressure high enough to keep the high-volume pump unloaded while the marking cylinder is being actuated. The marking-cylinder pressure must be maintained at a level suitable for the type of die being used. This value changes with each die used and is held at a suitable pressure setting by valve B. Regardless of the pressure on the other parts of the machine, the pressure at B will be constant during the marking operation. Valve B may be located convenient to the operator or provided with a remote-control adjustment as previously described, either by remote pilot valve or by an electrohydraulic remote-control device. Movement of the marking cylinder is slow so that the die will not be damaged by too rapid contact with the work. After the marking operation, the piece is unclamped rapidly and the ejection cylinder (which is a regenerative unit) rapidly moves the piece to the shipping station.

10-8 HYDRAULIC FUSES

A type of safety device that is not automatically reset is the hydraulic fuse. Fluid under pressure is held within the fuse by the prestressed disk in the valve. When the fluid pressure reaches the level for which the disk is prestressed, the disk ruptures, releasing the fluid to a lower-pressure area. Thus, the piping or machine to which the fuse is attached is protected from excessive pressure by selection of a suitable disk. When a disk ruptures, the piping or machine is out of service until the cause of excessive pressure is corrected and a new disk is inserted in the fuse. Hydraulic fuses are excellent protectors for jack rams and other devices which are in circuits subjected to high intensification of fluid pressure.

10-9 PRESSURE SWITCHES

A pressure switch is actuated by the fluid pressure; the switch converts this pressure signal into an electrical signal by closing the contacts in an electric circuit. Electricity flowing through the electric circuit is then used to perform some function in the hydraulic circuit, such as starting or stopping a pump, opening or closing a valve. The pressure signal, which can be either a rise or fall of fluid pressure, can be converted into switch actuation by using a small piston, diaphragm, or Bourdon tube.

Figure 10-18a shows a typical piston-type pressure switch which balances the input pressure against a spring. When the fluid pressure exceeds the spring setting, piston P is moved to the right by the fluid. The piston moves pin S to actuate the electric switch. A Bourdon-tube-type pressure

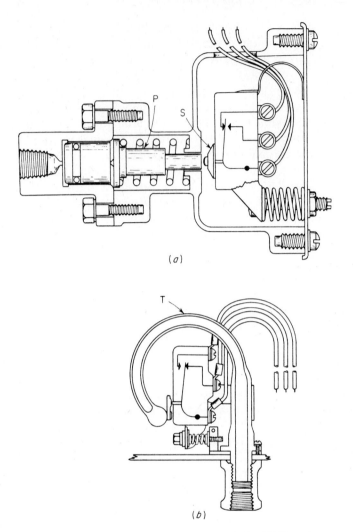

Fig. 10-18 (a) Piston-type pressure switch. (b) Bourdon-tube-type pressure switch.

switch with an electric switch directly coupled to the free end of the tube is shown in Fig. 10-18b. When the fluid pressure rises, it causes the end of tube T to move to the left, actuating the electric switch. As shown, the fluid under pressure enters the Bourdon tube, and any change in the pressure in the circuit is converted into motion of the tube end.

Pressure switches are used in hydraulic cylinder lines to sense a rising or falling pressure. They are also used in accumulator circuits and in filtering and cleaning devices. Where pressure surges may occur, the pressure switch must be *damped* (protected from the pressure surges). Various damping devices are available for this purpose.

SUMMARY

Pressure-control valves are built for a variety of uses. Each type of valve has distinct characteristics that make it suitable to the service for which it was designed. Pressure-control valves used in hydraulic circuits include:

Relief valve Normally closed two-way valve used to establish a pressure level by creating an orifice and bypassing the fluid to a lower-pressure level. This type of valve is usually internally drained.

Safety valve Normally closed two-way valve whose primary function is to establish a maximum pressure level. Functional dependability is of utmost importance; noise level is rarely important.

Hydraulic fuse Safety device equipped with a disk that will rupture at a fixed maximum pressure so that equipment will not be damaged by excessive pressures. A fuse will not automatically reset itself after functioning.

Pressure-reducing valve Normally open two-way valve that senses operating pressure from a low-pressure secondary area. The spool or control mechanism restricts the flow from the primary to the secondary port at a predetermined pressure level. The drain is usually external.

Sequence valve Normally closed two-way valve that may contain an integral check valve. Drain may be either internal or external. This type of valve is used to maintain pressure at a certain level in the inlet area prior to passing the fluid to the outlet area.

Counterbalance valve Normally closed two-way valve that usually contains an integral check valve for return flow through the unit. Drain may be either external or internal. Pilot pressure may be internal or external. This valve is used to maintain pressure within a specified area at a certain preset value.

Holding valve Counterbalance valve provided with a specific ratio auxiliary relief mechanism for thermal protection. Always has return check valve. Drain is internal. Pilot assist is always from remote point (usually opposite cylinder line).

Unloading valve Normally closed two-way valve that rarely contains an integral check. It is always externally piloted and usually internally drained.

Pressure switch Pressure-sensitive device that actuates an electric switch to transmit a pressure signal into an electrical control for an electric signal. This switch may require an external drain and damping devices to prevent unwanted surge signals.

Combination pressure-control valves These valves may contain relief valves or combinations of any of the above valves plus check valves for varying functions within hydraulic circuits. The most common type is the high-low pressure control for dual pump circuits.

REVIEW QUESTIONS

10-1 Explain the action of a compound relief valve.

10-2 When may a sequence valve drain be internally connected?

10-3 Why is an unloading valve externally piloted?

10-4 Is it possible to pass fluid through a reducing valve in the reverse direction? Explain.

10-5 Will a spool-type counterbalance valve hold a cylinder on a vertical machine slide in the up position indefinitely without other devices?

10-6 What feature is common to relief, safety, counterbalance, and unloading valves?

10-7 Why is a reducing valve normally open?

10-8 Why are solenoid-operated two-way valves used in conjunction with compound relief valves?

10-9 Why is the control for an electrohydraulic relief valve of the current rather than the voltage type?

10-10 What is a pressure additive?

10-11 Show both the simplified and complete symbol for a compound relief valve.

10-12 Is a noisy safety valve always objectionable?

10-13 What is the objection to using a sequence valve as a relief valve?

10-14 Can a relief valve be used as a sequence valve? If so, is it necessary to make any modifications?

10-15 Can a direct spring-loaded relief valve be remotely controlled?

10-16 What is the purpose of the dual valve unit shown in Fig. 10-14*b*? Why are two identical units used? How would you identify this type valve?

10-17 What is the purpose of the flat surface on the nose of the pilot cone shown in Fig. 10-15?

10-18 What is the purpose of the check valve in Fig. 10-14*c*? Why are four relief valves used? What are the relative pressure settings? How do the relief valves in the cylinder lines affect the valve adjacent to the pump?

10-19 Can a variable-displacement pump be used in a circuit such as that shown in Fig. 10-16? Explain.

10-20 What is the advantage of a regenerative circuit in the ejection cycle in Fig. 10-17? Are the force values different as the directional control is shifted? What are the speed ratios?

10-21 What is meant by a cascade circuit?

10-22 What is the purpose of the tube located in the face of the poppet in Fig. 10-14*b*?

10-23 Why is valve G included in the circuit of Fig. 10-13?

10-24 What is the purpose of the relief function of the holding valve of Figs. 10-10 and 10-11?

10-25 What is the purpose of the flow-control valve shown in Fig. 10-5*b*?

LABORATORY EXPERIMENTS

10-1 Secure a piloted relief valve over an open container so that the discharge can be observed. Pipe a hand valve in parallel. Connect the two valves to a pump capable of delivering a supply of fluid under pressure (usually more than 1 gpm). Completely relax the pressure adjustment on the relief valve. Close the hand valve. Start the pump with the full delivery passing through the relief valve. Gradually raise the pressure to the rated value. Slowly open the hand valve. Observe the time necessary to permit the relief valve to cease discharging fluid. Observe the pressure at which the relief-valve discharge closes. Slowly raise the pressure by closing the hand valve. Observe the pressure at which pilot fluid starts to pass through the relief valve. Note the pressure value when a major flow is established. Repeat this operation and check for repetitive accuracy.

10-2 Add a small hand valve to the vent connection on the relief valve. After adding the vent valve, close the large hand valve. Close the vent valve and raise the pressure on the relief valve to the recommended maximum. Open the vent valve. Observe the residual pressure value. Completely relax the pressure adjustment and observe the minimum pressure. Close the vent valve and reset the pressure to maximum. Observe the pressure values when pressure is slowly reduced by the vent valve. Note the point at which the valve starts to pass pilot fluid on a rising pressure. Observe the point when the pressure and flow are at "cracking" points (points at which major changes occur).

10-3 Repeat Experiment 10-2 with larger or smaller piloted valves and compare repetitive accuracies.

10-4 Repeat Experiment 10-1 with a direct spring-loaded relief valve.

10-5 Parallel an unloading valve in the circuit with the hand valve and relief valve. From another source of pilot pressure, raise the pressure to the pilot of the unloading valve and observe the point at which it starts to pass fluid. Observe the point at which it passes full volume. Gradually relax the pilot pressure. Note the point at which the flow through the unloading valve becomes insignificant. Repeat to check for repetitive accuracy.

10-6 Install a reducing valve in the discharge line of the large hand valve of Experiment 10-1. Install a second hand valve at the secondary or low-pressure side of the reducing valve. Pipe a small hand valve to the drain of the reducing valve. Leave this hand valve on the drain line open. Open the hand valve from the reducing valve to the tank. Close the hand valve at the input of the reducing valve. Raise the relief-valve setting to the maximum input pressure value allowed at the reducing valve. Open the hand valve to the input of the reducing valve. Relax the adjustment of the reducing valve. Gradually close the hand valve on the outlet of the reducing valve. Observe the pressure at the gauge connection of the reducing valve (low-pressure side). Gradually raise the pressure adjustment on the reducing valve in 100-psi increments. Open and close the hand valve on the discharge at each operation. Observe repetitive accuracy when the hand valve is closed each time.

10-7 Gradually restrict the drain connection with the hand valve. Observe the effect on the secondary pressure. (Make certain the valve is capable of containing the pressure in the drain area.)

10-8 Actuate the hand valve rapidly on the secondary port of the reducing valve to simulate machine action. Observe possible overpressure conditions. Note the speed of response.

10-9 Connect a pressure switch just ahead of the reducing-valve hand-valve control on the secondary. Set the pressure switch at a pressure equal to that on the secondary of the reducing valve. Connect an electric light or bell through the pressure switch. Observe the repetitive accuracy when the hand valve is opened and closed on both the pressure switch and reducing valve.

10-10 Connect a suitable pressure pickup connection to an oscilloscope and to the relief-valve gauge connection. Observe the pressure pattern on the scope screen as tests are conducted.

10-11 Repeat Experiment 10-10 with a micronic filter ahead of the relief valve. Observe the increased sensitivity and repetitive accuracy.

10-12 Cycle relief pressures by connecting a solenoid two-way valve to the vent connection and watch the pressures at each test station. (The vent should be connected to a two-way valve; discharge the two-way valve to the tank.) The solenoid may be actuated by a manual push button if cycle timers are not available. Observe overpressure conditions.

10-13 Pipe an internally piloted counterbalance valve or sequence valve in the discharge line of a piloted relief valve. Connect the discharge of the sequence valve to the tank. Set the relief valve at half pressure with the counterbalance valve or sequence valve completely relaxed. Gradually raise the pressure on the sequence valve. Note the effect on the relief valve. This simulates the effect of backpressure on the discharge of a relief valve. Vary the pressures in each valve to note the relationships.

10-14 Set the counterbalance valve in Experiment 10-13 at minimum pressure with some tension on the main-spool spring. Connect compressed air through a suitable reducing valve and safety valve to the drain connection of the counterbalance valve. Observe the increase in pressure in the main line as air pressure is added to the drain chamber. Reduce the air pressure. Observe the drop in the main-line pressure.

10-15 Make a compound relief valve from a direct spring-loaded counterbalance valve and a small, direct spring-loaded relief valve. Test its performance.

10-16 Connect a counterbalance-type holding valve into a circuit consisting of a mechanical boom arrangement similar to that shown in Fig. 10-12. Arrange a significant load at the end of the boom. Note the overcenter control attributable to the holding valve by relaxing the pressure adjustment and operating with the directional control alone. Gradually increase the pressure on the holding valve and note the control obtained.

10-17 Suspend a load in the air with the circuit of Experiment 10-16. Gradually relax the pressure adjustment of the holding valve. Note the rate of load descent.

10-18 Obtain a differential unloading valve similar to that shown in Fig. 10-15. Connect it into a circuit as a conventional pilot-type relief valve. Direct pressurized fluid through a check valve to a short length of pipe terminating in a plug-type needle valve. Tee the pilot-type relief valve ahead of the check and direct the return fluid to the tank without resistance. Connect the circuit pilot port of the relief valve to a tee between the secondary of the check and the hand valve. Close the hand valve. Start the pump with the relief-valve adjustment completely relaxed. Gradually build up pressure. Note the response. Open the hand valve slowly; then reclose. Note the pressure pattern.

10-19 Introduce a pneumatic accumulator beyond the check valve in the circuit of Experiment 10-18. Repeat the experiment and note the response time. Note the pressure pattern.

10-20 Build a pressure switch to handle low voltages by using contacts insulated and fixed to a Bourdon-tube pressure-gauge pointer. Note the effect of shock on the unit by connecting the contacts to a low-voltage signal-light circuit. Arrange the fixed contact so that it can be readjusted to provide a different pressure setting. Note the effect of minor shock loads at various pressure settings.

11
Flow-Control Valves

Flow-control valves are used to regulate the flow of fluid in hydraulic circuits. Control of fluid flow is extremely important because the rate of movement of machine elements depends on the rate of flow of the pressurized hydraulic fluid. A variety of flow-control valves are used in hydraulic circuits. Flow-control valves discussed in this chapter include sliding-spool, needle, globe, gate, noncompensated, combined, and compensated valves. Each type has important uses in hydraulic circuits.

11-1 BASIC TWO-WAY VALVES

These are sliding-spool, needle, globe, and gate valves. Sliding-spool valves [Fig. 10-3a and b] often combine flow and directional-valve functions. The spool, centrally located in a bore within a housing, alternately blocks flow through the passages in the housing and permits selected flow of fluid as the spool is moved within the bore. The corner of the major diameter of the spool as it joins the reduced diameter which creates the passageway for a fluid flow may be a right angle, as shown in Fig. 10-3a. This basically creates a sharp-edged orifice through the valve. The spool of Fig. 10-3b is provided with a gradually reduced diameter which provides a gradual flow change as the spool is moved in a linear pattern relative to the body with the integral passageways that are to be opened or closed.

Needle valves [Fig. 11-1a] have a pointed stem that can be adjusted manually to control accurately the rate of fluid flow (gallons per minute) through the valve. Needle valves are often made from steel-bar stock and are among the most common hydraulic flow-control devices. The needle

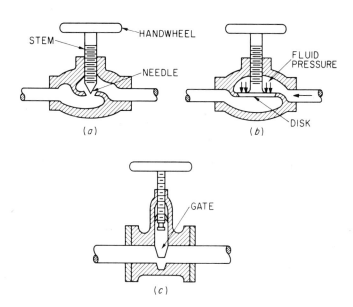

Fig. 11-1 Basic two-way valves. (*a*) Needle valve. (*b*) Globe valve. (*c*) Gate valve.

valve is also used as a *stop valve* in hydraulic circuits to shut off the flow of fluid from one part of a circuit to another part.

Globe valves [Fig. 11-1*b*] have a round disk to control or stop the fluid flow. The flow area through a globe valve is larger than through a needle valve. Hence, a globe valve will have a larger flow capacity at a lower pressure drop than a needle valve of the same size will have. Needle valves *throttle* the flow through them; that is, the flow area is slowly reduced as the valve is closed, gradually reducing the quantity of fluid passing through the valve. Globe valves are not so suitable for throttling service.

Gate valves [Fig. 11-1*c*] are not normally used as flow-control valves. The gate provides a large opening with minimum pressure drop. Most gate valves are used as *stops*—to shut off fluid flow or to open the line to full flow.

11-2 NONCOMPENSATED FLOW-CONTROL VALVES

Simple valves like those in Fig. 11-1 are not *compensated* for changes in the fluid temperature or pressure. Without compensation, flow through these simple valves can vary at a fixed setting if either the pressure or temperature of the fluid changes. Viscosity changes, which often accompany temperature changes of the fluid, can also cause flow variations through a valve. *Compensation* automatically changes the valve adjustment or pressure drop across the orifice to provide a constant flow at a given setting.

Noncompensated flow-control valves create an orifice in the pipe to restrict flow. The control element of the valve may be the needle or globe type or modifications of either type.

11-3 CHECK VALVES

This type of valve allows flow in one direction but prevents flow in the opposite direction. During flow through the valve, the fluid holds a flap-type gate, spherical ball, flat disk, poppet, spool, or some similar check element off the valve seat. When the fluid flow reverses, the weight of the fluid, a spring force, or gravity forces the check element onto its seat, preventing flow past the valve. Check valves may be built as part of other types of flow-control devices.

11-4 COMBINED FLOW-CONTROL AND CHECK VALVES

Figure 11-2 shows how a needle valve is used in a directional-control valve to control flow through the valve during part of the cycle. The major flow through the valve is controlled by the cam-operated, two-way valve spool below the needle valve. With the cam roller in its extended position (to the left), the fluid travels freely from the *in*-port to the *out*-port. As the roller is gradually depressed (moved to the right), the fluid flow is diverted through the needle valve. The flow through the needle valve is not compensated. Hence the quantity of fluid passing through the needle valve will vary with differences in fluid viscosity and pressure. Despite this limitation, this type of valve finds wide use in industrial hydraulics. It makes an excellent deceleration valve for predetermined minimum speed

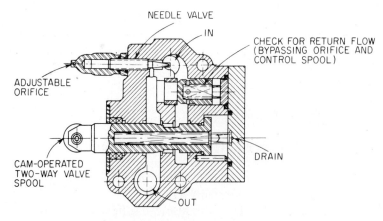

Fig. 11-2 Noncompensated adjustable orifice in a two-way-valve envelope.

288   *Industrial Hydraulics*

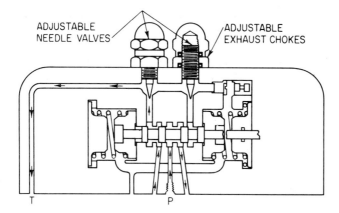

Fig. 11-3 Noncompensated adjustable exhaust checks in a directional-control envelope.

to position the mass prior to engaging locating pins or reaching positive stops.

Figure 11-3 shows how noncompensated adjustable needle valves can be installed in a pilot valve to meter fluid out of the tank line. This provides independent control of spool-speed movement of a machine or small-size cylinder that is being piloted. By adjusting the left-hand needle valve (turning it into or out of the fixed-size orifice), flow of fluid to tank line T is controlled. This flow is noncompensated.

The pressure inlet port P of Fig. 11-3 is tapped to provide a place to insert a nonadjustable orifice to prevent excessive flow of fluid through the valve. Fixed orifices to control fluid flow are used in many types of hydraulic control valves. The right-hand adjustable needle valve will control the flow to the tank line when the spool of the control valve is shifted to the other position, directing pressure to the alternate cylinder port. The needle at the left will then be inoperative.

The compound relief valve [Fig. 10-3] has a nonadjustable orifice between the input pressure line and the valve control chamber. Flow of fluid through this orifice can vary with changes in fluid density. The flow changes, however, are normally not large enough to require a more sensitive flow-control mechanism than the orifice.

Figure 11-4 shows how a fixed orifice is located in the poppet of a check

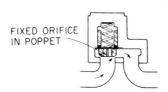

Fig. 11-4 Orifice-type check valve.

Flow-Control Valves **289**

valve. This valve permits a free flow of fluid in the direction indicated by the arrows. Flow in the opposite direction is restricted to the amount of fluid that can pass through the integral orifice. Orifice-type check valves are widely used in tilt mechanisms for metal ladles and chemical handling to prevent too rapid return movement, which might spill the ladle contents. Location of the orifice within the check-valve spool precludes tampering by unauthorized personnel. Many types of lift mechanisms and hydraulic jacks incorporate various forms of orifice check valves to prevent uncontrolled dropping of the load.

A method of providing adjustment for a noncompensated flow-control and check valve is shown in Fig. 11-5. This assembly can be sandwiched between a subplate-mounting-type, directional-control valve and the associated plate in which the connecting pipelines terminate. Two flow-control units are incorporated in this unit to control the speed of a cylinder or motor independently in each direction of travel. A single unit may be used if control is needed in one direction only.

Fluid entering at a metered-flow port contacts the face of the hat-shaped closure, urging it against the bias spring. The pressure required is relatively low so that little energy is consumed in the free flow through this assembly. When flow ceases, the spring urges the closure against the seat area. A plunger, seal assembly, and adjusting screw concentrically located

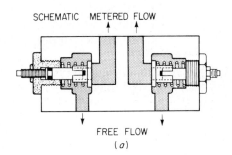

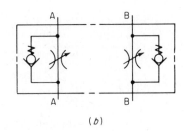

Fig. 11-5 (*a*) Schematic of noncompensated flow-control and check valve. (*b*) ANSI symbol.

within the hat-shaped closure can be adjusted to create an orifice at the machined notch in the portion of the closure within the bias spring. With this plunger adjusted to a position where the notch is completely blocked, a flow of a few cubic inches per minute may be expected. Because this is a noncompensated device, the leakage and adjusted flow will be affected by the pressure level and fluid viscosity. Flow entering at the lower port tends to urge the hat-shaped closure against the seat so that the flow restriction will be a function of the orifice created and external leakage will be prevented by the ring seal on the periphery of the plunger.

11-5 COMPENSATED FLOW-CONTROL VALVES

Accurate control of machine feed rates which are fixed by the fluid flow in fluid-power systems requires more than a simple orifice. A sharp-edged orifice is relatively more stable over a wide range of fluid temperatures. Since flow through a sharp-edged orifice varies with the fluid pressure, compensation mechanisms are built into the control-valve structure to maintain a uniform pressure drop across the orifice regardless of input or output pressure variations. Both adjustable and nonadjustable compensated flow-control valves are used. Either type may have an integral check valve for a return free flow through the control valve.

Figure 11-6 shows a nonadjustable compensated flow-control valve. Fluid entering the inlet port impinges on the full face of the poppet, urging it against the spring. The poppet can move only against the spring until it closes off the outlet port. Then a flow is established through the fixed orifice in the poppet. This flow equalizes the pressure on both ends of the poppet. The poppet then is in balance and will be urged by the spring back toward the inlet port. This movement opens the outlet port. A balance will automatically be established between the amount of fluid that will pass through the orifice and the upper lip of the poppet. Note that the open end of the poppet is attempting to shear off the flow of fluid to the outlet port. The flow rate through the fixed orifice is determined by the size of the hole and the value of the spring loading. The heavier the spring for a given-size orifice, the greater the fluid flow; the larger the hole with a fixed size, the greater the fluid flow. When a spring size is established and an orifice size chosen, the repetitive flow will always remain the same regardless of the pressure upstream or downstream because the spring and orifice size determine the flow. The area of the poppet is equal on both ends; so no differential is created. The poppet may be located within a body with other port configurations so that a return flow through the body can be established. This poppet mechanism may also be used in other valve structures to provide the compensated-orifice effect [Fig. 10-5b].

A valve of the above design is used to control the lowering speed of pallets on lift trucks and similar devices where the pressure values may be

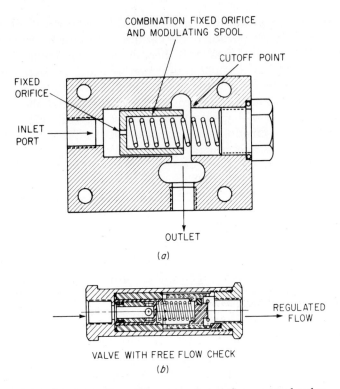

Fig. 11-6 Nonadjustable compensated flow-control valve. (a) Angle structure without return check. (b) Through-flow type with integral return-flow check.

different during each functional operation. The compensated flow control using the design in Fig. 11-6 must have a certain minimum flow of fluid to function so that the orifice will create enough restriction to move the poppet. The pressure required to move the poppet is usually a relatively low value and does not hinder effective operation of conventional machinery.

Figure 11-7 shows an adjustable compensated flow-control valve. The cross section of the valve shows its construction, and the symbolic drawing at the right shows the principle of operation.

The symbolic drawing shows the inlet connected directly to a reducing valve. A signal to actuate the normally open spool in the reducing valve is taken from the downstream side, just ahead of the adjustable-orifice spigot. Another signal is taken downstream from the adjustable orifice. This signal is reflected back into the spring cavity of the reducing valve.

Assume that the spring in the reducing valve can be compressed to a point where flow through the valve will cease at 50-psi pressure at the orifice. Then, when the adjustable orifice is completely closed, there will be no pressure above 50-psi at the input of the orifice. This assumes that no pressure is reflected into the spring chamber of the reducing valve.

Now, as long as there is a supply of fluid under more than 50-psi pressure at the inlet port and a volume of fluid exceeding that to be passed through the valve, the pressure drop across the orifice will always be 50 psi. The reducing spool will control the pressure at the orifice because the signal controlling the reducing spool is taken from the input to the adjustable orifice.

When a load is handled by metering fluid to a cylinder or other device, fluid pressure on the outlet of the valve can be expected. To compensate for this pressure, a control line is provided from the orifice outlet back into the spring chamber of the reducing valve. This line senses the value of the load and adds it to the spring, thus urging the reducing spool to the open position. Thus the signal pressure to one end of the reducing spool is taken from ahead of the orifice, and the energy available to balance the spool is made up of the mechanical spring and the reflected pressure from beyond the orifice.

In a typical installation, the load can consist of a dead weight such as an elevator mechanism that is lifting a part out of the machine. The part must be lifted a certain distance with a fair degree of accuracy of movement to clear the machine members prior to being ejected over and into a conveyor. A compensated flow-control valve at each cylinder used for ejection will ensure speed sufficiently accurate to prevent the product from skewing during the ejection portion of the cycle and will keep both ends traveling at nearly equal speed. In this installation the input pressure is not necessarily stable, but it is always adequate for the ejection operation. The weight of the part being lifted is constant, but the friction encountered may vary somewhat. Variations in input pressure and flow volume are caused by the actuation of several other parts of the machine concur-

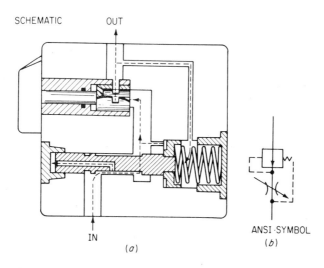

Fig. 11-7 Adjustable compensated flow-control valve. ANSI symbol.

Flow-Control Valves   **293**

rently with the ejection device. (These concurrent functions can include such movements as conveyor indexing, retraction of protective shields, and projection of slide arms to receive the finished product and guide it to the conveyor.) The variable inlet pressure is compensated for by the reducing valve, and the variation of outlet load is reflected back to the spring chamber of the reducing valve where it is compensated as described above.

Figure 11-7 shows that fluid entering the valve inlet is directed to a two-diameter reducing spool. Two diameters are used to provide (1) a precise control at the cutoff point and (2) the desired area to receive the operating signals for the reducing valve. The areas at the bottom of the spool and under the head opposite the spring cavity are ported to the outlet of the reducing valve prior to entering the spigot orifice. The spring cavity is connected to the outlet of the flow-control valve. In this valve the adjustable orifice is a spigot that varies the flow by use of a contoured notch over an arc of approximately 180°. A drain line is not necessary because of the use of the high-pressure, captive, O-ring-type seal on the adjusting shaft.

A frictional or toothed lock mechanism is often provided in flow-control valves to prevent vibration of the machine from causing the adjustment to vary. The lock mechanism may also prevent valve movement from some accidental cause. If the valve setting is critical (such as in valves used on precision machine tools), it may be desirable to use valves equipped with auxiliary covers over the adjusting mechanisms. The cover may be provided with a keyed lock, or the lock may be built directly into the adjusting-knob mechanism.

Modifications may be encountered in pressure-compensated flow-control valves which are provided to permit use in unusual circuit applications.

A stroke limit device may be provided at either end of the reducing spool in the valve of Fig. 11-7. When applied at the end where the bias spring is located, it will limit minimum flow through the valve for rapid response and eliminate cutoff on rapid operation. An adjustment at the opposite end of the spool controls the maximum opening of the reducing spool. This limitation of travel of the reducing spool will increase the valve sensitivity. Since the valve is normally open, it must travel a certain distance from the rest position before restricting the flow to the orifice. By limiting the travel, it is possible to adjust the opening to a point at which it will pass the maximum desired flow yet reduce the response time to an absolute minimum.

11-6 CONTROL-VALVE CIRCUITS

Flow-control valves are used in three different circuit configurations. The flow of fluid may be controlled as it enters the end-use device—a *meter-in*

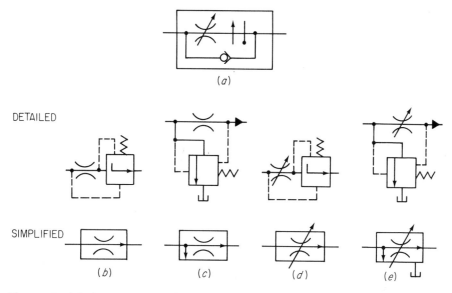

Fig. 11-8 (*a*) ANSI flow-control symbol showing pressure compensation plus temperature compensation with an adjustable orifice. (*b*) ISO symbol for nonadjustable orifice and reducing-valve-type compensation. (*c*) ISO symbol for nonadjustable orifice and relief-valve-type compensation. (*d*) ISO symbol for reducing-valve-type compensation with an adjustable orifice. (*e*) ISO symbol for relief-valve-type compensation with an adjustable orifice.

application. Control of the fluid flow as it leaves the end-use device is termed *meter-out*. A diversion of flow to the tank or some other lesser pressure area is a *bleedoff* function.

The flow-control symbol of Fig. 11-8a is drawn to ANSI symbol standards. The vertical arrow indicates pressure compensation. The vertical line with the bulb-like base indicates temperature compensation. The slash arrow across the orifice symbol indicates that the orifice can be adjusted.

These symbols will be found in many drawings originating in North America.

Fig. 11-8b through e shows the ISO flow-control-valve symbols. It is expected that the ISO symbols will ultimately supersede the ANSI symbols. ANSI is the United States representative to the ISO.

Figure 11-8b shows a nonadjustable pressure-compensated flow-control valve using a reducing valve for compensation.

Figure 11-8c shows a nonadjustable pressure-compensated flow-control valve using a relief valve for compensation.

Figure 11-8d shows the symbol of Fig. 11-8b with an adjustable orifice. Figure 11-8e shows the symbol of Fig. 11-8c with an adjustable orifice. The use of the appropriate ISO symbol is recommended if the drawings are to be used for devices that may be used worldwide.

Flow-Control Valves **295**

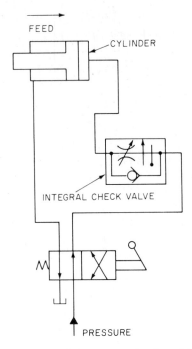

Fig. 11-9 Meter-out circuit with integral check valve. (ANSI flow-control symbol.)

METER-OUT CIRCUITS

The meter-out circuit is commonly used in machine tools that require precise control of the fluid on the discharge or exhaust side of the cylinder. Machine tools often require large restraining forces to prevent excessive pull on the cylinders. Mills and drills passing through the workpiece often tend to drag the entire tool unit forward. Considerable force may be required to push the tool during the cutting cycle. When the drill point gets to the breakthrough point in the workpiece, it tends to grab and drag the drill through. Metering the fluid out of the cylinder as in Fig. 11-9 prevents this breakthrough condition from affecting the speed of drill feed. The integral check valve connected around the compensated flow control permits a rapid travel of the tool to the start position, ready for the next cycle. Meter-out flow-control mechanisms are often incorporated in envelopes containing the directional-control valves. Some envelopes also contain diversion valves that direct the fluid around the flow-control valve for a rapid approach and then through the flow-control valve for the feed operation. The return check is usually included in the same envelope.

METER-IN CIRCUITS

Figure 11-10 shows how the pipes are connected to meter the fluid into the cylinder when the directional-control valve is shifted to direct fluid to the cylinder-rod end. Feed direction is shown by the arrow. The fluid must pass through the compensated flow-control valve before entering the rod end of the cylinder. Exhaust fluid is directed freely to the reservoir. When the lever of the directional-control valve is released, it permits the spool to return and the flow is directed to the head end of the cylinder. The fluid from the rod end will pass through the integral check valve in the flow-control mechanism, and the rod can move rapidly to the extended position.

Assume that this is the lift cylinder on a hoist used to lower baskets of parts into a quench tank adjacent to a heat-treating furnace. This cylinder must provide the power to lift the basket of red-hot parts gently out of the furnace without damaging the furnace walls. The basket is then swung over the quench tank and dropped quickly into the quench oil. As the valve is shifted to let the cylinder drop the load, the fluid can enter the head end of the cylinder without restriction. The pressure created by the load on the rod end of the cylinder can force the fluid through the integral check valve. There will be no cavitation on the head end of the cylinder because check valve X adds capacity to the pump to prefill the cylinder during the rapid drop.

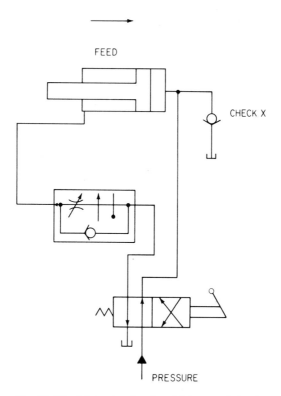

Fig. 11-10 Meter-in circuit with integral check valve.

Flow-Control Valves

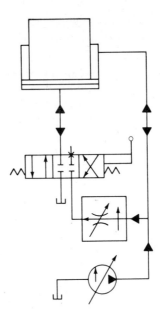

Fig. 11-11 Meter-in circuit.

As the operator shifts the valve again to bring the basket out of the tank, the basket will rise slowly at the controlled rate and the quench oil will drain out of the basket as it emerges from the tank. The descent of the basket to the unloading station can be controlled by the operator in the shifting operation of the four-way directional-control valve. A manual speed control is positioned by the operator using the directional valve, and a reasonable rate of speed can be established without the use of compensating devices or a feedback of information. Visual information is normally adequate. The cylinder may contain cushion devices at the bottom of the stroke to decelerate the load if shock conditions of any magnitude are anticipated. Check valve X does not affect the downspeed when the unit is unloaded because there is insufficient weight to cause the prefill actuation. During the empty downstroke the pump will force the cylinder rod out, and the check valve X will be held closed.

A large ram like that in Fig. 11-11 may create large forces in pressing operations. A draw operation may require an accurate movement rate of the ram so that the metal flows smoothly without tearing or cracking. The fluid cannot be metered out of the cylinder in many operations because of the possibility of pressure intensification. In this type of installation it is necessary either to meter the fluid into the head end or to use the bleedoff system explained in subsequent parts of this chapter.

Note the vertical arrow shown in the pump symbol of Fig. 11-11. This is a device used in ANSI symbols to indicate pressure compensation (reduced pump output flow as pressure increases). The ISO symbol will not include the vertical arrow because the assumption is that the variable pump will be compensated unless a manual adjustment is shown.

To keep the cylinder stable, it is common to use a reflected pressure on the lip of the ram [Fig. 11-11] from a source ahead of the flow-control valve, without any restriction in the line. When the four-way directional-control valve is shifted to permit an exhaust of fluid out of the ram, the pressure on the lip of the ram drives it down and forces the fluid out faster than would be possible with a gravity drop. The effective force is limited to the rod area or that extending through the seal.

METER-OUT CIRCUITS ON RECIPROCATING DEVICES

Reciprocating cylinders on grinders, hones, and similar machines may have equal speeds and feeds in both directions of travel. With double-rod cylinders there is no difficulty in providing a smooth, even speed in both directions with a solid table control by metering the fluid from the tank port of the four-way valve as in Fig. 11-12. Note, however, the external drains on the seals of the four-way directional-control valve. Most machine tool directional-control valves can be provided with external drains so that resistance created by the flow-control valve will not be reflected to the packing of the four-way valve. There are many different types of reciprocation control-panel valves that contain the directional-control valve, compensated flow-rate control valves, and perhaps other functional valves for stop-and-start or cross-feed functions.

BLEEDOFF CIRCUITS

The third basic type of control circuit uses the flow-control valve to divert the fluid to the tank or another lower-pressure area. The bleedoff function is not so widely used as the meter-in and meter-out types.

The bleedoff valve may be in the pressure line, as in Fig. 11-13, or this valve may be connected in either cylinder line, depending on whether it is

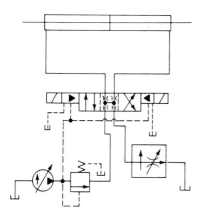

Fig. 11-12 Meter-out circuit.

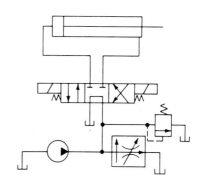

Fig. 11-13 Bleedoff circuit with emergency safety-type relief protection.

desirable to have the speed control (1) in both directions of flow as shown, or (2) so that the bleedoff fluid is limited to operation on a specific line and limits speed in only one direction of cylinder travel.

Bleedoff circuits are widely used in broaching machines, shapers, planers, and similar machines where a fairly large quantity of fluid (more than 1 gpm) is to be used. The degree of accuracy of bleedoff circuits is adequate for large flows. A bleedoff circuit may not be sensitive enough to compensate for very small flows such as those encountered in precise boring operations.

11-7 FLOW-CONTROL VALVES WITH RELIEF-VALVE COMPENSATION

Figure 11-14a shows by use of ANSI symbols how a relief valve can be the compensating agent for a meter-in flow mechanism. Figure 11-14b shows the full ISO symbol. Figure 11-14c shows the simplified ISO symbol for the flow control. Assume that the bleedoff flow control illustrated in Fig. 11-13 is actuated at the pressure level created by the work load of the cylinder. The line pressure is determined by the load; the pressure at the input of the flow-control valve will be the same. If the pressure rises to a

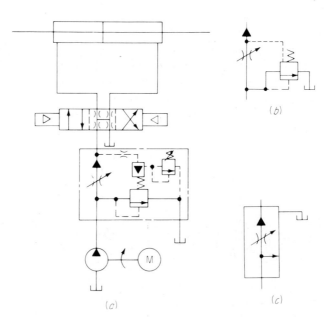

Fig. 11-14 (a) Pressure-compensated flow-rate control with integral relief valve (ANSI symbols). (b) ISO symbol for flow-control portion, full structure. (c) ISO simplified symbol for flow control.

maximum value because the cylinder encounters an unusual load, the relief valve will operate to prevent excessive system pressure.

Upon reaching an unusual pressure, the relief valve in Fig. 11-14 serves a dual purpose. It will provide the safety function as an auxiliary to the compensating function. Note that the entire orifice and compound relief mechanism are contained in a single envelope. The pump delivery is directed to the adjustable orifice. An internal tee connects the input of the main spool of the relief valve to this area. The balancing line to the control chamber of the relief valve is connected from a point beyond the adjustable orifice. This pilot sensing line is orificed to prevent an undue volume of fluid from entering the small integral pilot safety valve on the relief-valve control chamber. The location of the pilot sensing line at the point beyond the adjustable orifice means that the only pressure that can be developed by the main spool of the relief valve, until flow is established downstream, will be that corresponding to the spool spring setting. The setting of the spring determines the pressure drop across the externally adjustable orifice. The downstream load is reflected into the control chamber as an additive to the main-spool spring, so that the pressure drop across the adjustable orifice remains constant at the value of the spring, regardless of the downstream loading, until the sensing-line pressure becomes sufficiently high to crack through the safety relief-valve pilot. The small, integral safety relief valve determines the maximum pressure that can be built up in the control chamber. The maximum pressure built up in the control chamber determines the maximum hydraulic additive and the maximum system pressure.

Use of the relief valve and the downstream signal precludes using this device for anything but meter-in operation. Further, it can be used with only one functional unit and is usually considered as a complete circuit. It is not possible to connect two of these controls in parallel because the fluid will seek the course of least resistance and only one of the two controls would be effective.

Figure 10-3b and c shows a practical application of the meter-in flow-control system with relief-valve-type pressure compensation. In this application the major adjustable control orifice is created by the directional-control valve spool. Appropriate control channels are automatically provided as the directional spool is shifted to the various work positions. Maximum-pressure control is integral in the control-chamber pilot-relief mechanism that limits pressure additive to the set value established by the spring and cone assembly.

11-8 PRIORITY VALVES AND CIRCUITS

The use of two compensated flow-control valves to meter fluid into independent circuits is common; the resultant losses are not always as well known. To meter fluid into two circuits with conventional valves, the fluid pumped must exceed the combined flows. This excess fluid must pass over a relief valve to the reservoir at the maximum set pressure if constant-displacement pumps are employed in the circuit. Use of a variable-displacement pump will minimize power loss because the pump can be designed to supply only that flow necessary to supply the flow-control valves at their various adjusted positions. The pump compensating mechanism provides the needed control to adjust automatically the pump output to the needed flow through the flow-control valves.

The conventional meter-in flow-control valve for this service may have an integral pressure-reducing valve to control the pressure drop across the orifice. Use of a relief-valve-type compensator is precluded by the parallel path; the relief valve with the lowest setting would take all the fluid flow. Use of a reducing valve prevents this problem but introduces the problem of diverting excess fluid at maximum pressure.

A priority circuit plus two adjustable orifices in a single housing provides an answer to this circuit problem. To minimize a complex description, the circuit is explained in steps, rather than as an entire unit.

Figure 11-15 shows the symbol for a priority valve. The spring holds the valve spool in the position shown in the left flow block. Pressure is connected to circuit 1. Circuit 1 is also connected to the control area opposing the spring. As the pressure increases in circuit 1, the control signal tends to move the spool toward the spring and divert some of the fluid to circuit 2. This is the first portion of the priority circuit. Note that the entire flow will go to circuit 2 if circuit 1 is completely blocked.

Figure 11-16 shows how the knob assembly controls an orifice metering fluid into the load circuit associated with the spring cavity on the left-hand block of the priority valve. The knob also opens a pilot two-way valve to the tank when the flow is completely stopped to circuit 1. In the position shown, the fluid signal to the spring chamber of the priority valve is vented to the tank so that the fluid will pass to circuit 2 at the pressure setting of the priority-valve spring. This can be a relatively low pressure, usually well under 100 psi.

As the control knob is moved to meter fluid into circuit 1, a signal is reflected back to the spring chamber of the priority valve. This signal (pressure) will be the value of the load in circuit 1. The input pressure will be at least as much as the spring setting plus the load pressure in circuit 1.

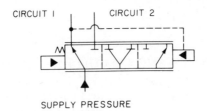

Fig. 11-15   Priority valve.

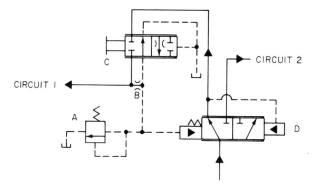

Fig. 11-16   Priority valve and circuit 1.

The spring value will control the pressure drop across the orifice. A secondary orifice in the control circuit limits the amount of reflected signal fluid to a quantity that will not overload the small signal lines or pass an excessive amount through the pilot overload relief valve connected to the spring chamber of the priority valve. This small, integral, maximum-pressure, pilot-relief valve prevents excessive line loads and subsequently controls maximum pressures in this area. It is adjusted to the maximum expected load value commensurate with the circuit components and the mechanical devices being operated. This valve usually serves as a safety relief valve rather than as a maximum-pressure valve. Maximum working pressures generally are a function of loads rather than of this valve setting.

In the circuit in Fig. 11-16 the priority valve serves both as a reducing valve controlling the pressure in circuit 1 and as a diversion valve passing the remaining fluid downstream to circuit 2. It bypasses fluid and maintains the correct flow to actuate circuit 1. Further, this valve can reduce pressure when necessary on circuit 1 and divert fluid to circuit 2 at full pressure when needed.

Figure 11-17 omits the controls in circuit 1 to avoid complexity. The

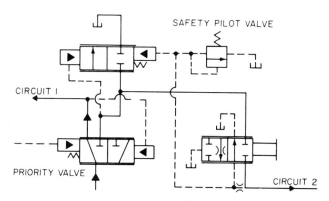

Fig. 11-17   Priority valve and circuit 2.

Flow-Control Valves   303

priority valve is shown to illustrate how it connects the two circuits. The valve shown immediately above the priority valve is a pilot-operated bypass valve or a relief valve. If the knob in circuit 2 is turned so that the flow through the orifice is completely blocked, the knob also actuates a portion that diverts the control fluid to the spring chamber of the relief valve back to the tank. In this position any fluid diverted from circuit 1 would be passed on to the tank at the value of the main-spool spring in the upper relief valve. The spring in the priority valve and the spring in the relief valve are in series so that the pressure drops are additive. However, the total value is relatively low, and little power is lost when both levers are in the closed position. The pump can be considered in the idling position.

As the control knob of circuit 2 is moved toward the open position, the vent line or line from the control chamber of the relief valve is blocked and the reflected signal from circuit 2 load is applied to the spring chamber along with the basic spring value. This permits the excess fluid passing back to the reservoir to do so at the circuit 2 load value. The safety pilot valve merely establishes a certain maximum pressure value that can be contained within circuit 2. This pressure value is usually much in excess of the work values normally encountered. Compensation for circuit 1 is by reduced pressure to the orifice controlled by the priority valve. Compensation for circuit 2 is by the input relief valve controlling the pressure drop across the orifice.

Figure 11-18 shows the complete circuit. The fluid under pressure from the pump is directed into the priority valve. This valve directs the fluid to the input of circuit 1. As the needs of circuit 1 are met by a portion of the fluid passing into this area, the excess fluid passes to circuit 2. The orifice established by the control knob or lever of circuit 1 determines the speed of operation of the mechanism fed by this circuit. Fluid being fed to the relief valve in circuit 2 meets resistance caused by a normally closed spool and is diverted to the orifice section controlled by the position of the knob or lever. As the pressure across the orifice establishes a maximum flow, the excess passes through the relief valve. This function occurs because the fluid in the spring chamber of the relief valve is at a controlled maximum pressure established by the work load in circuit 2. This work load plus the basic spring value determine the pressure value at which the spool can move to create an orifice of a size to bypass the excess fluid back to the reservoir.

Levers are provided to adjust independently the flow to circuit 1 or circuit 2. The ports consist of a pressure input, circuit 1 outlet, circuit 2 outlet, and tank outlet. As shown in Fig. 11-18, the drains from the levers and safety pilot valves are connected directly to the tank port. These units can be externally drained if the circuit is to be more sensitive. Usually there is a maximum-movement adjustment on the levers to establish preset maximum flows and calibrations for repetitive adjustments.

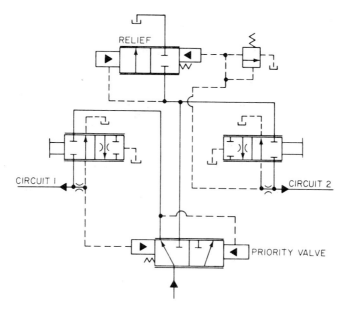

Fig. 11-18 Complete priority circuit.

One of the first uses for these valves was in the control of fluid motors driving salt and sand spreaders in highway ice-control work. Priority valves can also be used for balancing feed- and stock-removal conveyors, for steering tracked vehicles by changing the speed of fluid-motor turning drives, on ship drives using fluid-motor screw actuation, and in many other applications where a fluid motor or reciprocating device needs a diversion of the fluid through flow-control mechanisms for balancing purposes. Power losses are low, and efficiencies are high.

SUMMARY

Flow-control valves are used to control the flow rate of the fluid in a hydraulic circuit. These valves restrict the flow of fluid in the conductors or pipes.

The types of flow-control valves used include sliding-spool, needle, globe, gate, noncompensated, combined, and compensated valves.

Flow through orifices, either fixed or adjustable, is affected by pressure and temperature. Compensation is provided to ensure a uniform pressure drop across the orifice regardless of input or outlet pressures within the rated capacity of the valve.

Compensation can be provided by a reducing valve to control pressure drop or by a relief valve to provide the same function by diverting fluid to a secondary, lower-pressure area rather than restricting flow to the orifice.

Flow-Control Valves **305**

Control-valve circuits are of three basic types: meter-out, meter-in, and bleedoff.

Flow-control valves are often incorporated within envelopes that control a number of machine functions through a multiplicity of control valves.

Use of two or more flow-control valves with interrelated pilot controls to split flow can be structured to provide priority and/or predetermined flow patterns.

REVIEW QUESTIONS

11-1 Can a check valve be incorporated within a noncompensated flow-control-valve envelope structure?

11-2 What is the purpose of flow-control valves in hydraulic circuits? When is a gate valve used for flow control?

11-3 What is meant by a priority circuit?

11-4 Will the pressure value of the spring in Fig. 11-6 affect the rate of flow through the valve?

11-5 What type of compensation is used in the valve in Fig. 11-6?

11-6 What effect does the cutoff point have in the operation of the control valve in Fig. 11-6?

11-7 Why is the reducing spool sometimes restricted in movement?

11-8 Is an external drain essential in a compensated-type, flow-control valve?

11-9 Why is the four-way, directional-control valve used in the circuit in Fig. 11-12 shown with external drains?

11-10 What controls the normal pressure level in a circuit using a bleedoff flow-control valve?

11-11 Why is the control line provided from the outlet of the flow-control valve back to the spring chamber of the reducing valve in Fig. 11-7?

11-12 Does a beginning cut with a drill affect the feed rate in the same manner as the breakthrough does?

11-13 Is an integral check usually provided in a bleedoff-type flow-control valve?

11-14 Can a flow-control valve with a relief-valve compensation be used when only one pump and cylinder are provided in the circuit?

11-15 Why is a bleedoff flow-control circuit often used in broaching machines?

11-16 What is the purpose of the milled notch in the movable hat-shaped piece adjacent to the spring in Fig. 11-5a?

11-17 What is the significance of the slash arrow across the inverted parenthesis in the symbol of Fig. 11-5b?

11-18 What is the purpose of the check poppet in Fig. 11-6b? Why does it have an orifice in the nose?

11-19 Why is fluid metered from the blind end of the cylinder of Fig. 11-9?

11-20 What does the center block in the directional-valve symbol of Fig.

11-12 show? What do the hyphenated vertical dividing lines signify?

11-21 Describe the action of the valve of Fig. 11-15. What is the purpose of the parallel lines adjacent to the boxes in the symbol?

11-22 Is the center position of the symbol of Fig. 11-15 functional? Why is it shown?

11-23 What is the purpose of the orifice feeding from the outlet line to circuit 1 of Fig. 11-16 back to the priority valve?

11-24 What is meant by priority? What is the relationship to the term *sequence*?

11-25 What is the purpose of combining flow-control valves on a common manifold plate with other control valves? Is there more than one reason? What are the two most important reasons?

LABORATORY EXPERIMENTS

11-1 Connect two needle valves in parallel to the discharge of a hydraulic pump providing a constant flow of fluid in excess of 50 in³/min. Connect a relief valve set at 1000 psi in parallel with the needle valves. Open one needle valve to provide a path to the tank. Reduce the pressure with this needle valve to 500 psi. Reduce the pressure with the second needle valve to a value of 400 psi by passing fluid through it to the tank. Observe the flow through each valve. Reduce the pressure with the first needle valve to 300 psi. Observe the change in pressure and the effect on flow through the valves. Observe the relationship between pressure and flow.

11-2 Replace one needle valve in the preceding experiment with a reducing-valve compensated flow-control valve. Close the remaining needle valve. Observe the gauge reading as the compensated flow-control valve is opened and closed. Note the same readings when the pressure is raised and lowered with the needle valve.

11-3 Mount a cylinder in a vertical plane. Attach sufficient weight to the rod to move the cylinder freely. Build the circuit in Fig. 11-9 and observe its action.

11-4 Repeat Experiment 11-3 using the circuit in Fig. 11-10. Connect the components so that the prefill action of check valve X can be observed.

11-5 Oppose the cylinder in the circuit of Fig. 11-11 by the cylinder in the circuit of Fig. 11-13. Observe the pressures within the circuit by placing gauges in each working line.

11-6 Leave the four-way valve in the circuit of Fig. 11-13 centered. Observe the pressure in the head end of the cylinder of the circuit in Fig. 11-13 when pressure is applied by the cylinder in the circuit of Fig. 11-11.

11-7 Remove the line going to the rod end of the cylinder in Fig. 11-11.

Plug the tee going to this rod end and direct the rod-end fluid to the tank. Observe the difference in pressure on the head end of the cylinder in the circuit of Fig. 11-13. Insert a vacuum gauge in the rod end of the cylinder of Fig. 11-11 and carefully apply pressure on the cylinder of Fig. 11-13. Note the amount of vacuum at the gauge. Relax pressures and remove the gauge. Observe the difference in cylinder movement during application of force through the cylinder of Fig. 11-13.

11-8 Build the circuit of Fig. 11-14, using a needle valve for the adjustable orifice and commercial piloted relief valve. Observe the pressures in each portion of the circuit.

11-9 Obtain a spring-offset, oil pilot-operated, four-way valve with an open, or partially open, center spool. Plug the tank port. Connect the pressure port to a supply of at least 3 gpm at 1000-psi potential. Connect as in Fig. 11-15. At circuit outlets 1 and 2, tee in a pressure gauge which will read to at least 1000 psi. From the other outlet of the tee, connect a globe or needle valve with the discharge to the tank. Open these two valves to the tank so that no resistance will be encountered at either outlet 1 or 2. Supply oil at the pressure potential previously indicated. Gradually close the valve at circuit 1. Note the pressure at which flow is diverted to circuit 2.

11-10 Add the components of Fig. 11-16 tc the circuit in Experiment 11-9. Use a small pilot-relief valve for valve A. Orifice B can be a small needle valve. Valve C can be a globe and needle valve in parallel. The shutoff valves at the outlets of valve D may be left wide-open. Close valve B and open needle valve at C controlling the bleed to the tank. Open the globe valve at C to the tank or to operate a fluid motor. Relax the spring adjustment of valve A. Install a globe valve at the outlet of circuit 1 or use a fluid motor for a load.

Apply pressurized fluid to valve D. Close the outlet valve at circuit 2 and block the outlet to circuit 1. Close the needle valve within C. Open valve B to permit flow through A of approximately 30 in^3/min. Gradually raise the pressure at valve A. Note the pressure on the gauges. Open and close the valve at circuit 2. Note the pressures.

Continue to build circuit as shown in Figs. 11-17 and 11-18. Use fluid motors as the load driving a suitable pump. Note the pressure required to create the priority function. Experiment with different settings.

12
Directional-Control Valves

The primary function of a directional-control valve is to control the direction of fluid flow in a hydraulic circuit. A secondary function of the directional-control valve is to control, in some hydraulic circuits, both the pressure and rate of flow of the fluid. In this chapter a variety of useful directional-control valves are discussed.

12-1 CHECK VALVES

A *check valve* allows free flow of fluid in only one direction. Fluid flow in the reverse direction may be completely blocked, restricted, or piloted, depending on the valve design and use.

ANGLE CHECK VALVES
These valves [Fig. 12-1a] are built in the form of an elbow. The spring-loaded valve in Fig. 12-1a completely blocks flow in the reverse direction. During flow through the valve, the poppet is held off its seat by the fluid pressure acting against the spring tension. Because the spring exerts a downward force on the poppet, there is a drop in the fluid pressure as the fluid passes through the valve. The pressure drop varies with the rate of fluid flow and the type of fluid. When fluid flow through the valve ceases, the spring forces the poppet onto the seat. With the poppet held tightly against its seat, fluid cannot pass back through the valve.

The spring-loaded angle check valve in Fig. 12-1b has a removable pipe plug in the center of the poppet. If desired, this plug can be removed from the poppet or drilled to a specific size. With the plug removed, or with a hole drilled in the plug, a fixed-size orifice is formed for fluid flow through

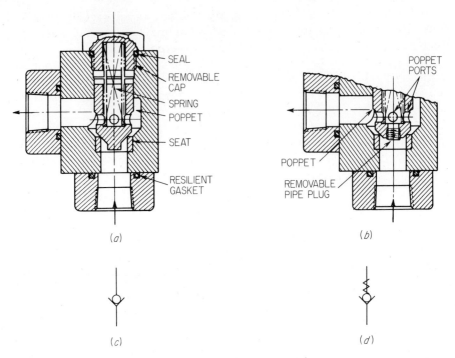

Fig. 12-1 (*a*) Poppet-type angle check valve for complete seal in reverse-flow direction. (*b*) Orifice-type angle check valve. (*c*) ISO symbol for check valve which is free to open if the inlet pressure is higher than the outlet pressure. (*d*) ISO symbol for a check valve which opens if the inlet pressure is greater than the outlet pressure plus the spring pressure.

the valve in the reverse direction—outlet to inlet—when the poppet is held on its seat by the spring. The free-flow condition is the same as that shown in Fig. 12-1*a*. Figure 12-1*c* shows the ISO symbol for a check valve which is free to open if the inlet pressure is higher than the outlet pressure. The weight of the poppet which will close the passageway by a gravity drop must also be considered. The poppet will be urged against the seat by pressure differential between inlet and outlet port.

Figure 12-1*d* shows the ISO symbol for a check valve which opens if the inlet pressure is greater than the outlet pressure plus the spring pressure.

Spring-loaded check valves are also built with the inlet and outlet connections in a straight line. In some designs the valve body is cored, so that the fluid makes two right-angle turns during passage through the valve. With this design the poppet can be serviced by removing the cover cap, as in the angle-type valve. Other designs have an in-line flow through the valve seat and through or around the poppet. This arrangement does not permit servicing without removing the entire valve from the line prior to disassembly. The in-line design does provide savings in manufacture while retaining adequate flow characteristics. Circuit requirements dictate the most advantageous design for specific application. Special considerations within a circuit may dictate the need for resilient seating action or auxiliary seals at the poppet area.

In evaluating the economics of circuit piping, it is apparent that an angle check valve provides a right-angle turn of the fluid conductor and can replace an elbow fitting. The valves in Fig. 12-1 have a flange union connection which makes the valves easier to maintain.

Spring-loaded check valves are often used to create a resistance to flow in order to provide pilot pressure or to establish a certain pressure value in one portion of a circuit. The spring-loaded check valve may be a terminal point in certain tank lines. It will provide enough resistance to flow to divert fluid through a heat exchanger or filtering mechanism. A spring-loaded check valve may also be provided on tank lines to make certain that the lines stay full of fluid when the machine is shut down. Minor leaks in the circuit normally might permit the fluid to return to the tank. The empty lines can create some erratic movements when the machine is started again. Keeping the lines full of fluid eliminates start-up problems of this type, and the machine is ready to operate immediately.

SWING-GATE TYPE OF CHECK VALVES

These valves [Fig. 12-2] are designed to permit fluid flow through the valve with minimum resistance, yet stop flow in the reverse direction. This type of check valve is generally used on low-pressure plumbing where a vacuum exists and the forces on the fluid are small, as in pump suction lines. Some pump designs and construction require an intake check valve on the intake line below the fluid level. Rotary pumps would not pressurize this intake check valve in the same manner that a reciprocating-piston pump would. If there is likely to be a relatively high head pressure on the discharge of the pump when the pump is not operating, it is usual to install an angle check valve [such as that in Fig. 12-1] on the discharge of the pump, beyond the point where the relief valve is installed. With this arrangement, the pump has the resiliency of the relief valve to start against instead of the head pressure on the check-valve poppet.

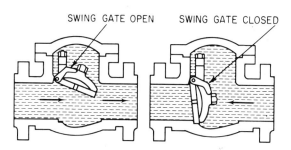

Fig. 12-2 Swing-gate type of check valve.

Directional-Control Valves

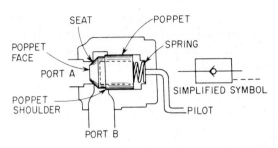

Fig. 12-3 Normally open pilot check valve.

12-2 PILOT CHECK VALVES

Four types of check valves are used for pilot service. These are the *normally open type, normally closed type, barrier type,* and *decompression type*.

NORMALLY OPEN TYPE
This type [Fig. 12-3] will permit flow in either direction until a pilot pressure of sufficient magnitude is placed on the pilot connection to force the poppet tightly against the seat. The poppet acts as a ram and closes off the line. The seat is smaller in diameter than the main bore. Pressure in one direction acts on the shoulder area of the poppet, consisting of the major bore diameter minus the seated area, and causes it to open if there is no pressure on the pilot line or if there is insufficient pressure to balance out the areas. Pressure in the opposite direction is effective against the face of the poppet.

NORMALLY CLOSED TYPE
Figure 12-4 shows a normally closed, pilot-operated check valve. This valve will permit free flow in one direction. In the other direction, the flow is stopped until a sufficient pilot pressure is applied on the pilot-piston assembly to force the poppet off the seat. Note the relative area of the pilot piston and the seated area of the check poppet. The ratio between the pilot-piston area and the seated area of the poppet will determine the pilot-line pressure necessary to upset the poppet and urge it against the spring. A flow through the valve from either direction will impinge on the face of the pilot piston adjacent to the poppet. If this flow meets resistance, it can reflect pressure against the pilot-piston face and negate part or all of the pilot pressure, so that the poppet will tend to close against the seat. If the valve is used to divert fluid to a low-pressure area, this may not be a problem. But if the valve is used to divert fluid into another high-pressure area, performance may be very unsatisfactory.

BARRIER TYPE

Figure 12-5 shows how the pilot check can be modified to minimize the effect of the pressure within the major passages on the face of the pilot piston adjacent to the poppet. A barrier sleeve is provided to minimize the difference in areas. The actuating piston is of relatively small diameter—just large enough to provide the needed column strength to urge the pressurized poppet toward the spring and open the passage between the ports. The pocket between the barrier sleeve and the piston face is drained to a low-pressure area. Valves of this type provide a control for the stored energy in an accumulator. The high-velocity flow from the accumulator is isolated from the actuating piston so that pilot pressures taken from the accumulator will not be nullified when the pilot valve opens the two main ports to each other.

DECOMPRESSION TYPE

Figure 12-6 shows a decompression-type poppet assembly. It has an intermediate poppet within the main poppet. Within this intermediate poppet structure, a ball poppet is concentrically seated with the other two poppets. A pin is provided to upset the ball prior to opening of the first poppet. The ball opens a small passageway to permit high static pressures to decay. After these pressures have relaxed to a certain point, the next step can be programmed. The small intermediate poppet extends slightly below the face of the main poppet. The pilot piston will contact the face of the intermediate poppet after the ball has permitted the locked static

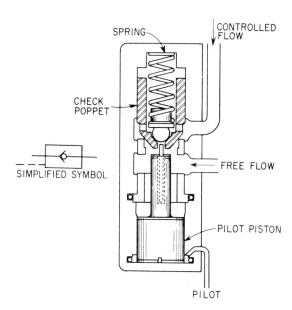

Fig. 12-4 Normally closed pilot check valve.

Directional-Control Valves **313**

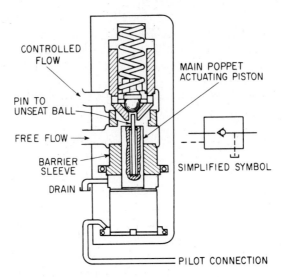

Fig. 12-5 Barrier-type pilot check valve.

pressure to decay to a point where sufficient pilot pressure is available to move the small poppet. This will provide an additional avenue to flow so that the relaxation of pressure will be hastened. When the pressure is sufficiently low, the main poppet can be unseated and the full maximum flow path will be established.

12-3 PILOT CHECK CIRCUITS

Figure 12-7 shows how two check valves, such as the unit shown in Fig. 12-4, can be combined in a single body. This assembly can be manifold-

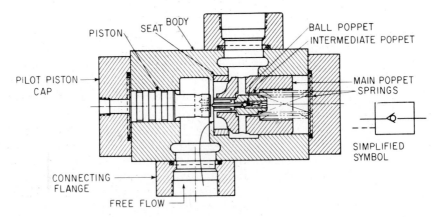

Fig. 12-6 Decompression-type pilot check valve.

314 *Industrial Hydraulics*

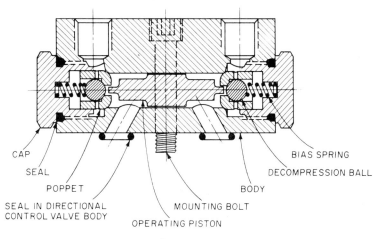

Fig. 12-7 Dual pilot check assembly.

mounted to a four-way directional-control valve. The actuating piston will operate in either direction according to the pressure level within the cylinder lines.

Figure 12-8 shows how the assembly of Fig. 12-7 is incorporated into a power-transmission circuit. The double-rod cylinder is used to position a chute on a concrete paving machine. When the chute is in position, it must be locked firmly to prevent variable loading from causing movement. The neutral position of the directional-control valve is designed to interconnect both cylinder ports and the tank conductor. Pilot pressure is relaxed in both pilot check valves. The check poppets lock fluid in the cylinder so that it is held rigidly. Pressure is blocked in neutral so that the

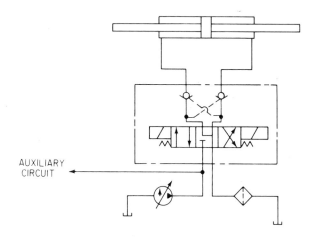

Fig. 12-8 Dual pilot-actuated check valve assembled to four-way directional-control valve.

Directional-Control Valves **315**

pump discharge volume is fully available to other functions within the machine. As the left solenoid is energized, pressurized fluid passes freely through the left check valve. Pressurized fluid urges the operating piston of the pilot check assembly against the right-hand poppet assembly, providing an opening for the return fluid to the reservoir. Reversal of the directional-control valve provides movement in the opposite direction. Release of either solenoid causes the valve to center and the check assemblies to lock fluid positively in the cylinder lines.

12-4 TWO-WAY VALVES

Many pressure-control valves are basically two-way valves. These pressure-control valves are sometimes used as directional-control valves by modifying the spring value or piloting the valve in both directions.

Figure 11-2 shows a cam-operated, two-way spool. The body also contains an integral check valve, to pass fluid back through the valve regardless of the position of the cam-actuated spool, and an adjustable orifice. This type of valve is widely used to divert control fluid through flow-control valves of either the pressure-compensated or the simple orifice type such as shown in Fig. 11-2. The machine member is permitted to travel at a high rate of speed until the cam encounters the roller on the spool of the two-way-spool-type valve. At this time the fluid is directed to the associated flow-control device. Note that the spool shown in this valve is normally open when the roller is not depressed. A normally closed spool, which opens as the roller is depressed, can be provided if preferable for *skip feeds* (moving through castings when cored passages are encountered which require no machining). This valve may be used to provide a mechanical and hydraulic interlock in a circuit. Flow is blocked to one portion of the circuit until a cam depresses the spool and roller at the desired location to permit flow into the secondary portion of the circuit.

Additional functions are provided in the valve assembly shown in Fig. 12-9. Rapid traverse is provided with the control spool in the extended position. As the cam encounters the roller, the control spool is urged against the bias spring. The direct path from *in* to *out* is blocked. Compensated flow then passes through *both feed spigots*. Further movement of the control spool blocks feed spigot 2. Fine feed rate is adjusted at dial 1; coarse feed rate is adjusted at dial 2. Coarse feed is a value established by dial 1 plus the additional capacity added by the spigot of dial 2. Both rapid-traverse and feed selections are compensated by the reducing-valve-type, two-way-spool mechanism. Because of this, the transition from one function to the next is smooth and free of erratic movements. The integral check valve of Fig. 12-9 permits free return flow of fluid.

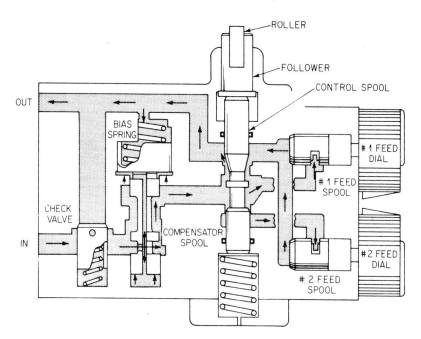

Fig. 12-9 Compensating-type two-way valve regulating flow to a three-way diversion valve and adjustable orifice assembly.

12-5 SHUTTLE VALVES

In certain types of machinery, the control must be from more than one point of origin in order to meet circuit requirements. The three-port shuttle valve in Fig. 12-10 can be used to provide a path for fluid from two alternate sources. The shuttle piston will stay in position, blocking the flow path to port 1, as long as pressures in port 2 and the out-port are greater than the pressure within the entry to port 1. When the pressure at port 1 is greater than that at port 2 and the outlet, it will urge the piston against the rest pin at port 2. This valve will permit flow in both directions when the piston is against a stop pin. The signal to shift the shuttle piston must come from either port 1 or port 2. The signal cannot come from the out-port. A definite pressure differential should be available between ports 1 and 2 so that the shuttle piston does not lodge halfway and block the out-port in certain types of circuit.

A ball-and-seat arrangement as in Fig. 12-10*b* may be used to provide a leak-tight seal. A fluid connection to all three ports will be encountered as the ball passes center in the crossover function. The shuttle piston of Fig. 12-10*a* is not completely dead-tight in the operating position. It does not, however, connect all three ports in the crossover position. The ball-type unit is widely used because of the economical construction and the tight

Directional-Control Valves **317**

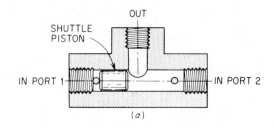
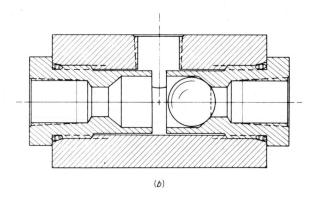

Fig. 12-10 Three-port shuttle valve. (*a*) With closed crossover. (*b*) With open crossover.

seal obtained as the ball abuts the seat. The small loss on crossover is usually of limited importance.

12-6 THREE-WAY VALVES

Cylinders or rams that do not require fluid under pressure to return the unit to a rest position can be operated by a three-way valve. The three-way valve may use a sliding spool or two poppets, such as those in Fig. 12-11, alternately to connect the cylinder to the pressure source and the tank. In Fig. 12-11 the pressure is blocked by poppet *A*. Poppet *B* prevents cylinder retraction. When poppet *A* is depressed, the pressure and cylinder ports are connected and the cylinder will be actuated. Closing poppet *A* will hold the cylinder extended. The cylinder will stay extended until poppet *B* is depressed, permitting the fluid to flow back through the tank port. The poppet-type construction provides a relatively tight closure. To make the valve completely tight, resilient static seals or ground poppet surfaces can be added.

The spool-type valve will have minimum flow conditions dictated by the tightness of the fit between the moving spool and the bore into which

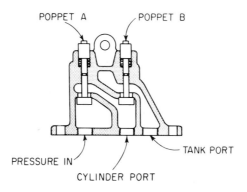

Fig. 12-11 Three-port, poppet-type, three-way valve.

it is fitted. Low-pressure valves of the spool type may have resilient seals fitted to the spool. This spool passes across small-diameter holes in a liner sleeve when passing from one port configuration to the other. The resilient seal also provides a completely tight seal with the sleeve construction. Pressures are generally limited to less than 150 psi when a resilient seal is used on the spool. Fluid pressures in excess of 10 000 psi have been contained by a metal-to-metal type of spool construction. Poppet-type construction can handle very high pressures if the mechanical strength of the valve is sufficient and if sufficient force is available to operate the poppets. Sliding-spool types of valves with metal-to-metal seals can be balanced so that operating forces are relatively low. Note in Fig. 12-11 that the energy necessary to depress poppet *A* will increase as the contained pressure within the pressure inlet port is raised. This factor is often the limitation imposed on the poppet design.

12-7 DIVERSION VALVES

The spool-type diversion valve in Fig. 12-12 is similar to a three-way valve. The basic difference is in the manner in which the ports are connected. In this valve the input port would be the cylinder connection if the valve were to be used as a three-way valve. Circuit 1 or 2 could be the pressure port, and the remaining would be the tank port.

When used as a diversion valve, the unit does not have a tank port. Pressure is directed to one of the two circuit connections in each shifted position. This valve is commonly used on mobile equipment such as multiple-purpose tractors that may have a front-end loader and follower equipment such as plows, disks, and other groundworking equipment. The diversion valve directs the hydraulic fluid to the front-end lift when the tractor is being used for this work. This diversion valve can also direct

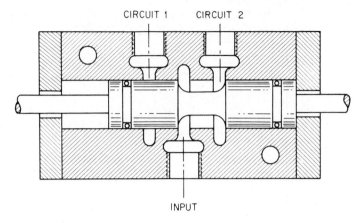

Fig. 12-12 Spool-type diversion valve.

fluid to the other circuits when the front-end unit is out of service. The circuits are usually so arranged that the diversion valve completely isolates one circuit while a second is functioning.

12-8 FOUR-WAY VALVES AND CIRCUITS

Figure 12-13 shows a rotary four-way valve controlled by the cylinder-rod movement. This rotary four-way valve, in turn, shifts the spool of the piloted four-way directional-control valve. The rotary four-way valve must be shifted well beyond center before the piloted four-way-valve spool can start to move. This will ensure a continuing supply of pilot pressure to the piloted four-way valve so that it will not stop on the crossover position.

Figure 12-14 shows how the rotary four-way valve works. Figure 12-14a shows how the fluid path is directed in one extreme shifted position; Fig. 12-14b shows that all flow ceases in the neutral crossover position. The other extreme shifted position is shown in Fig. 12-14c. From these it is apparent that the main spool of the piloted valve will be reversed when the dog connected to the cylinder reverses the rotary four-way valve. Note the coring in the piloted four-way valve to direct the return fluid back to the reservoir in both directions of travel.

Figure 12-15 shows how a valve uses the center of the spool to direct the fluid back to the tank connection in one direction of flow. Note the spring-loaded balls at the extreme left end of the main spool. This is a *balanced detent mechanism* with the springs loading a pair of steel balls which engage an annulus around the spool diameter. The retentive power keeps the spool in position, regardless of external vibrations, until a solid fluid signal source is applied to the end of the spool. A positive signal can provide sufficient force to move the two balls out of the annulus as the

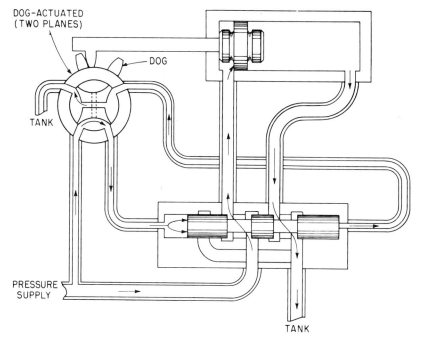

Fig. 12-13 Simple hydraulic reversal system.

spool is shifted. At the completion of the shift of the spool, the two spring-loaded balls will engage a second annulus at the other extreme shifted position.

The cylinder in Fig. 12-13 will continue to reciprocate as long as the pressure supply is maintained at a value sufficient to move the load. Speed of movement will be a function of the rate of flow of the supply fluid.

Actuation of directional-control valves can be by lever, pedal, stem, cam, dog, roller, fluid pressure (either pneumatic or liquid), solenoid, motor, or a combination of these devices.

Lever operation usually indicates complete control by an operator. Some

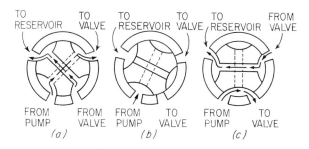

Fig. 12-14 Rotary four-way valve.

Directional-Control Valves

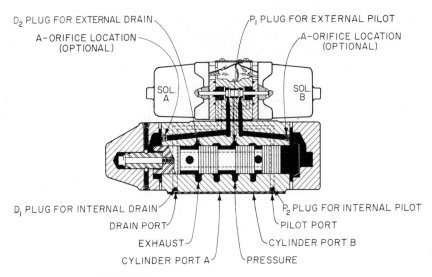

Fig. 12-15 Solenoid-controlled, pilot-operated, detent-held, four-way valve with hollow-main-spool design.

semiautomatic devices are provided with a pilot in one direction and lever actuation in the other direction. The operator can start the machine in motion, and a pilot source will shift the valve by fluid pressure in the opposite direction. Pedal operation is but a modification of the lever device. Stem operation may be an extension of the lever or an interlock with other movements within the machine. Cam operation is common in automatic machinery for interlocking functions. In Fig. 12-13, dog operation is used. Various roller actuators may supplement cam operation, or the valve may be actuated by the roller on the stem of the valve contacting a control slide with dogs of proper design. Rotary cams may abut the roller mechanism on the valve stem for certain types of programming.

The valve spool in Fig. 12-15 is shifted by an application of fluid under pressure directly against the end of the spool. Thus, the end of the spool is essentially a piston. The resultant forces available to shift the spool will vary according to the area of the end of the spool. Optional end caps may be fitted to a directional-control valve to permit use of compressed air or other fluids to shift the spool.

12-9 HYDROPNEUMATIC CONTROL CIRCUIT

One of the most popular of the less complex signal sources is the compressed-air system. Note in Fig. 12-16 that the hydraulic system operates at elevated pressures (3000 psi) to perform the direct labor on the machine. The pneumatic system actuates the control functions with a pressure less than 100 psi. The circuit in Fig. 12-13 is somewhat similar to that in Fig. 12-16 except for the pilot fluid and the use of cam-operated three-way valves in the latter circuit for pilot purposes.

Valve A in Fig. 12-16 is used to stop and start the reciprocating circuit. In the position shown, valve A relaxes all pilot fluid in the circuit and blocks the incoming source. The springs of valve C will center the main spool, and the cylinder will be locked in position. When valve A is shifted, it will connect fluid (compressed air) to valves B, D, and E in parallel. Valve B remains in a fixed position when air is applied to either pilot piston and the spool has made its prescribed movement. When the cylinder retracts and the dog depresses the cam on valve D, fluid is directed to valve B to shift the spool and cause the flow condition shown to the left. This will cause valve C to shift and reverse the cylinder, and the piston will move to the extended position. This will, in turn, depress the shifting mechanism on valve E. The spring in valve D causes the spool to shift to the relaxed position and block the flow to the left end of valve B. Valve E will then direct fluid to the right end of valve B. This will cause valve C to shift and start the retraction of the cylinder rod. The reciprocating action will be automatic and continuous until the cycle is stopped by blocking either the hydraulic supply or the pilot-air supply. Speed of reciprocation is basically controlled by the amount of fluid flow available in the hydraulic system. The length of stroke and size of cylinder will also

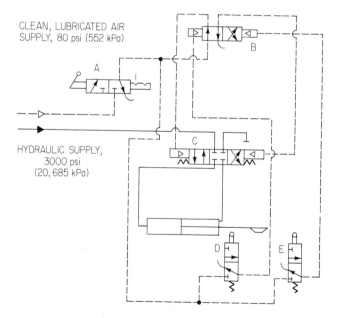

Fig. 12-16 Hydropneumatic-control circuit.

affect speed and force considerations. Valves D and E might control valve C independently of valve B if a spring-centered configuration were not used.

12-10 FLUIDIC CONTROL SYSTEMS

Compressed air at relatively low pressures can be used as a measuring device by a direct impingement against a machined part. The change in resistance to flow can be sensed by an appropriate valve or gauge mechanism which can be directly read or which can pilot a larger valve.

Low-pressure compressed air can also be controlled by special valves which can function by manipulating the air stream in a predetermined pattern.

This type of pneumatic control technology and the associated hardware, usually referred to as fluidic control systems, comprise the basis for special control systems used where explosives, adverse operating ambients, and unusual vibration preclude usual mechanical, electrical, or electronic controls. Because of the limited usage in industry, we are not attempting to cover the subject in this text.

Compressed-air systems, however, using valves similar to electrical devices, i.e., limit switch, pressure switch, relay, etc., have gained wide acceptance as signal sources and complete control systems.

In summary, compressed air, whether used as a complete system for controlling with fluidic devices or with moving part devices, offers many basic advantages as a signal-transmission agent. Some of these advantages are:

1. Response time may be somewhat slower than for electric signals but is usually faster than liquid hydraulic devices.

2. Repetitive accuracy, although it may be slightly less than with direct mechanical connections, is usually adequate for most machine tool requirements.

3. Extreme dependability can be built into the system with inexpensive components.

4. Air lines can be equipped with automatic lubricating devices for minimum maintenance.

5. Fire hazard is minimum.

6. Simple mechanical filtration of the air is usually sufficient. Intake areas can be conveniently located in clean or protected atmospheres.

7. Contamination of the product by nuisance leakage is minimized.

8. Relatively low pressures ensure minimum cost of plumbing.

9 Air consumption is low.

10 High operating forces are immediately available.

12-11 SOLENOID-CONTROLLED VALVES

Figure 12-17 shows a direct-solenoid, panel-mounted, four-way directional-control valve. The solenoid transmits force in a linear path through an extension of the fluid-directing spool. Fluid is isolated from the solenoid housing in this design by a dynamic-seal retainer plate. Full working pressure can be impressed on the tank port without malfunction. Because of the capability of using all ports at maximum working pressure, the unit may also be connected as a double two-way valve, a diversion valve, or a three-way valve. The detent mechanism prevents spool movement resulting from vibration caused by the machine on which the unit is mounted or by dynamic flow forces.

Note the port connections. The pressure port P is in the valve center so that the design is symmetrical. There is a cylinder line adjacent to the pressure port in each direction $C1$ and $C2$. The tank connections T abut

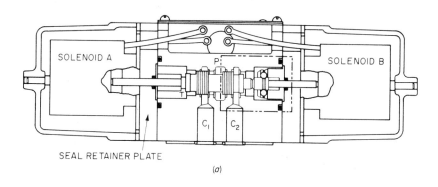

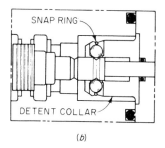

Fig. 12-17 Direct-solenoid, four-way valve. (*a*) Complete assembly. (*b*) Enlarged view of detent detail.

Directional-Control Valves **325**

the cylinder ports. These tank connections are ported together prior to leaving the valve housing. The pilot-valve tank port may be connected to a major return line in a piloted-valve section. Reflected pressures will not usually damage the dynamic seals in machine tool systems.

Oil-immersed solenoids or dc solenoids will usually fit in the same relative position on the valve. Oil-immersed solenoids are constructed with a cover that encapsulates the entire mechanism and permits operation in a bath of oil. This minimizes the hammerblow action on the anvil area of the armature and C frame when the solenoid is actuated and also provides lubrication and better heat dissipation. Direct-current solenoids of similar design are usually interchangeable. Direct-current solenoids of oil-immersed construction may require some modifications of the valve if the oil is common to the system fluid media.

The pilot valve assembled to the *stackable* four-way valve of Fig. 12-18 is equipped with dc solenoids. Because the pilot unit is designed for a centering-type valve assembly, the pilot spool is urged to the neutral position by compression springs. Shouldered centering washers determine the finite center position. No center position is intended in the valve of Fig. 12-17; thus, the detent mechanism of that unit establishes shifted positions. Energizing a solenoid associated with the pilot valve of Fig. 12-18 causes movement of the armature toward the anvil. This force is transmitted through a nonmagnetic pushpin to the end of the spool. The spool moves against a centering washer and spring. As the armature reaches the anvil, a shifted position is established. Pilot pressure is then

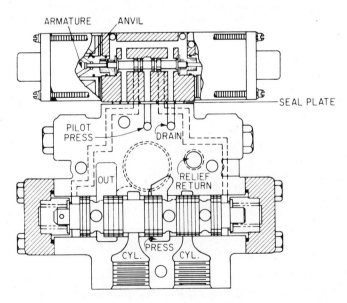

Fig. 12-18 Stackable, solenoid-controlled, pilot-operated, four-way valve.

326 *Industrial Hydraulics*

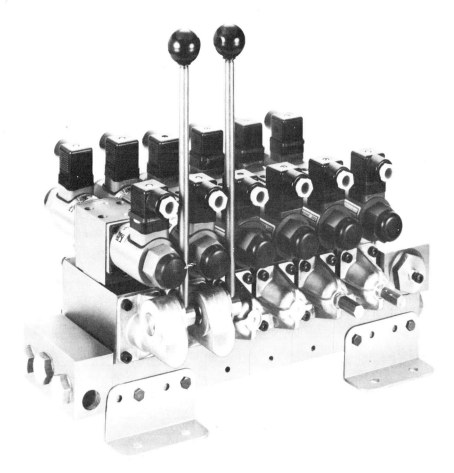

Fig. 12-19 Stackable valve assembled into a unified control center.

directed to one end of the main spool. Fluid displaced at the opposite end of the main spool is directed through the pilot valve to the tank or reservoir. In the neutral position, the pilot spool blocks pilot supply pressure, connects both pilot cylinder lines to the drain (from the end cavity of the main spool), and permits the spring on the main spool to center this flow-directing elements so that the desired major flow path is established.

Stackable valves are used widely in mobile fluid-power applications. Their use in general industrial fluid-power applications is predicated on the convenient-package concept. Space requirements are minimum and plumbing needs are reduced.

Figure 12-19 shows how several basic valves are incorporated to create a package that includes most of the components needed for a power-transmission control system. A relief valve and six directional-control valves are all incorporated in one compact assembly.

Large flows during retraction of a cylinder resulting from differential areas may be handled by an auxiliary two-way valve, sometimes referred to as a *dump valve,* manifolded to the cylinder ports of a stackable directional-control valve. The only special modification of the basic four-way valve is the addition of four tapped holes to accept the bolts used to hold the dump valve, shown in Fig. 12-22, to the main valve assembly. The O rings provide an automatic pressure-sensitive static seal between the interfaces.

The port B connection is piped to the large end of the cylinder. As pressurized fluid is directed through port A to the rod end of the cylinder, a pilot function can occur. Diameter $A1$ is less than diameter $A2$. The effective difference serves as an unbalanced area that can oppose the bias spring. As pressure in port A reaches the bias created by the installed load of the spring the poppet is urged from the seat. This creates a second path for return oil back to the reservoir. The seat area is equal to $A2$. Thus, the pressure at port B has no effect on the operation of the auxiliary return.

The only cost is associated with the basic unit price, plus the plumbing. Under most conditions the use of the dump valve doubles the capacity of the valving. This permits use of a ¾-in valve where a 1½-in size may have been suggested from preliminary analysis.

Single and double pilot-operated check valves, such as shown in Fig. 12-7, are packaged in a similar body. This unit mounts directly to the directional-control valve so that plumbing is at a minimum.

The main spool of the valve, shown in Fig. 12-18, connects the input port to the exhaust port in the neutral position through the hole in the center of the spool and the interconnecting holes to the surface of the spool. The cylinder ports are blocked in neutral. Different spool configurations will provide other conditions in the neutral position. Note in Fig. 12-21 how the first three valves use a spool as shown in Fig. 12-20. The

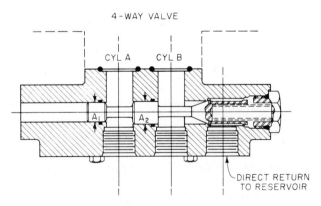

Fig. 12-20 Auxiliary two-way valve used to handle large flows.

328 *Industrial Hydraulics*

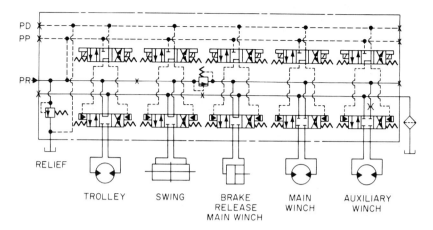

Fig. 12-21 Typical circuit for hydraulically actuated crane handling structural materials.

remaining two valves use a *closed-center* configuration wherein all ports are blocked in neutral.

Pilot pressure for the entire assembly is provided at the input to the stackable circuit. Pilot drain is combined with the relief-valve tank line. Filtering is provided in the major return to the tank conductor.

The circuit used to power a typical crane for handling structural materials is shown in Fig. 12-21. Pressurized oil enters the stack at the relief-valve segment. The relief-valve casting also serves as a connecting area for pilot pressure and pilot drain.

Trolley and swing may be required at any time. Maximum speed potential may also be needed. Force requirements are variable but usually comparatively low. Because of these conditions, the series circuit is used.

The main-winch brake release is actuated prior to operating either the main or auxiliary winch. Actuation of the brake release stops flow to the tank so that pressurized oil in parallel supply is available to both winches. A resistance cartridge valve between swing-and-release four-way valves guarantees minimum pressure for piloting purposes. The relief valve functions only in the event of circuit overload.

Pressurized fluid, entering the stack of Fig. 12-21, normally sees the pressure relief valve as a closed-branch tee. The plug (indicated by X in line) diverts pressurized fluid to the directional-control valve function associated with the fluid motor that powers the trolley. Because of the spool configuration, shown in Fig. 12-18, the fluid may be directed to the next valve series. If the trolley control valve is in neutral, the fluid passes through to the swing-control valve with the only resistance to flow created by the changes in direction and friction of conducting passages. If the trolley is moving, the energy necessary to create movement is absorbed prior to the fluid movement downstream to the swing cylinder. If the trolley is forcibly stopped so that the motor cannot function, all flow downstream will stop, and the pressurized fluid will be diverted over the relief valve at the set level.

Moving the swing cylinder to the end of the stroke will also stop all working flow and divert the pressurized fluid through the relief valve. The resistance valve in the line between the swing-and-brake-release valves guarantees pilot pressure to all units. The pilot-pressure connection is at the upstream position adjacent to the relief valve.

The brake-release four-way valve serves in several capacities. It must be held in the energized position for all downstream functions to be operable. It serves to pass oil through to the tank in a rest position. And it serves usual operating function by directing oil to actuate the brake release.

In other circuits, it is usually possible to find a clamp valve or outrigger circuit function that can be used for similar purposes in a combined series-parallel circuit. The main winch usually requires maximum power. Because of the lesser power requirements of the auxiliary winch, it is often supplied with a restriction in the supply line. This may be of the compensated type if wide pressure variance is common.

Cylinders with large differential area, such as those with unusually large rods, may need an auxiliary flow path to the tank to avoid an excessive pressure drop through the valve stack. The dump valve of Fig. 12-20 provides this additional flow capacity.

A circuit such as that shown in Fig. 12-21 can be assembled from threaded or flange-connected valves with interconnections made by use of tube, hose, pipe, or similar conductors. If many identical circuits are to be fabricated, it may be desirable to use subplate-mounted valves on a drilled or fabricated manifold plate wherein the major interconnections are through drilled passages. Subplate-mounted valves may be substituted for threaded or flange-connected valves with interconnections to the subplate for convenience in servicing the units. For major space and plumbing economies, the stackable units provide a well-accepted and proven interconnecting system. Service, if required, may be somewhat more difficult with the stackable unit.

Figure 12-15 shows a two-position, two-solenoid valve. The pilot spool is spring-centered, and the main spool is detented. The detent mechanism consists of dual, opposed spring-loaded balls that fit into a groove on an extension of the main spool. When the main spool has been fully shifted and the electric signal is relaxed on the pilot spool, there is no danger of vibration shifting the main spool. The pilot spool will remain in position, as shown in Fig. 12-15, since it is spring-centered and cannot be shifted by vibration.

Complete relaxation of a fluid-power system may be desirable. This can be accomplished by use of a spool such as that shown in Fig. 12-18. Further modification permits the relaxation of the cylinder-port connections in the neutral position as well as the pressure port. Unless the spool is directly operated by lever, solenoid, or other mechanical means, there must be pilot pressure available at a suitable value to pilot the mechanism to ensure desired major flow pattern.

Figure 12-22 illustrates schematically how check-valve units and a reversing pump can be used for directional-control service. The gear pump delivers pressurized fluid to the left-hand outlet. Flow enters through $T1$. $T2$ is closed so that the relief valve at $T5$ and the left-hand port provide the potential flow outlets. The relief valves, adjusted at B, are set to operate only in the event of circuit malfunction so that they would normally be considered a blocked port. Piston P is urged to the right, unseating a check poppet to permit return flow from the right-hand port through $T4$. Reversal of the pump will cause identical flow action in the opposite direction. Stopping the pump will lock the cylinder in the last

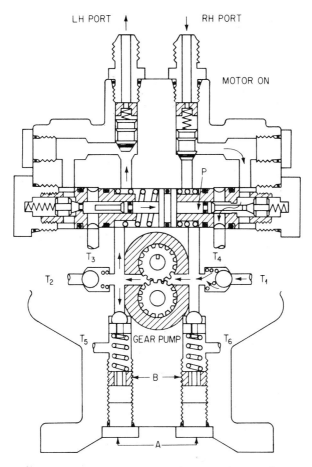

Fig. 12-22 Check-valve-type directional-control unit.

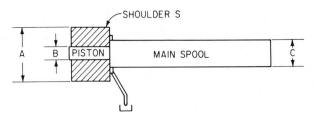

Fig. 12-23 Pressure-centering device for valve spools.

operating position. This type of circuit is primarily designed for small circuit components and compact power packages. Reversal or stopping of the pump is often accomplished by the use of a pressure switch, limit switch, or other automatic sensing device.

Springs may become bulky or difficult to actuate in the centering function of a four-way, directional-control valve. Hydraulic-pressure centering is fast and accurate. Hydraulic-pressure centering can be accomplished by providing a dual-piston assembly at the end of the valve main spool which is shouldered at the major diameter so that it determines the center position [Fig. 12-23].

The pressure port in the pilot valve is connected to both cylinder ports [Fig. 12-24] and is connected by suitable conductors to the end cavities of the piloted-valve section when the pilot valve is in the neutral or relaxed position. Pressure is held within the end caps of the main piloted valve. When a solenoid is energized or the pilot valve is actuated by any suitable means to direct fluid to the cavity adjacent to the end cap at C, piston B and sleeve A are directed to the tank. Pressure at the opposite end of the spool causes the piston and sleeve to move until they contact the end cap. This is the normal stroke. When the valve is again centered by deenergizing the solenoid, fluid is under pressure and directed to both end caps.

The combined diameters of piston B and sleeve A are greater than the diameter of the main spool. Fluid pressure, acting on the piston and sleeve, urges the main spool toward neutral. When sleeve A abuts shoulder S, it will stop. This is the neutral position of the spool, as shown. Piston B is smaller in diameter than the main spool and cannot force the

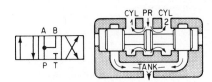

Fig. 12-24 Pressure to cylinder ports, tank port blocked in neutral position.

main spool across the valve center. When the opposite solenoid is energized, it will relax the pressure on the end of the main spool and direct fluid to the tank. Sleeve A is stopped by shoulder S, but piston B is free to carry the main spool on across the valve center until the main spool abuts the opposite end cap. This will connect the major pressure port with one cylinder port and the tank with the other cylinder port. Returning the pilot spool to neutral by deenergizing the solenoid or other pilot-control medium will permit oil to enter the cavity at the end of the main spool and force piston B back until the main spool abuts sleeve A, which has a larger diameter.

Large forces are available for positive centering of the main spool of the directional-control valve with this device. Note the port adjacent to shoulder S directed to the tank. This is a drain to care for the difference in areas between the spool, sleeve, and piston. When possible, the drain is usually connected above the fluid level in the reservoir to minimize the pumping action.

12-12 CARTRIDGE-DISK-TYPE SEALS FOR DIRECTIONAL-CONTROL VALVES

Figure 12-25 illustrates a ball-type, directional-control valve with a cartridge disk that provides a slight imbalance of areas so that the metallic seal is urged against the spherical surface. Various flow configurations can be programmed by the positioning of the holes in the ball element.

SLIDING-DISK TYPE OF CONSTRUCTION
Directional-control valves are also actuated by rotary electric motors and suitable cam mechanisms. Torque motors are used in the actuation of servo-type directional-control valves. These units are discussed in another chapter.

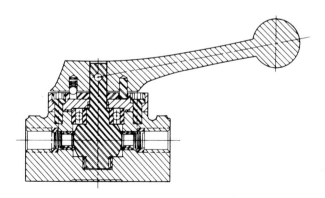

Fig. 12-25 Ball-type, directional-control valve.

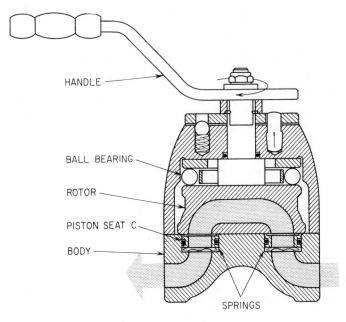

Fig. 12-26 Rotary-disk, seal-type, four-way valve.

There are many types of directional-control-valve construction other than the spool type shown earlier in this chapter. Many aircraft-type valves are designed with a multiplicity of poppets. These poppets are mechanically forced from their seats by suitable cam action so that the desired flow is produced. The poppet construction provides a relatively tight seal as opposed to the needed clearance flow with the spool-type construction. The poppet may not provide the smooth acceleration and deceleration characteristics that are possible with the spool-type construction. Sliding-plate and disk construction with various types of balance factors can provide less clearance flow than spool-type units.

The balanced-disk-and-plate construction may have a fixed amount of unbalance to ensure a satisfactory seal. In Fig. 12-26, piston seat C abuts a plate containing passages. When the seat is in alignment with certain of these passages, a flow is directed with a minimum of resistance through the valve. Seat C is initially urged toward the mating plate by a spring device. This provides a seal at relatively low pressures. As the pressure increases, the area created by the outer diameter of the seat minus the inner diameter of a lip on the seat will be effective in holding the seat against the plate. The area of the lip will determine the amount of force holding the two surfaces together. The sliding parts are provided with a fine finish to minimize flow between the sealing surfaces. Seals of various types may be used to prevent loss of fluid between the piston seat and the bore in which it is located.

12-13 POPPET-TYPE VALVES

Figure 12-27a and b shows the action of a three-way, poppet-type valve. This valve has two orifice-type seats. One will be closed and the other open in functional operation. Figure 12-27a shows the condition with the valve deenergized while Fig. 12-27b shows the energized position. Figure 12-27c and d shows how the three-way poppet valve can be used with a suitable spool-type valve to provide a four-way action.

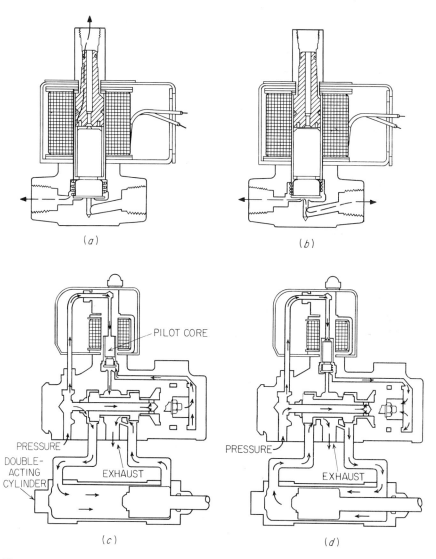

Fig. 12-27 Poppet-type pilot valves: (a) and (d) deenergized, (b) and (c) energized.

Directional-Control Valves **335**

DESIGN CONSIDERATIONS FOR DIRECTIONAL-CONTROL VALVES

Key design factors for directional-control valves can be summarized as follows:

1 Desired function
   a Permit flow in one direction only (check).
   b Accelerate or decelerate a machine member (2, 3-, or 4-way valve).
   c Change direction of flow (3- or 4-way valve). (Multiple flows in special designs.)
2 Working and maximum pressure levels
3 Maximum flow rate (usually dictates physical size).
4 Type of construction
   a Sliding spool (metal-to-metal)—maximum versatility with predetermined operating losses.
   b Sliding spool—elastameric seals—limited pressure to prevent extrusion of elastomeric seal.
   c Poppet type—minimum loss of flow with limited acceleration and deceleration capabilities.
   d Disc type—good seal—good acceleration and deceleration. Less versatile control capabilities.
5 Signal source
   a Manual (all functions controlled by operator).
   b Cam (valve responds to cam action associated with machine movements).
   c Pilot (pneumatic or hydraulic pressurized fluid actuates directional-control valve).
   d Electrical (solenoid, linear force motor, or rotary electric motor control directional valve control element).
   e Combination (more than one signal source controls directional-control valve element).

Once design decisions are reached, the directional-control valve can be integrated into the circuit as explained in the circuits shown throughout the text.

SUMMARY

Directional-control valves may control two or more fluid passages. The control of a multiplicity of passages may be programmed so that the fluid under pressure will follow a definite predetermined pattern causing the desired machine movement. Directional-control valves may also control both pressure and flow rate as a secondary function.

PRESSURE DROP THROUGH VALVES

Pressure-drop values or the energy lost in passing through a valve will differ with different types of construction. A two- or three-way valve usually involves one passage through the valve in each functional position. Four-way and multipurpose valves may have several passages functioning simultaneously. For example, a four-way valve will direct fluid to the cylinder or motor and at the same time will direct fluid from the opposite side of the motor or cylinder back to the reservoir. When calculating pressure drop, it is important to include all passages involved.

OPERATING CLEARANCE FLOWS

Valves with sliding members, such as spool-type units, will have a definite clearance-flow value. Clearance flow may be dependent on the type of fluid, temperature of the fluid (for both viscosity and expansion of metals), and whether auxiliary seal mechanisms, such as piston rings, are employed in the design. Calculations as to potential clearance flows may be of little importance in many installations. Vertical loads without mechanical counterbalancing and potential differential areas in cylinders may make the clearance flow of vital importance. Circuit characteristics will determine this.

DRAIN CONNECTIONS

The plumbing associated with the drain connections on directional-control valves is important. A high flow rate resulting from the shifting of a four-way valve will create a high flow rate in the drain or pilot-tank line for a relatively short period of time. If other drains are also connected into this line, there will be momentary reflected pressures that may upset other portions of the circuit. If the drain line can empty to the tank in the interim between operations, there may be little effect. A long drain line may be vented to the atmosphere with a suitable air breather at the extreme end away from the reservoir to ensure maximum gravity flow.

Most drain lines are connected above the fluid level in the reservoir to minimize resistance to flow. The exception to this may be certain motor or pump drains that may be subject to vacuum during some portions of the cycle where air inclusion would result.

Sight-flow devices in the drain line may be provided as visual aids in checking functional operations. The flow or lack of flow in a drain line will visually indicate whether a directional-control valve is being piloted at the correct time and will indicate the speed of actuation by the quantity of fluid coming through the drain.

PILOT-PRESSURE SOURCES

Piloted valves in many installations require provision for either internal or external pilot-pressure sources. An internal source of pilot pressure indicates that the potential for piloting is the same as that within the major valve segment. External pilot sources may provide economy in overall operation because of the small quantities of fluid normally required. The pneumatic system always provides an external source of potential for piloting hydraulic valves. A similar system can employ inexpensive pumping equipment which has the sole function of supplying fluid under some fixed pressure value to the units that are to be piloted. With an external and independent source of pilot pressure, it is possible to provide much more stable operation in many circuits. If the fluids differ, it may be necessary to establish certain barrier seals between the two fluid media. In many installations the pilot pump is driven from the common source of rotary power. The pump may even be an auxiliary part of the main pumping system.

A good knowledge of directional-control valves is a basic need in the understanding of hydraulic power-transmission circuits. Careful attention to the manufacturer's recommendations is advisable. Unusual operating conditions must be carefully analyzed.

REVIEW QUESTIONS

12-1 What is the purpose of the check valves shown in Fig. 12-8? Why are both cylinder ports within the four-way directional-control valve interconnected in the neutral centered position?

12-2 What would happen if both poppet A and poppet B were depressed at the same time in Fig. 12-11?

12-3 Describe the operation and use of a shuttle valve.

12-4 What is the purpose of check valves $T1$ and $T2$ in Fig. 12-22?

12-5 What is the purpose of the detent mechanism shown in Figs. 12-15 and 12-17?

12-6 What causes the poppet to open in the valve of Fig. 12-20? Why?

12-7 How does pressure to port B affect the valve of Fig. 12-20? Why?

12-8 What is the purpose of piston P in Fig. 12-22?

12-9 Are the relief valves of Fig. 12-22 intended for continuous operation? What type of reversal is provided in this circuit?

12-10 Describe how a valve spool may be centered by pressurized fluid.

12-11 Why is pressure directed to both cylinder ports in neutral in the unit shown in Fig. 12-24? What is its usage?

12-12 Can compressed air be used to pilot a valve? Why?

12-13 What is the purpose of the cartridge in the ball valve of Fig. 12-25? How does it differ from the cartridge of Fig. 12-26?

12-14 Why are poppet-type, directional-control valves used?

12-15 What makes a disk-type valve virtually leak-tight after it is shifted?

12-16 What is the purpose of the holes in the check poppet of Fig. 12-1? Why is the tapped hole provided in the face of the poppet in Fig. 12-1b?

12-17 What is meant by the term *normally open two-way*? How would you identify the spool of the valve shown in Fig. 11-2?

12-18 What is the purpose of the compensator spool shown in Fig. 12-9?

12-19 Why is nonmagnetic stainless steel used in the fabrication of the spool assembly of Fig. 12-17?

12-20 What is the difference between a series and a parallel circuit?

LABORATORY EXPERIMENTS

12-1 Build the following circuit. From a variable-displacement pump of approximately 5-gpm flow capability, direct fluid to a pilot-operated four-way valve. Connect a cylinder or motor to the cylinder ports of the pilot-operated four-way valve. Connect each pilot connection of the four-way valve to the outlet of a shuttle valve, as shown in Fig. 12-10. Tee a pressure connection ahead of the four-way valve. Direct this to the pressure port of two small, lever-operated, four-way valves. Connect the cylinder ports of the first small four-way to port 1 of the two shuttle valves. Connect the second small four-way to port 2 of the shuttle valves. Start the pump after checking to see that all tank lines are connected back to the reservoir. Alternately pilot the main four-way from the lever-operated valves. Note the combinations that are possible.

12-2 Build the circuit of Fig. 12-13. Restrict the supply of fluid to the rotary valve. Observe the change in reversal point. Restrict the fluid flow to the main four-way valve. Observe the change in reversal point. Place flat washers in the cavity between the end of the main spool and the cap. Note the difference in speed of cylinder travel as washers of various thicknesses are inserted.

12-3 Build the circuit of Fig. 12-16. Reduce the air pressure to zero. Gradually build up the air pressure until the unit starts to operate. Observe pressure values. Note the difference in response with the air circuit and the fluid circuit of Fig. 12-13.

12-4 Build the circuit of Fig. 12-8. Install the cylinder in a vertical plane. Attach a dead weight to the lower end or raise a platform with the upper end. Apply a significant weight. Stop the cylinder part way and check for drift.

12-5 Replace the four-way spool with a closed-center spool and repeat Experiment 12-4. Note the variations in drift. What is the cause?

12-6 Connect a closed-center four-way valve to a source of fluid under pressure. With the valve in neutral, check the flow through the valve with 250-psi pressure increments through each cylinder port. Measure leakage the first minute after applying pressure. Compare this with the leakage after the second and third minutes. Observe the change in rate. Put gauges in the pressure line and in each cylinder port. With the valve in neutral, observe the pressure at the inlet compared with the pressure on the cylinder port gauges. Observe the pressure drop with 500-psi increments until the rated pressure on the valve is reached. Note the clearance flow from the tank port during the last test after the first, second, and third minutes.

Install an oscilloscope pressure pickup at the cylinder port of the valve in the preceding test. Note the response time from energizing the solenoid until the pressure builds up to maximum value. Deenergize the solenoid and observe the decay time.

12-7 Repeat Experiment 12-6 with a solenoid-operated valve such as that in Fig. 12-19. Compare the reaction times. Note the reaction time of the dc solenoid compared with the ac solenoid.

12-8 Repeat Experiment 12-6 with a pressure-centered valve such as that described in the text. Compare the centering times for spring and pressure centering.

12-9 Block the drain on the pressure-centered four-way valve between the end of the spool and the centering pistons. Observe the time necessary to permit leakage to allow the valve to shift. (*Note:* This may require several minutes.) Connect this drain below the fluid level and observe speed times. Repeat with the drain above the fluid level and observe the shifting time.

12-10 Repeat Experiment 12-9 with thinner fluid. (If 300 SSU is used in the first test, use 100 SSU in the second test.) Repeat the test using kerosene or mineral spirits as the fluid. Note the different time values.

12-11 Use a reversible gear motor as a pump and build the circuit of Fig. 12-22. (Separate pilot check valves can be substituted for the dual unit.) Connect to the cylinder of Experiment 12-4. Note the drift.

12-12 Operate the motor of Experiment 12-11 with a hand crank. This is a similar circuit to the safety circuit on merchant ships for bulkhead door controls.

12-13 Connect the cylinder of Experiment 12-4 with a ball- or disk-type four-way valve. Note the load conditions as the valve is reversed.

12-14 Repeat Experiment 12-13 with a poppet-type solenoid valve such as that shown in Fig. 12-27.

12-15 Install an orifice check as shown in Fig. 12-1*b* in the lower cylinder line of Experiment 12-4. Note the controlled drop.

13
Electrical Devices for Hydraulic Circuits

Complete hydraulic circuits can operate effectively without electrical devices. For example, many mobile graders, earthmovers, shovels, and similar machines depend entirely on mechanical and hydraulic interlocks and actuation. However, effective use of electrical devices can provide more effective control, less expensive interlocks having many additional safety features, and simplified automatic sequencing. Even in mobile-equipment applications, use of electrical controls makes it possible to actuate the unit remotely. When the machine must operate in a hazardous area, remote actuation is sometimes desirable; the operator can provide satisfactory control through electrical devices from a remote point within a safe area. In this chapter you will study the various electrical-control mechanisms used in modern hydraulic circuits.

13-1 SOLENOIDS

Push or pull actuation can be provided by solenoids. They can also provide a limited rotary actuation. Solenoids actuate valves, pump controls, latch mechanisms, and similar elements in hydraulic or pneumatic circuits.

Solenoids fall into two broad classifications, those designed for use with alternating current and those for use with direct current. Modifications of each find wide usage in hydraulic power-transmission circuits.

The unit illustrated in Fig. 13-1a employs a laminated plunger and field assembly to provide quiet, dependable alternating-current service. This is referred to as an air-gap-type solenoid. The armature floating in the magnetic flux path to the seated position is not restricted by damping

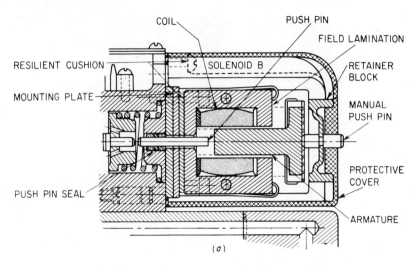

Fig. 13-1 (*a*) Alternating-current solenoid applied to four-way directional-control-valve pilot. (*b*) Direct-current solenoid with encapsulated armature. (*c*) Direct-current solenoid with encapsulated armature and rectifier and clip-cell assembly to permit use with alternating current. (*d*) Oil-immersed, ac solenoid.

agents other than the valve spool and spring assembly. Inertia can be high when used with *detent* or *no-spring* model valves. The armature must seal or seat properly when energized, or the coil will be destroyed by rapid heating. The pushpin assembly must pass through a *dynamic seal*. Nonmagnetic materials are usually employed in the pushpin assembly so that the solenoid force is not dissipated in the critical flux path.

The dc solenoid of Fig. 13-1*b* is fabricated with a nonmagnetic tube separating the armature and coil winding so that the armature is within the fluid medium and the coil is dry. A damping action is created which provides extended plunger life with some loss in response time. Dynamic seals are not needed; therefore, the associated frictional factors are virtually eliminated. Complete seating of the armature against the anvil is not critical other than for proper spool positioning. The coil will not overheat if properly sized.

Direct-current solenoids with dual windings are provided with a limit switch that automatically cuts one winding out of the circuit when the solenoid closes. This provides a high energy potential for the longest stroke available, or needed, in the application. The remaining coil, left across the line in the closed position, provides sufficient resistance and holding power to keep the device in position without overheating. However, if a solenoid with a dual winding is prevented from actuating a switch at the end of the stroke, it will overheat and be destroyed.

The solenoid shown in Fig. 13-1*c* employs the same general components as those shown in Fig. 13-1*b*. A rectifier, and a clip-cell to protect the rectifier, are added so that it will function with alternating current. The advantages of the armature damping and elimination of dynamic pushpin

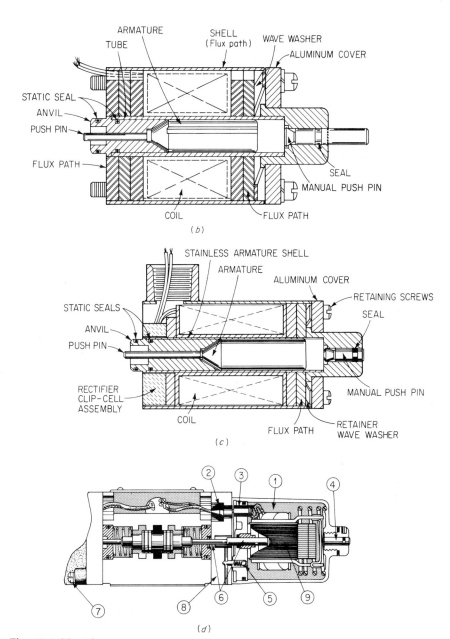

Fig. 13-1 (*Cont.*)

seal provide a considerably extended life. The clip-cell limits high-voltage peaks created by external apparatus. The manual pushpin, in the solenoids of Fig. 13-1, permits setup operation or emergency operation of the valve.

The solenoids of Fig. 13-1b and c may be provided with coil modifications and changes in the amount of iron in the washers at either end of the coil to provide differing force values. Heating factors are directly affected so that the ratings differ. The assembly can be designed to stay across the line indefinitely and be rated *continuous duty*. Other units may be rated for *intermittent duty* where the cycle of energization is of short duration, such as the swing cylinder on a snow blower or controlling latches on truck-body operating devices. Economy in coil size and original cost can be provided if the intermittent-duty units will provide the needed circuit function.

The solenoid assembly shown in Fig. 13-1*d* isolates the fluid within the hydraulic circuit from that within the cover assembly. It provides the advantages of oil immersion without potential contamination from the circuit fluid. It does require a dynamic seal as shown at 6. The fluid in case 1 is heated to some extent from radiation emanating from lamination 9 and the coil so that a thermal relief 5 is provided. The manual pushpin 4 is the same diameter as the internal pushpin at 6; so the only displacement is the electrical actuation when the plunger causes the pin to extend. This displacement is accommodated by a small amount of encapsulated air in the housing which provides the needed capacitance. Plug 3 and receptacle 2 provide convenient service capability.

Choice of solenoids results from circuit needs. Each basic design offers advantages in specific circuit applications.

Solenoids have their greatest holding power in the closed position. As the length of armature travel increases, the force at the start position is minimized. Because of this there is a tendency to limit solenoid stroke length to obtain the greatest force in the smallest physical package.

Solenoids shown in Fig. 13-1 are all push-type units. They are applied so that they push directly against a sliding-spool mechanism. The mechanism to be actuated by a *pull-type* solenoid is usually fastened to the armature at the large end opposite the seated area. A pull-type solenoid is shown in Fig. 13-2*a*. The valve member is fastened to the armature of the solenoid. Note that the armature is within the fluid path. The solenoid coil, however, is isolated from the fluid passing through the valve, much as the units illustrated in Fig. 13-1.

Modification of a pull-type solenoid can provide both push and pull functions from a single armature. The upper unit of Fig. 13-3 provides a single pull function. The lower solenoid is a dual unit used to provide both push and pull service so that three spool positions are possible.

Many types of solenoid coil fabrication are needed because of differing service requirements for fluid-power devices.

Class *A* coils are supplied in all standard valves that handle gases or liquids up to a maximum temperature of 212°F. They are suitable for a maximum room temperature of 40°C (104°F) and are designed for a maximum temperature rise to 80°C (176°F) by the resistance method. This construction consists of a varnish-impregnated, paper-intersection, layer-wound, enamel-insulated wire.

Class *A* molded coils are paper intersection, layer wound. Coil wind-

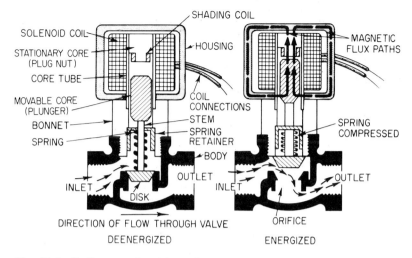

Fig. 13-2 Pull-type solenoid used to actuate a poppet-type valve.

ings are fully encapsulated with an epoxy resin, resulting in a homogeneous watertight cover. They are recommended for installations where considerable moisture and humidity are or may be present.

Class H coils are supplied in valves handling gases, liquids, or steam at temperatures above 212°F or valves installed where there is a high ambient temperature. The coil insulation is rated by AIEE for 356°F (180°C) "hottest spot" temperature, which includes the normal temperature rise of 117°F (65°C), by thermocouple method. High-temperature coils have been used in valves controlling gases and steam up to 450°F (232°C) without failure.

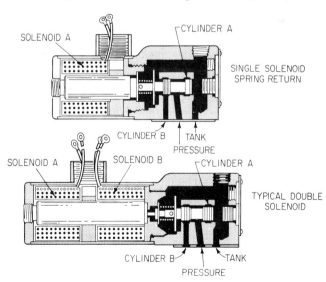

Fig. 13-3 Pull-type solenoid used to actuate a sliding-spool valve. Upper unit—single pull; lower unit—push-pull.

Electrical Devices for Hydraulic Circuits **345**

Extensive use of solenoid-operated fluid-power devices in critical machine tool circuit applications creates a need for indicating devices and various types of protectors for the solenoid assembly. A signal light is often wired in parallel with the solenoid coil to indicate whether energy is available to the solenoid. A suitable fuse or circuit breaker can be added to provide further protection for the unit.

The assembly shown in Fig. 13-4a contains a suitable fuse in series with the solenoid coil so that it will blow if a short circuit is encountered as a result of solenoid or valve malfunction. A white light is wired in parallel with the solenoid coil indicating normal function and availability of signal energy. A red light is wired in parallel with the fuse. If the fuse blows, a voltage differential is created across the red light and it is actuated.

The fuse circuit is shown in Fig. 13-4b. The circuit of Fig. 13-4c substitutes a circuit breaker and fuse for the fuse alone. Figure 13-4d shows this assembly in a housing.

If the white light is not lit, it indicates component failure or no power to the valve. A quick check of wiring to the valve will determine power availability. Most signal lights are designed to operate on voltages in excess of 80 to 90 V. Voltages below this representative value would damage the components, so that the initial check can show this condition. Component failure can be determined by a continuity check with a resistor and lamp.

A double white light indicates improper wiring or high-cycle application. By checking the wiring it can be determined if both solenoids are being energized at the same time. A red light appears when the solenoid does not complete the full stroke. On high-cycle applications it is desirable to check the limit-switch wear. Uneven wear on limit switch can cause solenoids to be energized at the same time unless interlocking relays are provided.

If the Red Light is Lit, the Following Causes and Remedies are Suggested:

	Cause	Remedy
1	Valve will not shift.	Clear valve with manual pushpin.
2	Overvoltage.	Check line voltage. Overvoltage for long duration will cause fuse to blow or trip circuit breaker.
3	Improper voltage applied.	Check to ensure proper voltage to solenoid. Low voltage will cause solenoid to short-stroke. In turn, it will blow a fuse or trip the circuit breaker.
4	Component failure.	Check fuse continuity. Reset circuit breaker. Retest to see if spool will shift.
5	Improper wiring.	Check wiring. Both solenoids may be energized at the same time. The red light appears on the solenoid which did not full-stroke.

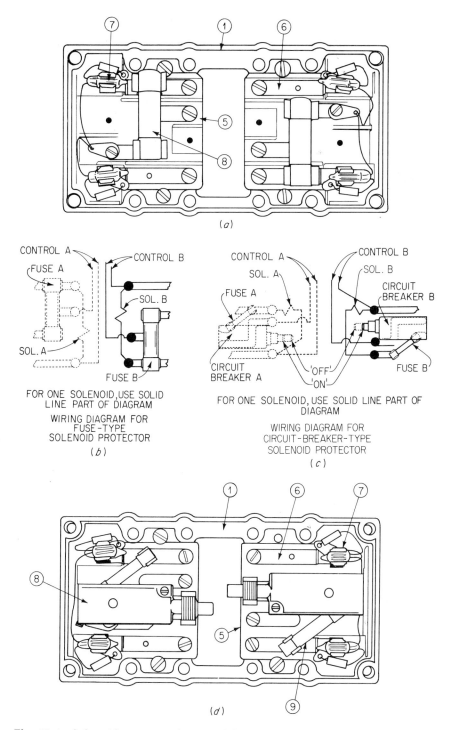

Fig. 13-4 Solenoid protector devices. (*a*) Housing to manifold unit to directional-control valve has fuse protection. (*b*) Circuit for use with fuse. (*c*) Circuit for use with circuit breaker. (*d*) Housing of manifold unit of directional-control valve to provide circuit breaker.

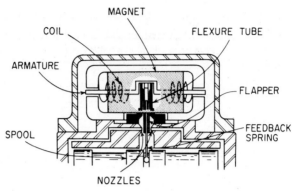

Fig. 13-5 Torque motor applied to flapper-type servo valve.

13-2 TORQUE MOTORS

Limited stroke and force equal to the applied current are provided by a torque motor. Figure 13-5 shows how a torque motor is applied to a servo valve. In this type of construction the motor actuates a flapper that changes the resistance to flow at the point of the opposed nozzles. There are two windings within the motor. Movement of the armature follows the pattern of the current impressed on the winding. The stroke of a solenoid is the same for each actuation. The stroke of the armature of the torque motor may be infinitely variable, according to the current impressed on the magnet winding. The torque motor can be mechanically attached to a small directional-control-valve spool for direct actuation. The armature is limited by the flexing of the integral spring design.

13-3 OTHER ELECTRICAL CONTROLS

The electrical components most applicable to control of hydraulic equipment are motors, both rotary and limited-motion torque motors (used on servo valves), motor controls (starters), push buttons and selector switches, limit switches, pressure switches, timers, relays, solenoids, and transformers. Figure 13-6 shows the symbols for many of these devices. Solenoids and torque motors are the devices most often incorporated directly within a hydraulic component.

Electric motors are incorporated in a device used to actuate the adjustment of pressure-control valves and flow-control valves. These control assemblies incorporate limit switches, an electric motor, and the necessary cams and screw devices to provide the desired mechanical motion. The mechanism may provide a rotary motion for a pressure-control valve adjustment actuation; or the control can incorporate a screw mechanism that produces a linear output force by varying the tension on a spring

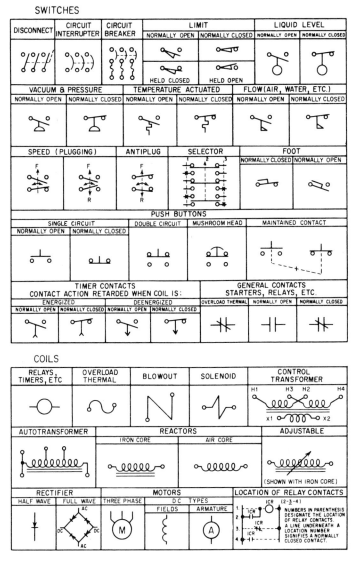

Fig. 13-6 Electrical symbols for hydraulic circuits.

within the motor and speed reducer housing. The motion used to change the flow through a metering-type valve is usually rotary but might be linear in some configurations.

13-4 CIRCUIT DIAGRAMS

Electrical diagrams for hydraulic controls are of the *ladder type*, in which the left side represents the power lead and the right side represents the

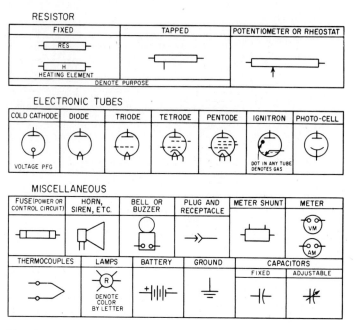

Fig. 13-6 (*Cont.*)

ground lead. Components are shown connected across these two leads. When components have more than one position—such as relays, limit switches, pressure switches, and timers—they are usually shown in the deactuated or reset position. Contacts of relays are designated as 1-*CR*; all other contacts of this relay are also labeled 1-*CR*. The common reference is used because all contacts function simultaneously.

When designing the electric circuit for control of a machine's hydraulic system, one must thoroughly integrate all mechanical and hydraulic factors. In many instances the mechanical and hydraulic devices will actuate the electric components which, in turn, will actuate solenoids or motors on the valves.

The series of steps involved in designing the electric circuit corresponds to the phases in the machine sequence. In designing the circuitry to accomplish each succeeding step, the previously designed circuit must be considered and care must be taken to avoid disturbing previous functions. To illustrate this procedure, consider the circuit in Fig. 13-7, showing a machine having two cylinders designated as *A* and *B*. Cylinder *A* is mounted horizontally, and cylinder *B* is mounted vertically. *A* must move to the right, from position 1 to position 2, and stop. Cylinder *B* must then move up from position 3 to position 4. Cylinder *A* exerts increasing pressure against the workpiece until the desired pressure is reached. Then *A* and *B* retract together.

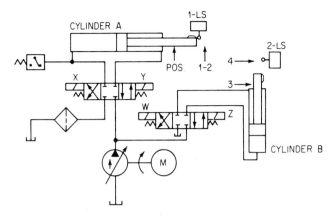

Fig. 13-7 Electrohydraulic circuit.

The circuit of Fig. 13-8 shows the arrangement necessary to move cylinder *A* to the right. The solenoid must be energized through the relay contact so that it remains energized after the momentary start switch is released. Another contact of the relay locks in to maintain the relay in an energized state.

In Fig. 13-9 the circuit has been supplemented with components to stop cylinder *A* and extend to cylinder *B*. Relay 2-*CR* has been added; it is energized by limit switch 1-*LS,* which is actuated as cylinder *A* extends the rod to the point where the limit switch is located. Relay 2-*CR* deenergizes solenoid *X* to stop cylinder *A* and energizes solenoid *Z* to extend cylinder *B*.

To complete the cycle, the necessary components are added as in Fig. 13-10. When cylinder *B* is fully extended, limit switch 2-*LS* is actuated, energizing a third relay 3-*CR*. Relay 3-*CR* deenergizes relay 2-*CR* to energize solenoid *X* and deenergize solenoid *Z*. This allows cylinder *A* to exert pressure against the work and also stops cylinder *B*. A pressure switch has been added in series with relay 1-*CR* so that when the pressure reaches the preset value, the pressure switch opens, deenergizing relay 1-*CR*. Relay 1-*CR* then deenergizes solenoid *X* and energizes solenoids *Y*

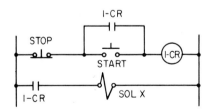

Fig. 13-8 Starting operation of control relay.

Electrical Devices for Hydraulic Circuits **351**

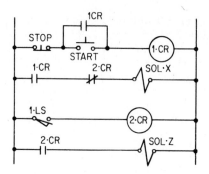

Fig. 13-9 Circuit with components to stop cylinder A, Fig. 13-6.

and W to return both cylinders. The circuit represents only the total basic operation and would be further modified to incorporate safety and interlock provisions.

13-5 CHECK POINTS IN CIRCUITS

To make a fluid-power system and its electric circuits fully compatible with existing components, the designer must be aware of the electrical devices used, understand what they can do, and know how they can

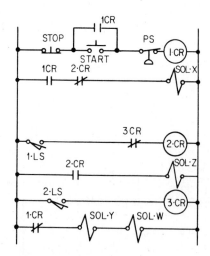

Fig. 13-10 Circuit with all components for completion of cycle shown in Fig. 13-7.

352   *Industrial Hydraulics*

accomplish their functions. Some basic steps necessary to check electrical-control systems used in a fluid-power network are reviewed here:

1 Study the mechanical and hydraulic operations required and the sequence of these operations. Make certain these cannot be further modified for simplification of the electrical control.

2 Starting with the first operation in the sequence, build the electrical schematic, making certain not to disturb any previous function as additional controls are needed.

3 After completion of the electrical schematic, prepare a written sequence of operation of the electrical portion, using care in identifying the sequence in relation to the hydraulic circuit.

4 Check for the possibility of component failure and how it might affect safe operation of the circuit or any part thereof.

5 Determine the characteristics of each component and prepare information for ordering.

6 Identify power sources adjacent to the machine and be sure components are compatible.

7 Check the proposed location of electrical components and make certain that there will be no interference with other mechanical parts.

13-6 CHOICE OF CONTROL METHODS

In some cases an electrical device may be the only mechanism that will provide the desired degree of control. When there is a choice among electrical, mechanical, hydraulic, and pneumatic controls, it is necessary to consider safety, dependability, flexibility, maintenance, atmospheric conditions, and initial cost. Electrical controls frequently offer a flexibility and convenience that are unobtainable with other control devices. With electrical controls, all functions of a multipurpose machine can be conveniently controlled from one or more positions. Safe operation of a press, for example, can be easily provided with duplicate controls requiring the operator to use two hands to start the machine. Machine positions are easily monitored by limit or pressure switches that can start one phase of machine operation after completion of another. The location of a malfunctioning element can be determined easily and repairs made with little loss of time.

Initiation of a signal to unload pumps and similar devices and to sense a pressure upon completion of a work stroke in coining and shearing is an easy task for a pressure switch of suitable design. Electrical timers can become the sequence controls of the machine, either alone or in conjunction with limit switches to integrate both time and position. The timer also may regulate the period of each operation. The speed and variety of electrical arrangements make them a valuable asset to any machine where accurate *dwell* time is required.

13-7 SAFETY CONSIDERATIONS

Safety is always a prime consideration to the hydraulic-system operator and designer. The means of starting the cycle is important in both the installation and design of the electrical system. Where a clamping action is involved, two switches are often installed so that the operator must actuate both simultaneously, thus keeping hands away from the vicinity of the clamping mechanism. The system may be designed so that the operator must keep switches manually actuated until the clamping process has proceeded to a point where it no longer endangers the operator. To avoid unintentional repetition of the cycle, the design may require that the operator release the start switch and reactuate it before the cycle can be repeated. Emergency means should be provided for stopping or reversing a moving piston. Where two operators are working at a single machine, duplicate emergency controls may be desirable.

Proper steps must be taken to safeguard against damage to the machine or operator in case a limit switch or other control device does not operate or remains actuated. Possibilities of injury can often be minimized by proper selection of either normally open or normally closed contacts of the switch and suitable interlocks.

13-8 INTERLOCK FUNCTIONS

Some machines have auxiliary motors for oil cooling pumps, pilot-pressure sources, etc.; these motors may have to be in operation before the main drive motor starts. To accomplish this, *interlock contacts* from the auxiliary motor starters can be placed in the main-motor start circuit. If a certain pressure level is necessary in these auxiliary circuits, a pressure switch also may be placed in series with the control of the main-motor starting circuit. In stopping, it may be required that the auxiliary motors remain running until the main motor stops. For this a timer can be used to delay the stopping of the auxiliary motors a predetermined length of time.

Electric circuits should be so designed that motors will not restart automatically and controls will not become energized when electric power returns after a power interruption. In a clamping circuit, for example, a hazard would exist if the clamp could close unexpectedly. To prevent this a manually operated spring-return two-way valve might be used. In the normal unoperated position of the valve, clamping pressure is diverted back to the reservoir. When the valve is operated, fluid is allowed to operate the clamp. For normal operation, a limit switch is then actuated to block the manual valve so that the operator may remove his hand from the valve.

13-9 PROTECTION OF COMPONENTS

Protection against overloads and malfunctioning of the electrical components is easily provided. The entire electrical system is protected by fuses or overload relays so that it is immediately deenergized when trouble occurs.

Industry recommendations require certain protective measures in electrical systems. A disconnect switch capable of interrupting the maximum operating overload current must be installed to disconnect all lines of the power circuit. This disconnect should be interlocked with the control-box doors, so that power is disconnected when the doors are opened. Undervoltage protection must be provided for all motors or equipment that might be damaged if the motor should start on return of power after interruption.

Certain other requirements, including mounting of components, are outlined in industrial standards for protection of electrical controls and safety to personnel. The components should be protected from passing vehicles, splashing liquids, or moving objects. The location of the control equipment should not interfere with machine operation or maintenance. Equipment such as push-button controls and limit switches should be rigidly mounted in as clean a location as possible and should be protected against accidental actuation. Limit switches should be mounted so that accidental overtravel will not damage them.

13-10 MAINTENANCE OF ELECTRICAL DEVICES

Preventive maintenance of electrical control equipment can consist of a monthly inspection and use of the following routine precautions:

1 Check proper voltage of the control circuit.

2 Keep the equipment free of dirt or foreign material.

3 Check for dust within the housings. Blow out or vacuum clean at stated intervals, depending on atmospheric conditions.

4 Check for excessive heating of components.

5 Remove corrosion products.

6 Watch for erratic operation.

7 Check for loose connections.

8 Note worn or broken parts (broken barriers between terminals may permit short-circuiting, etc.).

9 Note the condition of gaskets in oil-tight, dust-tight, and watertight equipment.

10 Watch for the collection of moisture in any enclosure.

11 Note whether timing devices, sequencing of operations, etc., function properly.

12 Remove excess grease and oil with suitable solvent or cleaner.

13 Remove fuses from fuse holder and polish ferrules (except silver-plated holders and ferrules).

14 Use the proper amount of grease for the motor; too much can be as detrimental as too little and can cause heating and other unwanted conditions.

SUMMARY

Electrical devices provide accurate, inexpensive control for hydraulic pumps and for pressure-, flow-, and directional-control valves.
   Control may be by the use of either direct or alternating current.
   Alternating-current solenoids are used for predetermined repetitive-stroke applications requiring fixed-force values.
   Torque motors and dc solenoids are used to provide movement of the armatures (against a spring force) at a value proportionate to the applied current.

REVIEW QUESTIONS

13-1 What happens to an ac solenoid if the armature does not complete its stroke when energized?

13-2 What is a double-wound dc solenoid? Why are two windings provided? Are all dc solenoids double wound?

13-3 What is the operational difference between a double-wound dc solenoid and a dc solenoid with a single-wound coil?

13-4 Must a torque motor travel the same distance at each actuation? Explain.

13-5 Illustrate a normally open limit switch by an electrical symbol.

13-6 At what point is the holding force greatest in an ac solenoid stroke?

13-7 Describe the use of a pressure switch in a hydraulic circuit.

13-8 What is the purpose of a limit switch?

13-9 What types of controls are needed for a motor starter? What is a disconnect switch?

13-10 Can direct current be used to actuate hydraulic directional-control valves with solenoid actuators?

LABORATORY EXPERIMENTS

13-1 Insert solenoids of various capacities in a standard spring-testing machine. Note the value at which an ac solenoid will start vibrating. Increase the force until the solenoid will no longer stay energized. Compare this value with that at which the solenoid became noisy.

13-2 Install a standard hydraulic cylinder in a circuit with a three-position, solenoid-controlled, pilot-operated, four-way directional-control valve to direct fluid to the cylinder. Install a limit switch at each extreme of cylinder piston travel. Operate the solenoid of the valve with the limit switch. Observe the repetitive accuracy and the degree of dependability.

13-3 Repeat Experiment 13-2 with the addition of relays interlocked so that when the limit switch energizes one it will automatically deenergize the other. Note the degree of repetitive accuracy.

13-4 Repeat Experiment 13-2 with an orifice that can be adjusted to supply pilot fluid to the directional-control valve. Note the point at which the valve "dies" on center.

13-5 Repeat Experiment 13-4 with the relays of Experiment 13-3 added. Note the high degree of dependability.

13-6 Check the response time of the four solenoid types shown in Fig. 13-1. Install the solenoid on a similar four-way valve. Plug one cylinder port. Insert a pressure switch or pressure transducer in the other line. Use an oscilloscope or elapsed-time clock to record actuation time.

13-7 Repeat with a recording-type pressure indicator. Check for repetitive accuracy.

14
Servo Systems

Servomotor systems (termed *servo systems* for briefness) are power-transmission devices used to transform low-power signals into motion or force with a high degree of precision as to position, speed, or both. The term *servo* is often associated with four-way directional-control valves. A four-way valve of suitable design can be an *integral* part of a servo system, but is not always an *essential* part of such a system.

14-1 MECHANICAL-LINKAGE SYSTEMS

All servo systems have parts that perform certain basic functions. First, a signal is impressed on a control linkage. This linkage may be a lever, a rod, an electronic device, a fluid column, or any similar source of initial movement. Servo systems also contain a device to provide a reference value which will establish a pressure level or a stop position. The spring setting of a relief valve can maintain a pressure level; a closed-center four-way directional-valve spool can establish a stop position. The impressed signal on the control linkage starts a sequence of events. A power mechanism (such as a cylinder or motor) is caused to move by the control that received the initial signal. Movement of this motor or cylinder provides a signal to the initial control device to cancel out the original impressed signal.

The four-way directional-control valve in Fig. 14-1 is mechanically fastened to the cylinder-barrel assembly. It is drawn with ISO symbols to show the direction of flow through the valve. The box around the valve and cylinder indicates that the unit is in a single package. As the *input displacement rod* is moved to the right, it will cause the valve spool to move

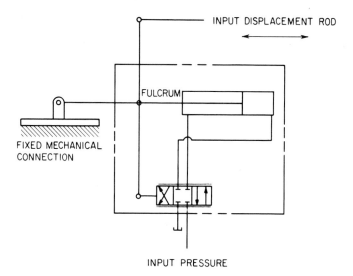

Fig. 14-1 Four-way directional control used in a mechanical-linkage system.

to the left. This directs fluid to the rod end of the cylinder and causes the cylinder to move to the left. Since the valve body is also connected to the cylinder assembly, the body will also move to the left, moving the spool back to the neutral position to stop movement. There is a direct relationship between the element being moved by the cylinder and the input-displacement rod. Movement of the rod to the left will cause a similar major movement of the cylinder to the right and another cancellation of movement. The movement of the input-displacement signal is at a fixed ratio with the cylinder movement. This will affect both the speed of movement and the amount of movement as well as the direction.

14-2 SERVO SYSTEM COMPONENTS

Figure 14-2 shows certain functions of a servo system by the use of simple box symbols. The system components will vary widely in method of construction and degree of control. The components and their basic functions are as follows:

PRESSURE PICKUP
A pressure pickup consists of a device that accepts a pressure rise or fall within a contained tube or cylinder and translates this information into a proportional electrical signal. The unit may be similar to the Bourdon tube used to actuate the indicating needle of a pressure gauge. This tube actuates a *potentiometer*. Another type of pressure pickup has a wire-

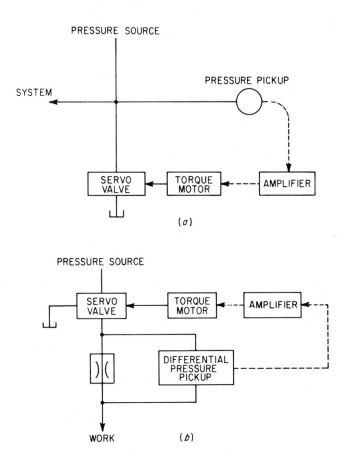

Fig. 14-2 (a) Relief function performed by a servo system.
(b) Metering function performed by a servo system.

wound tube whose resistance varies as the contained pressure fluctuates. The varying resistance of the wire, as the tube deflects, provides a small magnitude signal. Likewise, any potentiometer actuated by a Bourdon-tube type of mechanism is normally expected to create only a tiny signal. These signals must be enlarged to some minimum value before they are usable in conventional mechanical-movement devices.

A *comparison value* may be integrated in the pressure pickup device so that the signal forwarded through the system will show how much the measured pressure differs from a desired value. In Fig. 14-2a, the desired pressure is established by the amount of signal current furnished through the circuit by the pressure pickup. Any adjustment for pressure readings would be made at the pressure pickup; the signal strength through the pickup can be varied so that the resulting signal provides the desired motion at the servo valve.

360 *Industrial Hydraulics*

AMPLIFIER

The desired increase in signal strength from the pressure pickup is provided by a device that can multiply the power of the signal. This multiplication can be in the voltage, amperage, or fluid signal. Regardless of the type of signal being magnified, the result is an exact copy of the initial signal, with the additional power as needed. The amplifier, as the name implies, increases (magnifies or multiplies) the magnitude of the signal to a satisfactory working level.

TORQUE MOTOR

A torque motor is a *low-displacement* electric motor. Movement of the armature proportional to the direct current applied to the windings of the motor is produced by restraining the motor armature with a mechanical spring. The torque motor actuates the servo valve.

SERVO VALVE

The servo valve is a sensitive device used to control the direction of flow of the fluid within the system.

DIFFERENTIAL-PRESSURE PICKUP

A differential-pressure pickup consists of two basic pressure pickups. They are so arranged that they can provide to the amplifier and servo system a signal that reflects the *difference* in pressure between two stations within the hydraulic circuit. This signal can provide the information needed to maintain either a desired pressure or a desired flow, in answer to changes fed into the comparison device.

14-3 RELIEF-VALVE FUNCTION

Relief valves, reducing valves, and certain metering valves may be regarded as self-contained closed-loop systems. The relief valve continuously compares the output or pressure load on its spool with the input or spring load and opens its passage according to the pressure difference until the output and input are equal. It is possible to duplicate the characteristics of these valves by means of electrically controlled servo valves, pressure pickups, and an electric circuit.

Figure 14-2a shows how a relief valve can be controlled by a servo system. The pressure pickup would be downstream in the circuit of a reducing valve. The servo valve is a two-way valve that creates an orifice or blocks the flow path in accordance with the signal fed into the device. Note that the pressure pickup accepts a signal from the input-pressure source and compares it with an established pressure value in the system. The pressure pickup will not cause movement if the input-pressure signal equals the established pressure value. However, if the input-pressure value is low, the pickup will present the proper signal to the amplifier to cause the torque motor to actuate the servo valve, diminishing the flow to the reservoir. This will increase the pressure in the system. If the pressure is higher than the reference-signal value, the pickup signal to the amplifier will cause the torque motor to move in the proper direction to increase the flow to the tank and diminish the system pressure. The servo valve is the equivalent of the main spool of a relief valve. The torque motor and amplifier can be compared to the working areas of the relief valve, while the spring and input-pressure orifices can be compared to the pressure pickup and reference mechanism.

Figure 14-2b shows how the pickup for a metering function is sensed by a *differential-pressure pickup*. In this circuit the pressure drop across the orifice is to be maintained at a uniform value. The differential-pressure pickup is provided with a reference device equal to the desired pressure drop across the orifice. The servo valve can be a normally open two-way valve. When the pressure to the orifice builds up to a point higher than the pressure downstream plus the reference value, a signal is presented to the amplifier and then to the torque motor. The torque motor causes the normally open servo valve to move in a direction to restrict fluid flow to the orifice. If the work system calls for a higher pressure through the servo valve and across the orifice while maintaining the same differential pressure, the differential-pressure signal will be reversed. This will cause the servo to open again and permit flow of additional fluid to the work system, with a resultant rise in pressure.

The metering function can be directly compared to the reducing valve and adjustable orifice in a pressure-compensated flow-control valve. The differential pressure in the flow-control valve is a function of the main spring of the reducing valve. This function is provided in the servo system by the differential-pressure pickup [as shown in Fig. 14-2b].

Servo valves and additional components are not always used in place of general-purpose valves. Economic factors preclude the use of the more complex servo system unless requirements within the transmission system indicate the need. Difficulties like response lag can cause instability or can permit high transient pressures or flow peaks with the electronic servo system. High transient pressures or flows may be avoided by using hydraulic accumulators [Chap. 6] or dampening devices at certain points in the system.

To provide the desired reference flexibility from the signal source, a servo system for flow and pressure control may be necessary. Extreme accuracy with flexible remote adjustment is most practical with electronic mechanisms. The alternative means of providing relief, reducing, or metering action by use of servo controls is valuable in installations requiring a degree of pressure or flow control that is unattainable through use of standard general-purpose automatic-acting valves.

14-4 SERVO-TYPE DIRECTIONAL-CONTROL VALVES

Four-way directional servo-type valves may be classed as mechanical, pneumatic, or electric in operation. The mechanical type [Fig. 14-1] is commonly used to amplify signals in the stroke control of a piston-type pump, automotive and mobile off-the-road power-steering devices, and many aircraft devices.

The *nozzle valve* in Fig. 14-3a is electrically operated. Kinetic energy of the fluid at the nozzle exit is reconverted to pressure in the tapering passages of the receiving block. Pressure reconversion is virtually complete. Input energy from the torque motor moves the jet pipe with a minimum of friction. Electrically operated servo valves are widely used in process-control applications requiring low power output.

Flapper-type valves [Fig. 14-3b] can also be electrically operated. The valve design shown is normally used as illustrated, in conjunction with a fixed orifice to produce variable pressure rather than variable flow. As the input signal to the torque motor causes the flapper to approach the variable orifice, it creates additional restriction to free flow to the atmosphere. This will increase the pressure to the work. As the torque motor causes the flapper to move away from the orifice, it will reduce the restriction at the variable orifice and the pressure to the work will decrease.

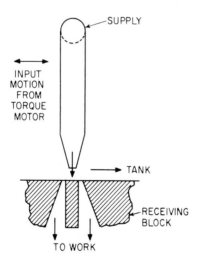

Fig. 14-3 (*a*) Nozzle valve (Askania type) converts pressure energy of the fluid to kinetic energy of the jet. Energy is reconverted to pressure in tapered passages of receiving block in accordance with load resistance.

Servo Systems 363

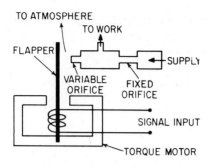

Fig. 14-3 (*b*) Pneumatic flapper-type valve provides variable pressure at work connection as determined by position of the electrically operated flapper with respect to the orifice.

Pneumatic two-stage circuits use a jet of air against the workpiece to establish a relationship of position. The flapper is the workpiece. The valve in Fig. 14-3c translates the movement provided from the torque motor alternately to affect the air flow at the nozzle; the main spool is moved by the difference in pressure at the nozzle. The difference in pressure will always cause the spool to move if the pressure values are within the frictional and flow-force characteristics of the valve.

Pneumatic-follower systems, using a jet of air against a workpiece such as a template, can also establish a relationship of position. If the jet comes close to the workpiece, an increase in pressure results. This pressure

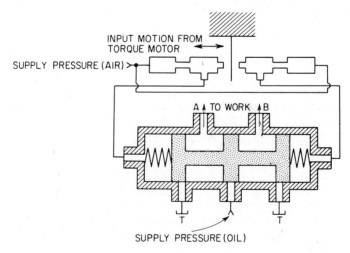

Fig. 14-3 (*c*) Pneumatic two-stage servo circuit.

364 *Industrial Hydraulics*

increase or signal value can actuate a four-way valve that is offset in one direction by a spring. The opposite end of the four-way-valve spool is equipped with a pneumatic-piston assembly. The signal from the jet, applied to this pneumatic piston, indicates movement away from the template to establish correct position. If the piece being followed moves away from the jet, it causes the pressure in the air system to drop. This pressure drop provides a signal to the control valve to shift in the other direction and move the jet closer to the piece being followed. Close-contour relationships can be followed by devices of this type. They are used on several different types of copying lathes and in similar applications.

14-5 OIL-PILOT TWO-STAGE ACTUATION

Electrically operated servo valves, actuated by some form of torque motor, are used frequently in elaborate systems such as those needed for tape-controlled machines. The torque motor used in servo actuation is provided with an armature between magnetic fields. Without energy being fed into the magnet windings, a static neutral position of the armature is obtained. A spool can be mechanically fastened to the armature in the same manner as the flapper used in the pneumatic valve previously described. Modifications of the flapper design can be used in fluid-actuated valves.

Figure 14-4a shows a single-stage servo-valve spool directly fastened to a torque motor. The relative porting of the first stage and how it actuates a cylinder mechanism are shown in Fig. 14-4b. The ends of the main spool form this cylinder mechanism. The second stage, containing the main spool, receives all energy for movement from the primary or first stage actuated by the torque motor. Energy fed into the appropriate winding will cause the armature to move in direct relationship. The winding into which the energy is fed will determine the direction in which the torque-motor armature will move. If a small flow is adequate, the armature of the torque motor can be directly connected to a pilot-spool mechanism or other device that can actuate a cylinder directly. This pilot spool can shift another spool in an amplification circuit.

Armature movement is relatively small and is linear. The forces developed are small enough so that most systems use this as the first stage of energy transfer and depend on piloted systems to amplify the signal. Deflection of the armature directly actuating a spool requires construction different from that of the usual general-purpose, four-way, directional-control valve. Cylindrical slide valves used in servo systems resemble general-purpose directional-control valves; there are, however, several important differences.

Commercial directional-control valves have a clearly defined neutral

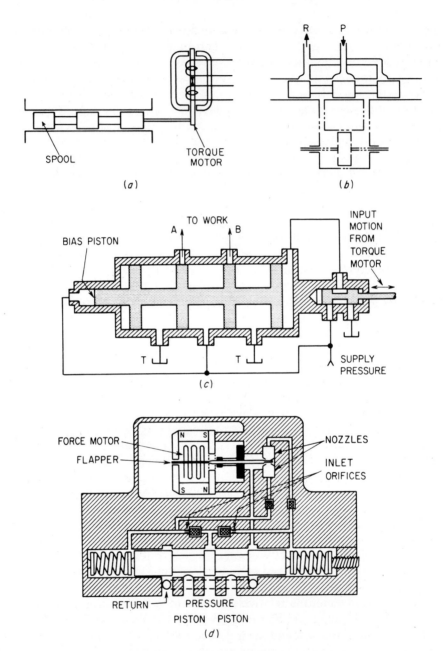

Fig. 14-4 (a) Single-stage servo valve directly actuated by torque motor. (b) Spool of torque-motor-actuated pilot valve. (c) Oil-pilot two-stage actuation. (d) Two-stage servo valve with open center, first stage using a flapper-type pilot.

position which extends over an appreciable portion of the valve-slide stroke. In closed-center valves the neutral position is an *overlap;* in open-center versions it is an *underlap.* In servo valves the lap is always less than in general-purpose valves; servo valves are designed to act almost as soon as the valve slide is moved away from neutral. This prevents the servo valve from having a large *dead zone,* or area of *reduced system gain,* which would mean reduced accuracy. Dead zones are permissible, however, in certain noncritical servo applications.

A second important distinction between general-purpose directional-control valves and slide-type servo valves is in their respective flow characteristics. In most circuit applications, general-purpose directional-control valves are designed to be ON-OFF devices, with a very fast shift from full-closed to full-open and vice versa. The spool may have a speed of spool-shift control that will soften acceleration and deceleration functions, but it is a fixed value, usually established by some type of restriction in the pilot lines. Servo valves may also respond quickly to a signal and open wide for a period of time, but they generally function by permitting progressively greater amounts of flow as the valve slide is moved away from neutral.

The fact that the port opening in a commercial valve may be excessive need not rule out the possibility of its use in simple servo installations. Rate of opening is a factor, and even though satisfactory flow control may be obtained by interposing a suitable lever arrangement or other mechanical gearing down, this may prove inconvenient. There are power-steering control valves used in commercial applications, and modifications of some normal general-purpose, spool-type valves provide adequate servo control in simple systems. Stringent performance requirements may modify the design of the valves. "Laps" may have to be held to accuracies under 0.001 in. Variations of port openings with valve-slide displacement may be nearly linear. This condition can prohibit the use of valves having liners with large round holes for porting.

Military requirements, such as those for missile systems, have caused the development of the high-power, high-performance servo system. To overcome some of the difficulties presented by high force requirements, the servo control chain started by the torque motor is augmented by pilot devices so that the torque motor is strictly a control on the first-stage mechanisms. The actual work is then performed by the second stage at high working-force potentials. The first stage can be regarded as a preamplifier. As such it can be designed to have low friction and hydrodynamic forces, needing only sufficient power to operate the second stage.

The unit illustrated in Fig. 14-4c uses pressurized fluid to actuate a small piston in much the same manner as a wire spring. A small three-way valve is directly actuated by the torque motor. It, in turn, acts to apply fluid to the large area at the spool end opposite the *bias piston* to move the main spool in its directional function. Figure 14-4d shows a two-stage servo valve with an open-center nozzle-type control for the second-stage spool-type assembly.

Figure 14-5 illustrates a pilot-valve system that uses direct-current solenoids. The pilot valve of Fig. 14-5 is not a conventional four-way three-position valve, but a double-pressure regulator. When one solenoid is energized, the armature exerts a force on the pilot-valve spool which is proportional to the solenoid current. This force is balanced by the output pressure, which is directed to the opposite end of the pilot spool. A given solenoid current will therefore generate a certain secondary pressure (a common value is usually up to 250 psi). It is this secondary pressure which is applied to one end of the spool of the main valve to displace the spool against an opposing (centering) spring, this displacement being proportional to the applied pressure. The net effect is that the user speed may be controlled by varying the solenoid current—independent of the load and any load fluctuations.

The solenoid pilot valve [Fig. 14-5] can actuate a major valve and system, such as that shown in Fig. 10-3b and c, replacing the lever actuation or functioning as an alternate to the lever actuator, as shown in Fig. 12-19.

An important feature is that the pilot choke used to control spool shifting time operates under constant-pressure conditions. For a chosen valve opening, the secondary or piloting pressure is constant and independent of the load. Switching times are, therefore, constant, without resorting to external pilot lines.

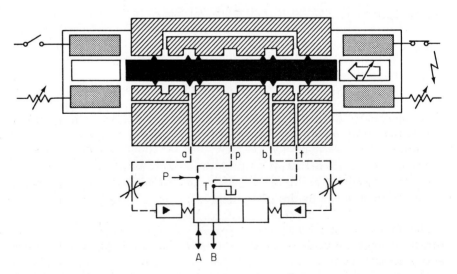

Fig. 14-5 Solenoid-operated double-pressure regulator used to control piloted spool position.

The relief-valve bias spring in the control chamber of the manual valve structure of Fig. 10-13b and c usually provides approximately 50 psi. This is the pressure drop across the directional spool and the value at which the pump unloads.

To improve stability, the unloading pressure of solenoid valves is usually increased to approximately 100 psi by fitting a stronger spring in the relief-valve control chamber.

For simple applications, a straightforward potentiometer or rheostat in series with the solenoid coil may be used. The spool position is determined by the solenoid current and with increasing temperature the current is reduced because of the change in coil resistance. Fortunately, this is compensated by decreased oil viscosity, and within reasonable limits constant speed is obtained.

Using transistor controls not only permits a reduction in the physical size and a wider choice of potentiometers, but also facilitates superimposing a low-frequency dither to reduce hysteresis. These devices control solenoid current and lend themselves to programming of more sophisticated applications.

Solenoid valves may, of course, be used without potentiometers or similar controls, and connected directly to an electrical supply. The flow obtained will then be determined by the setting of the end stops.

The flow regulation of this type of valve begins with approximately 25 percent of the nominal voltage, and the maximum flow is obtained at approximately 85 percent of the nominal voltage [Fig. 14-6]. The maximum flow is determined by the ground angle of the profile part of the spool. The accuracy of control is determined by the following factors:

1 The friction of the spools (pilot-valve spool, distributor spool, and unloading-valve spool) cannot be exactly determined. It depends upon the manufacturing tolerances. The friction is also dependent upon the load. When the spool remains in the same position for a certain time, the friction increases because of the ultrafine particles which are always present in the oil.

2 An increase or decrease of the applied voltage has therefore to overcome the friction hysteresis. By using pulsed direct current the hysteresis is reduced. The solenoid resistance increases with increasing temperature, and consequently the current is reduced and with this the flow. On the other hand, the viscosity of the oil is reduced and a partial compensation takes place.

Diagram 1 of Fig. 14-6 shows the control of a double solenoid-operated directional-control valve by means of a potentiometer with two windings and a neutral point.

Diagram 2 of Fig. 14-6 shows a connection for obtaining three advance speeds for the direction "*a*," and return at maximum speed for the

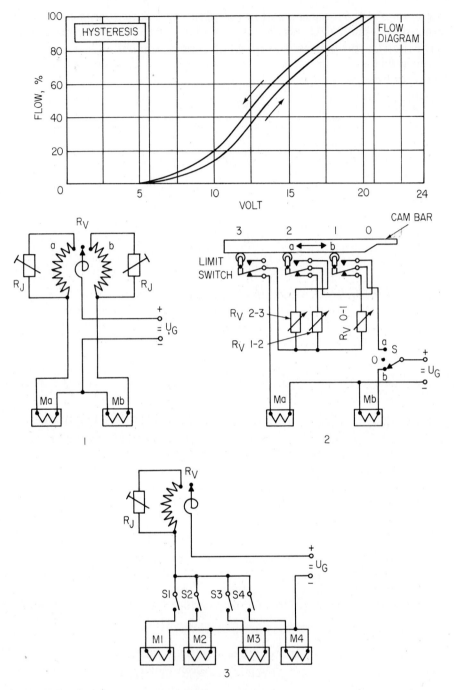

Fig. 14-6 Flow diagram (1) Double-wound potentiometer control. (2) Limit-switch controls. (3) Control of four solenoids by a single potentiometer and four switches.

direction "b." By means of this circuit high speeds can be obtained between 1 and 2, with reduced speed from 2 to 3. Heavy loads can be cushioned in this way, and brought accurately to a halt.

Diagram 3 shows the control of four solenoids by a single potentiometer and four switches.

Electrical diagrams R_v control potentiometer
R_j trimming potentiometer
U_g direct-current source
M solenoid
S switch

14-6 FEEDBACK CIRCUITS

If accurate proportionality is required between a velocity-demand signal and the resulting output-member velocity, the servo can be equipped with a *velocity-feedback transducer* and an amplifier. Figure 14-7 shows how a tachometer generator is used as the transducer. If the requirement is for a given displacement rather than velocity, the output member is usually equipped with a position-feedback device. Figure 14-8 shows a position-feedback device consisting of a synchro or a potentiometer feeding the information to the amplifier.

Feedback information is provided by a linear variable differential transformer (LVDT) for the control circuit associated with an overcenter-type piston pump.

Figure 3-19d shows a schematic view of the relative position of the cam mechanism that transmits the signal information from the tilt box to the LVDT. Figure 14-9 shows an actual cross-section view of a typical axial-piston pump that can be used for overcenter service to provide both speed and directional control. The position of the tilt box determines pump delivery and the ports through which the fluid will be delivered. The

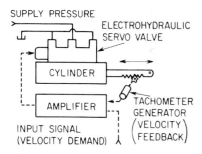

Fig. 14-7 Velocity feedback circuit.

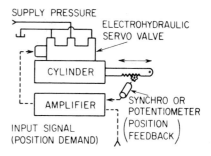

Fig. 14-8 Position feedback circuit.

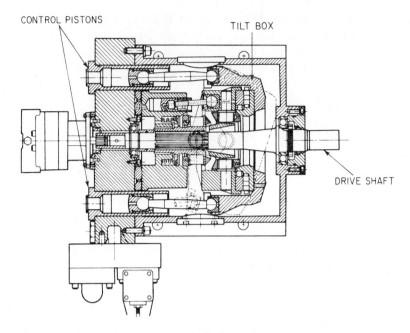

Fig. 14-9 Reversible, servo-controlled, axial-piston pump.

control pistons are supplied by the hydraulic, servo-type, four-way control valve shown in Fig. 14-10. Note the cam attached to an extension of the pivot pin on which the tilt box swings. The valve assembly of Fig. 14-10 is directly attached to the pump shown in Fig. 14-9.

The circuit of Fig. 14-11 controls this close-coupled assembly. Note the relationship of the tilt box, LVDT, and servo valve. Various command signal sources can be applied to provide the desired action.

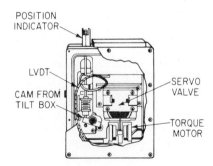

Fig. 14-10 Servo valve and linear variable differential transformer.

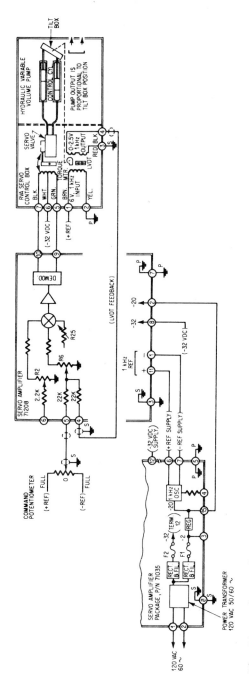

Fig. 14-11 Circuit for servo-controlled, axial-piston pump.

Limitations on servo circuits often rest in the mechanical motion required from the associated mechanical devices. Response capabilities of the servo devices can usually be tailored to most industrial requirements with adequate speed and power capabilities with almost infinite control over speed, force, and direction. The most intricate acceleration and deceleration programs are possible. Circuits may be designed for either linear or rotary power transmission or combinations thereof.

SUMMARY

The electrohydraulic servo element may be the last link in a complex system involving a number of measuring and computing operations for which only electrical means can furnish the needed speed and precision. Process-control systems, computer-controlled machines, and aircraft autopilots are examples of such systems. In these systems the input to the hydraulic actuator is predetermined, and the output member is a device which converts an electrical signal into mechanical motion, usually at relatively high power.

Essential elements for the hydraulic portion of the servo are a power source, an electrohydraulic servo valve, and a cylinder or motor. Some systems may use mechanical or pneumatic servo valves. The servo valves are usually made with symmetrical port configurations so that the speed and movement in each direction will be the same. For easier transmission of rotary motion, a rotary hydraulic motor may be used in place of cylinders. The rotary motor will also minimize elasticity effects due to fluid compressibility, since the volume of fluid contained in the motor can be relatively small. However, if the motor operates at high pressures and speeds, inertia effects can be serious.

This book is intended to give only a broad knowledge of the field of servos as used in hydraulic power-transmission devices. One important point to remember is that the input to a hydraulic actuator may be a demand for a given displacement or for a given velocity, while the input to a servo valve of the usual type will produce a corresponding and roughly proportional velocity of the cylinder or motor. If no feedback is present, the actuator will act as an integrating stage.

The small orifices and relatively close fit within the servo-valve structure make it essential that proper filters be included in the system. Chapter 8 on filtration pointed out that most of the needs for exceptionally low particle size hinge on the use of servo controls. A thoroughly clean circuit is imperative if stickiness and speed-response losses are to be avoided. Some mechanical-feedback servo systems have sleeves in the main control area that oscillate at a fairly high rate over a very short stroke to prevent particles of foreign matter from settling and forming banks of silt within the valve structure.

REVIEW QUESTIONS

14-1 Describe the function of a torque motor.
14-2 Why is the reducing-valve pilot signal taken from the outlet of the valve?
14-3 Where does the signal to actuate a relief valve originate?
14-4 How does a flapper-type servo pilot differ from a spool type?
14-5 What is a differential-pressure pickup?
14-6 What is the effect of a large overlap on a servo-valve spool?
14-7 Describe the components necessary for a velocity feedback circuit.
14-8 What type of signal is impressed on a torque motor?
14-9 What is meant by *position feedback?*
14-10 Describe the operation of a jet-nozzle follower system as used on tracer lathes.
14-11 What is an LVDT? How is it used?
14-12 How are rotary speed and motion monitored?
14-13 How is linear position indicated? What is a common signal source for linear position?
14-14 Can power steering be considered a servo system?
14-15 What supplies the control pistons of Fig. 14-9?

LABORATORY EXPERIMENTS

14-1 Duplicate the circuit illustrated in Fig. 14-1. Observe the relationship of movements with different positions of the fulcrum. Note the speed relationships.
14-2 Build the servo-valve type of relief-valve circuit shown in Fig. 14-2a. Observe the speed of response on an oscilloscope screen. Make a direct comparison with a similar-capacity, commercially available relief valve of poppet-type construction. Repeat with a sliding-spool, single-diameter, piston-type valve. Note the response time and overpressure conditions.
14-3 Compare the relief-valve response time to the relief action of a compensating-type pump. Try axial-piston, vane, and radial-piston pumps.
14-4 Build the metering valve in Fig. 14-2b. Observe the output speed regulation while rotating the shaft of a fluid motor carrying a known load. Vary the input pressures and observe the corrective action of the servo valve.
14-5 Replace the valve of Experiment 14-4 with a comparable general-purpose compensated flow-control valve. Observe the response characteristics.

14-6 Construct the nozzle valve shown in Fig. 14-3a. Note the small signal needed to provide a high degree of control.

14-7 Duplicate the device in Fig. 14-3b. Observe the different pressures obtained by changing the fixed-orifice size.

14-8 Build a model elevator using potentiometer feedback to indicate position. Program the potentiometer to stop at predetermined positions. Observe the accuracy.

14-9 Observe the action of an automotive power-steering system. Attempt to turn the steering wheel without the engine running.

14-10 Observe the action of a water-supply valve used to control the power-unit temperature by immersing the thermal control bulb in hot water. Immediately remove it to chilled water. Note the movement of the flow-restricting element within the valve as a result of thermal changes.

15
Industrial Hydraulic Circuits

A circuit of some type is needed whenever hydraulic power is applied in industry. In this chapter you will study a variety of industrial hydraulic circuits and learn how they are arranged to give best performance. By acquiring a knowledge of hydraulic circuits, you will prepare yourself to handle many of the problems arising in the use of hydraulic power in any industry.

15-1 CIRCUIT COMPONENTS

Hydraulic circuits use a variety of components. These include pressure, flow, and directional controls; pumps; hydraulic motors; pipes; hoses; tubing; reservoirs; and accumulators. These components are arranged in various ways to produce a desired output from the circuit.

15-2 PRESSURE-REGULATING CIRCUITS

Use of a relief valve permits easy control of the maximum pressure in the fluid-power circuit. In Fig. 15-1a, two relief valves give two working pressures in the circuit. A dual-relief-valve circuit like this is often used on a press which needs high pressure on the work dies on the downward stroke but little pressure to return the press ram. The low-pressure relief valve A also saves power and reduces pump wear if the ram must be held retracted for any length of time and no pump unloading device is used. With only the high-pressure relief valve B in the circuit, the pump motor would deliver large amounts of power at the top of the upstroke. High-

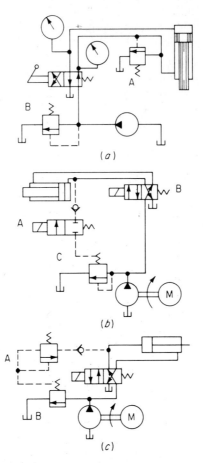

Fig. 15-1 (*a*) Two-relief-valve circuit. (*b*) Venting circuit. (*c*) Remote pilot circuit.

pressure fluid passing through the relief valve would overheat because of the energy being dissipated at this point; this could cause packing failures, excessive leakage, and fluid breakdown. With two relief valves, the high-pressure valve controls the system on the downward stroke, and the low-pressure valve limits pressure in the product loading position and during retraction.

Venting of the relief valve is another method of unloading the pump or changing the secondary pressure in accordance with the circuit needs. In the circuit in Fig. 15-1*b* the relief valve is vented to atmospheric pressure when the small solenoid-operated dumping valve *A* is open and the two-position, four-way valve *B* is in the cylinder *retract* position. Pump volume then unloads at low pressure through the open relief valve *C*. When the cylinder is fully retracted, a limit switch is tripped to energize the dumping-valve solenoid.

The dumping valve could be piped directly to the tank without going through the four-way valve. However, the relief valve might be slow in building up pressure at the desired time because of possible small leakages through the two-way valve. By making the vent connections as shown, the system operates properly because the directional valve closes the vent when the cylinder is to be extended. Although the line from the dump valve is open to the tank on the retracting stroke, the relief valve does not vent until the two-way-valve solenoid is energized. The slight leakage through the closed two-way valve does not unbalance the relief valve after pump pressure is built up. Complete opening of the two-way valve is required for unloading the pump through the piloted relief valve at the value impressed by the main-spool spring in the relief valve.

If the dumping valve of Fig. 15-1b is replaced by a remote pilot-relief valve A as in Fig. 15-1c, the pressure will automatically be reduced to a secondary lower value as the directional-control valve is shifted to retract the cylinder. For example, the main relief valve B can be adjusted for 2000 psi and the pilot A for 150 psi. When the solenoid is energized to retract the cylinder rod, the line to the head end will be connected to the tank as well as the discharge from the pilot-relief valve. This will establish circuit pressure at 150 psi. The connections on the four-way valve can be reversed if it is desired to have the circuit at 150 psi when the machine is idle and the cylinder retracted.

15-3 REDUCED PRESSURE

A reduced pressure is often required in one branch of a circuit which has only one pressure source. For example, the control of pressure on oil-driven spot-weld guns is important for good welds. If the guns are powered by the same pump that powers the work clamp [Fig. 15-2], a pressure valve placed in the pressure line between the welding and clamp

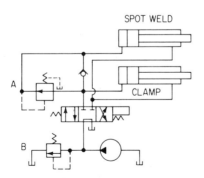

Fig. 15-2 Reduced-pressure circuit.

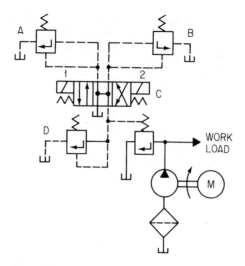

Fig. 15-3 Remote control of fluid pressure.

cylinders will regulate pressure on the spot-weld gun. Clamp pressure is determined by the pump relief-valve setting at B. Weld-gun pressure can be adjusted at any value less than the relief-value setting by adjustment of the pressure-reducing valve A. With the four-way control valve in neutral, the pump volume is unloaded to the tank. On the forward stroke, the clamp gets full pump pressure while the weld gun gets reduced pressure according to the needs of the work being processed. On the return stroke, pressure to both cylinders is controlled by the relief valve. A check valve is provided to bypass the reducing valve on the return stroke. Certain types of reducing valves can be drained into the opposite cylinder line so that the spool is held open on the return stroke and the check is then not required to provide the needed flow path.

The circuit segment shown in Fig. 15-3 provides two-station remote control of pressure with a safety maximum-pressure adjustment at the main relief valve D. The integral pilot relief within the head assembly of the main relief valve is set to the maximum safe operating pressure. With the directional valve C in neutral, the pump is vented with the open-center spool as shown.

If the pilot directional-control valve were provided with a closed-center spool, the adjustment on the main relief would be the controlling value when the unit was in neutral position. As solenoid 1 is energized, the pressure level will be adjustable at pilot valve B at pressures equal to or lower than the maximum established by the pilot on the main relief valve. Energizing solenoid 2 directs the control fluid flow to pilot valve A, which will then control the pressure adjustment up to the value established by the main-valve pilot. Either of the remote pilots can control the pressure through a range from the vented value of the main valve to the maximum established by the main relief-valve integral-pilot section. The component parts of the main relief valve are shown in Fig. 15-3 to illustrate the parts that affect the remote function.

15-4 VARIABLE PRESSURES

Figure 15-4 shows a circuit using a cam-operated relief valve to vary pressure values. Note the air-oil circuit used to provide an independent signal for the pressure-control function. This portion of the circuit may be replaced by a cam fastened directly to a machine member; or a rotary cam may time the pressure sequence.

Release of pressure is as important in most hydraulic circuits as the building up of pressure. Severe shock can result when a four-way valve is shifted after a ram has completed its work stroke. After completion of the work stroke in the circuit in Fig. 15-5, and with the head end of the cylinder under pressure, the operating valve is reversed. Then the only escape path for pressurized fluid from the cylinder is through the adjustable orifice, regardless of how fast the four-way valve may be shifted. During this decompression period, fluid from the pump will circulate through the unloading valve. This valve is held open by the pressure in the blank end until the decompression cycle has reduced the pressure to a desired point and a rapid return is desired. Decompression is generally necessary on systems using cylinders larger than 8 in in diameter operating at pressures in excess of 1000 psi and with large-capacity pumps.

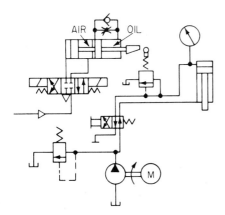

Fig. 15-4 Variable pressure controlled by auxiliary equipment.

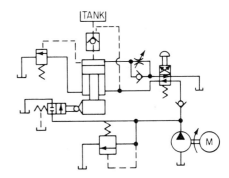

Fig. 15-5 Release of pressure at a controlled rate.

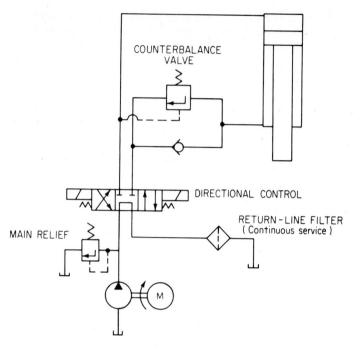

Fig. 15-6 Counterbalance circuit.

15-5 COUNTERBALANCE CIRCUIT

A normally closed two-way valve in the rod-end cylinder line of Fig. 15-6 is used to maintain pressure within this cavity to hold the piston in position. In this circuit the pilot pressure is taken from the line that directs fluid to the blind end of the cylinder. No motion of the cylinder is encountered until the four-way directional-control valve is actuated. Pressurized fluid directed to the blind end of the cylinder is also available to the pilot connection of the counterbalance valve. Load on the cylinder will not affect normal operation of the counterbalance valve because the pilot signal comes from the opposite cylinder line. Use of the counterbalance valve prevents the cylinder from lunging when the work load is suddenly removed and the fluid under compression in the blind end of the cylinder is free to expand at a rate equal to the flow being supplied by the pump. Pilot pressure to the counterbalance valve will decay, causing the spool to restrict or stop flow if the cylinder tries to "run away." This valve provides a constant resistance to flow; it does not act to meter the fluid flow except to the value of the pump delivery as reflected through the pilot signal. The counterbalance valve, connected in this manner, will support a negative work load. On a circuit having a vertical ram, the counterbalance valve will prevent the ram from dropping rapidly because of gravity when the system is not in operation. The holding-type counterbalance valve with a poppet-type seat minimizes potential *drift* due to valve-spool clearances.

15-6 SEQUENCE-CIRCUIT FUNCTIONS

Large presses require a large fluid flow to the main ram and low pressures during the rapid movement. When this volume exceeds practical pump sizes and capacities, prefill or surge circuits are used. During rapid closing of the press in Fig. 15-7, fluid from the pump flows to the small internal ram. Closing speed is determined by pump delivery and the bore of the internal ram. Downward motion of the large ram creates a vacuum satisfied by atmospheric pressure forcing fluid through the prefill valve; when

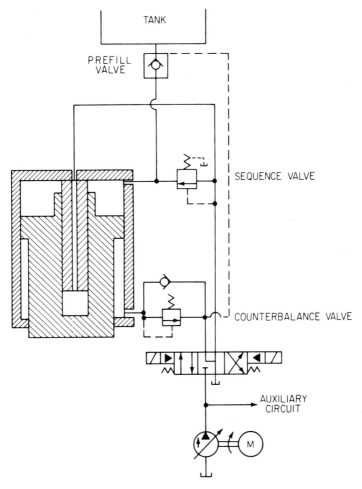

Fig. 15-7 Sequence, counterbalance, and prefill circuits.

Industrial Hydraulic Circuits **383**

the work is contacted, pressure builds up. The sequence valve then opens, and pump pressure is exerted on the large main ram to provide the required working force. The prefill valve is held closed by pump pressure.

The internal ram design might be replaced by auxiliary rams fastened externally to the press. The reduced area shown in Fig. 15-7 is for rapid closing. Area reduction is provided by boring out the main ram and fastening the smaller ram at the top of the outer cylinder casing.

15-7 SPEED-CONTROL CIRCUITS

When a constant-displacement pump is used and when it delivers a fixed volume of fluid to the circuit, feed control must be accomplished with metering valves. Pressure-compensated variable-displacement pumps may also be used with flow-control valves to ensure repetitive flows in an economical pattern. When the load is reasonably constant, the metering valve can be a fixed or adjustable orifice. Such an orifice will maintain a constant flow rate only if the load is constant. To overcome this limitation, pressure-compensated metering valves are used. The pressure compensator maintains a constant pressure drop across the orifice regardless of load.

METER-IN CIRCUIT
Placing the flow-control valve in the line to the blank end of the cylinder shown in Fig. 15-8 makes this a meter-in circuit. Since the pump delivery is variable, no relief valve is required. The compensating mechanism within the pump structure limits the output to the volume passing through the flow-control valve. Had this been a constant-displacement pump where the delivery is greater than the metered flow out of the flow-control valve, excess fluid would have to flow through the relief valve at full pressure. Consequently, the constant-displacement pump is always operating at relief-valve pressure.

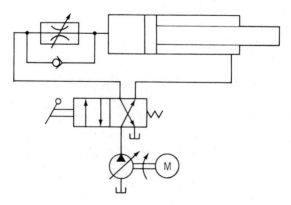

Fig. 15-8 Meter-in feed-control circuit.

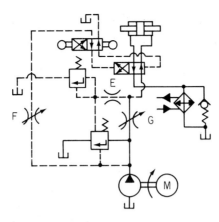

Fig. 15-9 Meter-in flow control with relief-valve compensation.

Metering fluid into the head end of the cylinder, as in Fig. 15-8, may be necessary because of the ratio between the piston and rod diameters. If the rod is large, the resulting area after deduction of displacement will be such that the intensification value will exceed the working-pressure capabilities of a flow-control valve used to restrict the flow from the rod end.

A reducing valve in the flow control maintains the desired pressure across the orifice in Fig. 15-8. The main relief valve is used to establish the pressure drop across the orifice in Fig. 15-9. Orifice G controls the flow from the fixed-displacement pump into the directional-valve portion of the circuit. The mainspring on the relief valve establishes the pressure drop across orifice G. The pressure in the control chamber can be no higher than the circuit requirements because of the connection through E to the downstream side of the main orifice. Needle-valve F determines rate of four-way directional-spool movement.

The meter-in circuit, using a relief valve for compensation, is usable only on circuits with one basic cylinder function. Auxiliary functions such as cross-feed on surface grinders (which operate at the end of the table stroke) may be integrated, but only one cylinder can be expected to use the precision flow-control function.

METER-OUT CIRCUIT

On many applications, such as drilling and milling, it is desirable to use a meter-out circuit which provides a resistance to flow. This resistance to flow helps minimize uncontrolled forward movement caused by sudden release of work loads. The pump pressure or forward pressure is constant regardless of work resistance. As the work load increases, the back pressure drops. With a single-rod cylinder, the backpressure exceeds the pump pressure at negative and low positive work loads. This intensified pressure can reduce the life of the rod packing. The load at which the

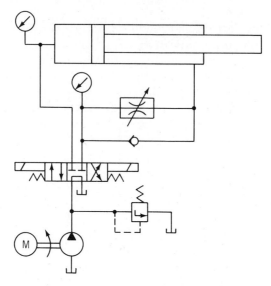

Fig. 15-10 Meter-out feed-control circuit.

backpressure starts to fall below pump pressure is determined by the ratio of head and rod areas.

With a double-acting cylinder having equal rod areas extending through the packing at each end, the speed will be equal and backpressure will exceed the constant forward pressure only when there is a negative work load. The maximum allowable system pressure (which occurs at maximum negative load) limits the forward pressure, thus limiting the capacity of the machine. Efficiency of a meter-out circuit is low; however, the advantage of supporting a negative work load has been obtained.

A check valve is provided for a return flow around the valve in the circuit in Fig. 15-10. With a double-rod cylinder and an installation where the feed is to be uniform in both directions, the meter-out flow-control valve may be installed in the tank line of the four-way directional-control valve.

BLEEDOFF FUNCTION

Efficiency of operation is improved by bleeding off excess fluid at actual working pressures. As shown in Fig. 15-11, pump pressure then varies with work load instead of being continuously fixed at the relief-valve setting. Although efficiency is improved, this circuit does not provide the same accuracy of feed control as do the meter-in or meter-out circuits. This is partly because of pump slippage. Fluid flow to the cylinder is the flow from the pump that is left after bleedoff. This flow will vary as pump slippage changes with load. Efficiencies cannot compare with variable-delivery pump circuits where the duty cycle is extensive. Economy of first cost or intermittent service usually dictates the bleedoff usage.

The bleedoff flow-control valve in Fig. 15-12 is used in a circuit having a fixed-displacement pump and motor. This usage results in a constant-torque, variable-horsepower circuit. Although this drive costs less than a circuit having a variable-volume pump, it is not as efficient.

The fixed-displacement fluid motor often carries variable loads which can result in high inertia at the deceleration period. Because of this, *crossover reliefs* and a makeup check (to compensate for circuit leakage) are integrated into the motor portion of the circuit beyond the four-way directional-control valve. Note the air-to-oil cooler on the discharge of the four-way valve just beyond the return-line filter. Mobile equipment systems often include an air-to-oil cooler adjacent to the radiator used for engine cooling. It may share the same fan and appear very similar to the engine radiator. The circuit shown [Fig. 15-12] is used to drive a cement mixer on a truck with power taken from the truck engine. The fluid motor speed can be varied by changing the speed of the engine driving the pump. This changes pump delivery. When engine speed is constant, the bleedoff valve is used to vary the motor speed.

Figure 15-13 shows a bleedoff flow-control valve used in a hydraulic brake circuit. When pump delivery is cut off from a fluid motor, the motor will continue to rotate because of its own inertia and that of the attached load. The motor then acts as a pump; therefore a source of fluid must be available to prevent sucking of air; if a coasting stop is undesirable, a brake device is required.

The circuit in Fig. 15-13, with the bleedoff flow-control valve, is set up to provide a constant-torque drive which may be brought to a coasting stop or a quick stop, depending on the position of the four-way valve. With the valve in neutral, the motor coasts to a stop. In the brake position of the four-way valve, the pump unloads and the fluid discharged from

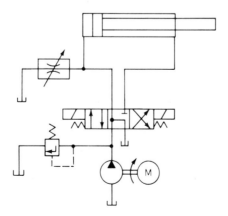

Fig. 15-11 Simple bleedoff flow-control circuit.

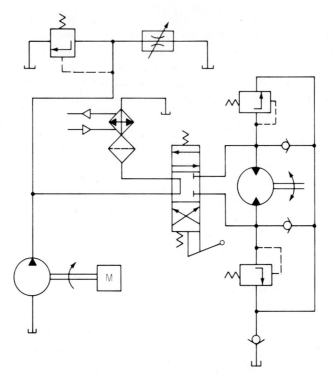

Fig. 15-12 Bleedoff control in two directions of motor operation.

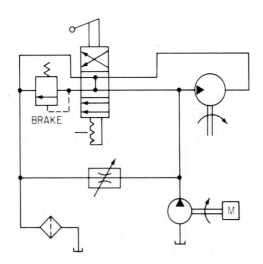

Fig. 15-13 Bleedoff flow control used in a brake circuit.

388 *Industrial Hydraulics*

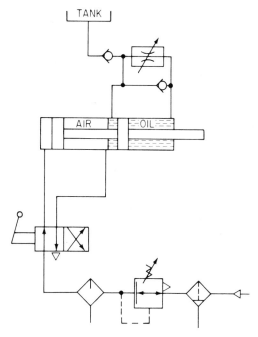

Fig. 15-14 Hydraulic feed control for pneumatic-powered cylinder.

the motor must overcome the adjustable spring setting in the brake relief valve; this creates a backpressure which stops the fluid motor.

CLOSED-CIRCUIT METER-OUT FUNCTION

Figure 15-14 shows how accurate control of feed can be obtained with a combination cylinder. Two cylinders are coupled to function in the same direction. This circuit can give, at speeds of an air circuit, feeds as accurate as those obtained with a full hydraulic circuit. The double-acting air and hydraulic cylinders have a common piston rod. Air directed by the four-way, hand-operated valve powers the piston in either direction. Speed control of the piston in the forward direction is obtained in the closed hydraulic circuit by adjusting the flow-control valve. A fluid filler tank keeps the hydraulic system full by supplying makeup fluid. Where force requirements can be met by air pressure, this circuit is usually quite satisfactory and may cost less than the full hydraulic circuit.

INTERMITTENT FEED

On some machines it is desirable to have a skip-feed arrangement. This is readily accomplished, as shown in Fig. 15-15, by having a cam-operated shutoff valve in parallel with the meter-out flow-control valve. Rapid forward movement will take place any time the shutoff valve is open. Cams on the cylinder-powered slide act to close the shutoff valve. Flow is

Industrial Hydraulic Circuits **389**

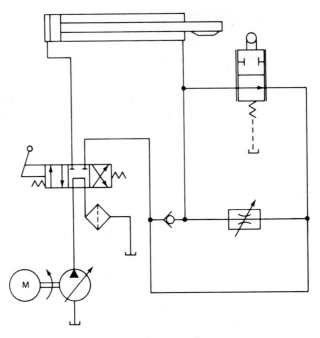

Fig. 15-15 Intermittent feed-control circuit.

then diverted through the flow-control valve to slow the slide to the set feed rate. Return speed is at the maximum as flow from the pump to the cylinder-rod end bypasses the flow-control valve by flowing through the check valve.

DECELERATION
Cylinders moving at high speed can be decelerated to a slow smooth stop in either direction by use of deceleration or cam-operated shutoff valves. Normally the valve is held open by its spring. Note in Fig. 15-16 that as the cylinder approaches the end of its stroke, a cam contacts the roller on the shutoff valve slowly to reduce flow from the cylinder. This action creates a backpressure to slow the cylinder travel. Cushions built into the end caps of a cylinder can perform the same function with an established internal mechanical signal. With the integral control the cylinder must be stopped within the length of its maximum stroke. External deceleration valving permits easy adjustment. An integral needle valve may also be incorporated into the cam-operated deceleration valve so that the valve will block off all flow and divert it through the needle valve at an established rate.

VARIABLE VOLUME
Control of speed can easily be obtained by using a variable-delivery pump. The initial cost of the pump is usually greater than the cost of the components used in a constant-displacement pump circuit with a flow-

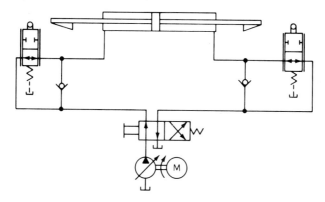

Fig. 15-16 Deceleration circuit.

control valve, particularly if heat generation does not pose a problem. The pump delivery can be varied manually, as shown in Fig. 15-17, or with an electric motor or servo control. The minimum flow delivered by the pump shown in Fig. 15-17 is determined by pressure values established by the balancing pistons and the pressure additive created by the auxiliary pilot-relief valve within the pump control circuit. Usually this minimum flow is that required to make up for leakage and control functions at maximum pressure. Maximum flow for this pump is determined by a manual adjustment as shown by the symbol. An adjustable bias spring is also provided for improved response characteristics. Pump controls may include other stop mechanisms to restrict overtravel and provide damping functions.

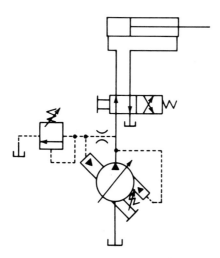

Fig. 15-17 Variable-volume flow control.

Industrial Hydraulic Circuits **391**

Use of a pressure-compensated pump permits high-volume delivery for rapid traverse and low-volume for feeding, holding, or squeezing. Pressure in the system acts to change pump delivery. Efficiency is excellent because the pump delivery is governed by load. Figure 15-17 shows a number of controls, including a pressure-compensating mechanism.

ROTARY HYDRAULIC TRANSMISSION

The circuit shown in Fig. 15-18 drives a machine used to coat, size, and laminate cardboard. Two pumps are used—a fixed-displacement unit and a variable-delivery, radial-piston pump having an electric pilot motor to change pump delivery. The jackshaft of the machine is driven by two constant-displacement fluid motors. Each motor can transmit up to 20 hp continuously or 25 hp intermittently at 850 rpm. By varying pump delivery, the drive speed can be varied from 120 to 850 rpm in 45 s. This system provides smooth high-speed acceleration and fast start-stop control.

The transmission in Fig. 15-18 has an adjustable hydraulic brake which will exert any desired braking force between 5 and 40 hp. This brake is operated from an overload pressure switch, from limit switches which guard against overrunning, or from push-button stations for manual control. Any of these electric controls will deenergize the solenoid of the

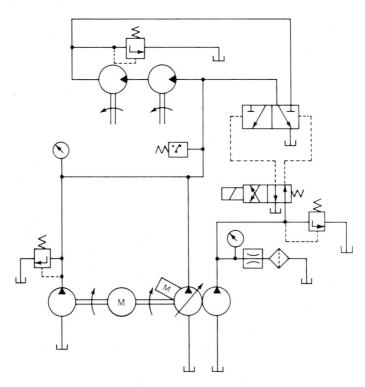

Fig. 15-18 Rotary hydraulic transmission.

392 *Industrial Hydraulics*

two-position, three-way valve. The valve shifts to direct pilot fluid to move the two-position, three-way valve to the indicated position. Pressure fluid from the pump is dumped to the tank through the three-way valve. The motor return port is blocked. At this time the inertia of the motors causes them to act as pumps, but fluid discharge against the spring load of the brake relief valve acts to brake motor rotation. Pilot pressure for the three-way valve is supplied by a gear pump built into the variable-delivery pump housing.

15-8 SPEED CONTROL FOR CONTINUOUS PROCESSING

In Fig. 15-19 the variable-displacement pump on a wire-covering machine is driven from the main extruder drive. This pump provides power to drive the capstan and reel in synchronism with the extruder. As the speed of the extruder is increased or decreased, the volume of fluid delivered by

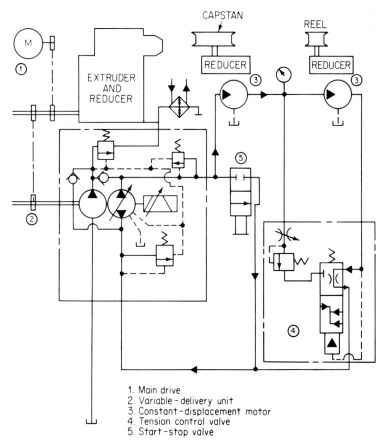

1. Main drive
2. Variable-delivery unit
3. Constant-displacement motor
4. Tension control valve
5. Start-stop valve

Fig. 15-19 Speed control for continuous processing.

the variable-displacement pump also increases or decreases, thereby causing a corresponding change in speed of the capstan and rewind hydraulic motors.

Displacement per revolution of the variable-displacement pump is changed by means of a solenoid with double winding. The solenoid operates in a variable way progressively in opposite directions to actuate a servo pilot valve which in turn controls pump displacement. The adjustable displacement of the pump permits the ratio of capstan and rewind to be varied with respect to the extruder speed. Volume of fluid delivered to the rewind motor is in fairly direct proportion to the speed of the capstan. This means that the reel speed is synchronized with the speed of the capstan by hydraulically connecting the motors in series. As wire is wound on the reel, the diameter increases, requiring a lower speed; therefore, an excess quantity of fluid is delivered by the capstan motor to the reel motor. This excess volume of fluid passes through the constant-tension-control valve.

15-9 BLEEDOFF ON DIFFERENTIAL CIRCUITS

Differential circuits are frequently used on high-speed broaching machines. In the typical circuit shown in Fig. 15-20, the working area (piston annulus) is three-quarters the return area (piston head). The ratio of areas used with this differential circuit will theoretically give a return speed three times the work or broaching speed.

During the broaching stroke, pressure fluid acts on the rod end. A bypass flow control regulates speed and a counterbalance valve creates a backpressure on the broaching tool. On the return stroke, pressure fluid from the rod end is returned to the pressure line. Use of a differential circuit is advantageous on an application where the required returning force is small; the penalty of less available force paid for higher return speed is not important.

15-10 BASIC AUTOMATION CIRCUIT

Figure 15-21 shows the four basic functions incorporated in an automated machining function circuit. These functions consist of (1) an indexing table mechanism to feed part to the machining unit, (2) a shot-pin locating device for precise positioning, (3) a clamp to hold the part to be machined, and (4) a unit to feed the cutting tool or machining function to the work.

The unit illustrated is part of a high-production machine. The circuit is split into two separate functions. System A connections are involved only in the drill unit. System B connections are used in all the remaining functions.

Limit switches 1, 4, and 7 must be engaged before valve G can be

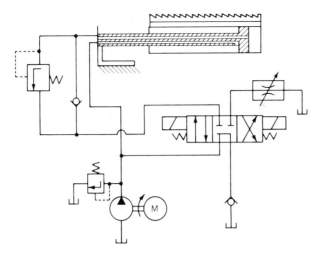

Fig. 15-20 Bleedoff on differential circuit.

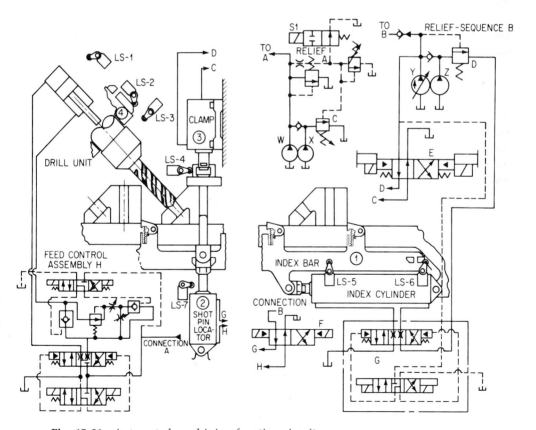

Fig. 15-21 Automated machining function circuit.

Industrial Hydraulic Circuits

actuated to index the push bar to move the part into position. This will ensure that all of these functions are clear. Pump Z moves the index cylinder while pump W, at low pressure, holds the drill unit in position and pump Y holds the clamp cylinder open and the pin cylinder down. A pressure signal from the outlet of pump Y opens relief sequence unloading valve D. This valve serves a multiple function. With valve G in neutral and pump Y deadheaded at maximum pressure, it serves as an unloading valve for pump Z. If pump Y should become inoperative, the flow from Z passing through the check valve could reflect to the pilot of valve B. At this time it will function as a safety relief valve. During rapid traverse movement of the clamp or locating pin cylinder, it blocks flow to the index cylinder so that it performs a sequence function. When pump Y is deadheaded and valve D is wide open because of the signal at point B, full flow of pump Z is available to move the index cylinder. When valve G is in neutral, the pump Z is relaxed to the reservoir.

As limit switch 5 is engaged, valves E and F are shifted, first to locate the shot pin by an electrical time sequence, and then to clamp the part to be machined. When the part is clamped, the feed unit is activated. The directional-control valve in feed-control system H is activated to provide rapid traverse. Limit switch 2 provides a signal to direct the rod-end fluid through a feed mechanism. Limit switch 3 diverts the flow through a second feed mechanism in series for a clean-up feed, after which a timer returns the unit from the bottomed position. During the feed operation, the index cylinder is retracted. The latch mechanism permits the return function without affecting the machined parts. Limit switch 1 must be contacted before the locating pin retracts and the clamp returns to the out position.

Pumps W and X both can be relieved through valve A if solenoid S1 is deenergized. This relieves the pressure down to the value established by the main-spool spring in valve A. When pressure is required for the drilling function, solenoid S1 is energized. Pump W is sized to handle the maximum feed rate. Pump X in combination with pump W provides the rapid traverse forward and return flow. When the unit is feeding during the normal cycle, valve C is used to unload pump X back to the reservoir. The excess of fluid from pump W not being used in the feed function is passed over valve A. The check valve in the line between pumps X and W provides the connection for combining the flow of the two pumps and prevents the loss of fluid from pump W when passing fluid to the tank when pump X is unloaded by valve C.

Pumps Y and Z can combine their flows through the integral check valve for rapid movement of the clamp or shot-pin locating cylinder in either direction. At the end of the stroke, pump Y is permitted to deadhead, pumping only sufficient pressurized fluid to compensate for circuit leakage. Pump Z is the only source of pressurized fluid for the index cylinder. The movement of the index cylinder is then isolated from the circuit so that it cannot affect the clamping force provided by pump Y. Pump Y is of small capacity and is primarily used for holding pressure with minimum horsepower need and resulting low heat generation.

15-11 ACCUMULATOR CIRCUITS

BALANCING LOADS

Figure 15-22 shows an accumulator used to hold a specific loading on a pair of rolls. The accumulator provides a live, resilient load as a result of the pressurized nitrogen contained within the shell, so that variations in product can be accommodated without breakage.

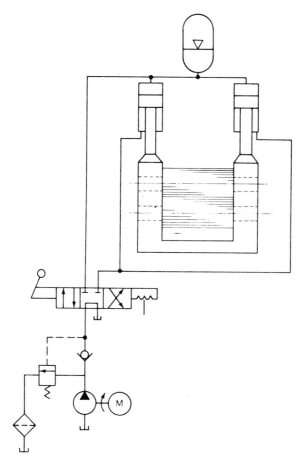

Fig. 15-22 Load-balancing circuit with accumulator.

Industrial Hydraulic Circuits 397

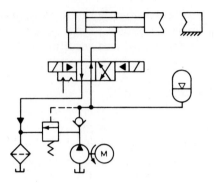

Fig. 15-23 Holding pressure (fluid) with an accumulator.

HOLDING PRESSURE

Leakage compensation and power savings are obtained by using the accumulator shown in the vise circuit of Fig. 15-23. When the vise jaws are in the clamp position, pressure builds up and is held by the accumulator. During holding, the pump volume is bypassed back to the tank through the unloading valve. Any leakage past the piston packings in the cylinder is compensated for by the volume from the accumulator. The pump can be operated intermittently because the accumulator holds pressure and makes up leakage. When the holding pressure drops to an allowable minimum, the unloading valve closes and pump pressure again builds up.

STORED ENERGY FOR STARTING DIESEL

Figure 15-24 shows how a circuit used to start a diesel engine can be provided with the needed auxiliary functions and emergency devices. In

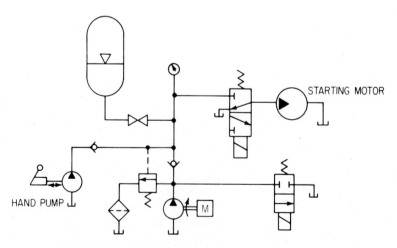

Fig. 15-24 Stored energy to start a diesel engine.

398 Industrial Hydraulics

starting a diesel engine, the instantaneous power required is high, but the period between operations is long. Power for the starting operation is taken from the accumulator. Between operations, the main pump charges the accumulator while the engine is running. A hand pump in the circuit serves as an emergency unit to charge the accumulator in the event of leakage. A solenoid-operated, two-way valve is connected to the main pump pressure line to permit relaxation of pressure and diversion of pump flow during the start cycle to minimize energy needs of the engine during this critical period. After starting, the solenoid two-way is deenergized so that the accumulator can recharge. Solenoids are operated simultaneously so that energy flow is correctly apportioned.

REDUCTION OF CYCLE TIME

This is often made possible by the use of the added volume available from an accumulator. While the press circuit shown in Fig. 15-25 could be designed with larger pumps, the initial cost would be higher. In this circuit the volume from the accumulator adds to that of both pumps to speed ram travel. When the ram meets sufficient resistance to build up fluid pressure and actuate the pressure switch, the solenoid valve shifts to direct fluid from the large pump to recharge the accumulator. The small pump supplies high-pressure fluid to the ram for the pressing operation. When the manually operated four-way valve is shifted for the return

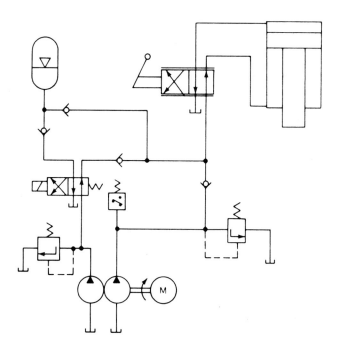

Fig. 15-25 Reduction of cycle time by use of an accumulator.

stroke, pressure is relieved, the solenoid valve is deenergized, and both pumps and the accumulator deliver fluid for rapid return.

15-12 BOOSTER AND INTENSIFIER CIRCUITS

Use of a booster or intensifier has the advantage of keeping high-pressure fluid in short lines near the work cylinder. This reduces maintenance of equipment and eliminates the cost of high-pressure pumps and valves. The intensifier may be of the simple *one-shot* design or a continuous-pressure booster [see Fig. 4-15a to e]. Where space prevents the use of a single large-diameter cylinder, a low-pressure circuit having a tandem cylinder can provide the large work force. The use of this force multiplication may cost less than pressure intensification.

FORCE MULTIPLICATION
The tandem cylinder circuit shown in Fig. 15-26 is designed to give rapid closing action and then a large clamping force, without developing very high fluid pressures. Shifting of the three-position, four-way valve directs fluid to the small cylinder, causing rapid closing. When the clamp contacts the work, pressure builds up to open the sequence valve and apply pressure to the large cylinder. Clamping pressure builds up to the pump-compensator value, and the clamp force becomes equal to the pressure times the combined area of both cylinders. This permits use of a smaller-capacity pump than would be needed in a circuit using a large cylinder and requiring greater pump capacity to provide the needed speed. A replenishing line to the blank end of the large cylinder prevents air from being pulled in behind this cylinder and supplies fluid during rapid advance. The rod end of the large cylinder may be connected to tank (above the oil level) rather than to the pump line to assist rapid return of the small cylinder. A pilot check valve may be added at the blind end of the large cylinder to provide an additional path to tank for the fluid during the retraction stroke. The valve can be piloted from the rod-end line of the small cylinder.

PRESSURE INTENSIFICATION
Figure 15-27 shows how pressure intensification is obtained by supplying low-pressure fluid to the large cylinder that is mounted integrally in tandem with a smaller-diameter cylinder. The pressure developed will be inversely proportional to the piston areas. High pressure is obtained at the expense of reduced volume, the volume of high-pressure fluid being inversely proportional to the ratio of pressure output to pressure input. Therefore, for short-stroke pierce cylinders, where little volume is needed, this circuit is ideal.

Shifting the four-way valve puts pump or air pressure, which is limited

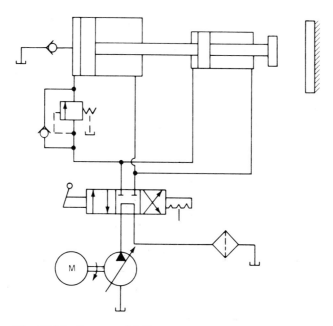

Fig. 15-26 Force multiplication with tandem cylinder.

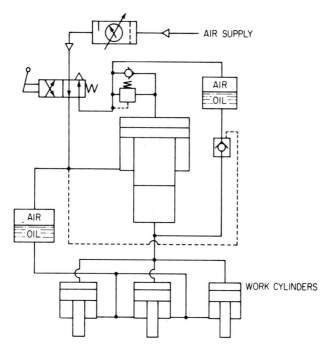

Fig. 15-27 Pressure intensification with air-oil circuit.

by the relief-valve setting or a pneumatic reducing valve, on the blank end of the large cylinder. High-pressure fluid from the output end of the intensifier powers the piercing cylinders. After piercing, the manual valve is shifted to the position shown to direct pump or air pressure to the rod end of the intensifier. The spring-loaded pierce cylinders retract. A filler unit on the small diameter of the intensifier makes up any fluid lost by leakage in the high-pressure circuit.

15-13 MOTION SYNCHRONIZING CIRCUITS

TIE CYLINDERS

Figure 15-28 shows two cylinders synchronized by the use of mechanical linkage. Getting two cylinders to move in unison is difficult because of leakage, friction, variations in cylinder-bore size, and contaminated fluid. The circuits that are considered here have been used on existing machinery; however, any of these variables can upset the operation.

Cylinders tied together for mechanical synchronization are perhaps the most positive in action. This arrangement works well on heavy equipment when the mechanical linkage is rigid. Rack-and-pinion connections are good if the mesh is proper and backlash is not important. This method is

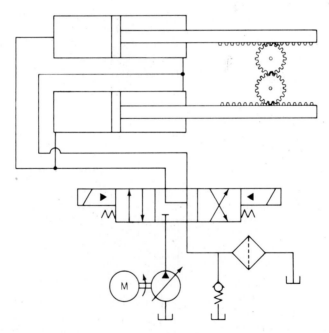

Fig. 15-28 Interlocking cylinder motion with positive gear tie.

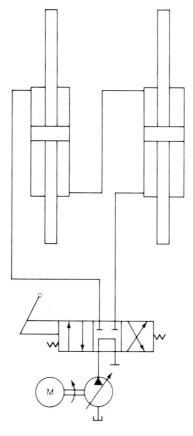

Fig. 15-29 Series piping to provide synchronous cylinder movement.

limited to cylinders which move in the same direction. A cable-and-pulley hookup between pistons is sometimes used in place of racks and pinions. For simplicity, in all circuits shown here the cylinders are identical in size.

SERIES PIPING

Figure 15-29 shows series piping of cylinders. This arrangement will give reasonably accurate synchronism. Double-ended cylinders must be used to keep fluid deliveries constant; their use, however, may be objectionable because of the space required for rod extensions. The cylinders will stay in step only if each cylinder receives the same amount of fluid. Any leakage will upset operation. To reduce leakage, new spools should be installed in worn valves; leaking gaskets and fittings should be eliminated. Air in the system will also cause trouble, as will thermal expansion of the fluid. In some installations, valving may be needed to compensate for leakage. A

Industrial Hydraulic Circuits **403**

cam-operated valve, or possibly a solenoid-operated unit, could be installed to pressurize the locked fluid at the end of the stroke in a predetermined program.

MATCHING PUMPS

Two pumps which have identical delivery [Fig. 15-30] will move the cylinders in unison. But leakage must be minimum, and other variables must be considered. If the loads are unequal, one pump may deliver a little more fluid than the other, depending on the volumetric efficiency of each pump. Unequal packing friction is also a factor. If friction on one piston is greater, it may not move until the other is well under way, thus destroying accurate synchronization. A common-shaft drive for the pumps in Fig. 15-30 might eliminate the losses or inaccuracies caused by the motor-coupling slack.

SYNCHRONIZING WITH FLOW-CONTROL VALVES

Metering an equal volume of fluid to or from each cylinder, as in Fig. 15-31, should result in synchronized movement. Each valve is adjusted independently. In this installation, too, leakage and friction will cause

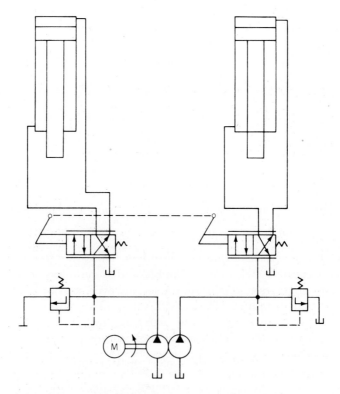

Fig. 15-30 Synchronization with matched pumps.

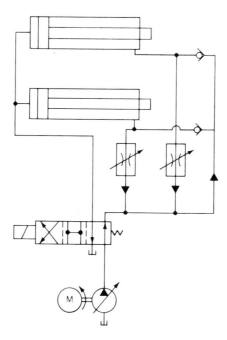

Fig. 15-31 Synchronization with compensated flow-control valves.

potential inaccuracies. Contaminated fluid must not be used, since it might contain particles that could interfere with the proper operation of the flow-control valves. With simple flow regulators, loads have to be equal if the cylinders are to move in unison. If reducing valves within the compensators of the flow-control valves stick in the open position, the valves will function as simple adjustable orifices and the synchronization will suffer consequent inaccuracies.

Figure 15-32 shows how accurate synchronization of movement to the right is provided by the use of accumulators and flow regulators. The stroke is limited to the available volume from the accumulators on the forward stroke. The accumulators are recharged on the retraction stroke. Good performance from this circuit requires minimum leakage in the valves.

SYNCHRONIZING MOVEMENT WITH FLOW-DIVIDING MOTORS
An effective flow divider consists of two hydraulic motors, each delivering fluid to a cylinder [Fig. 15-33]. The motors are mechanically coupled so they must rotate at the same speeds and therefore will deliver equal volumes. (Both motors must be of identical size.) This flow divider will act in both directions of motion. With this system, variables in load or friction do not greatly upset the movement. There are many installations where two carefully matched double pumps have been coupled together and

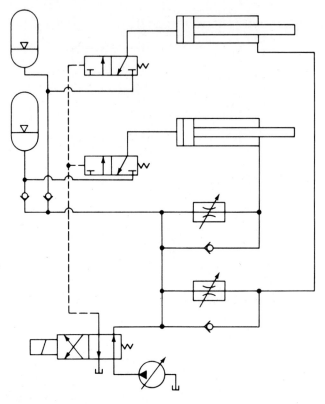

Fig. 15-32 Synchronization with flow-control valves and accumulators.

used as motors for dividing the flow to as many as four separate cylinders. Both piston- and generated-rotor-type motors have proved very effective in synchronizing circuits.

15-14 SERVO CIRCUITS

TRACER CIRCUIT
Contour milling is an important application of hydraulic servo systems. In the circuit in Fig. 15-34, the rotation of a fluid motor which drives the vertical feed of a milling-machine table is related to the contour of a master cam. The master cam moves with the table. The finger of the tracer valve rides on the master cam to follow its contour. This causes the tracer valve to move and direct fluid flow to shift the directional-control valve of the fluid motor. When there is no movement of the tracer finger, the fluid motor is stopped. Upward or downward tracer movement causes clock-

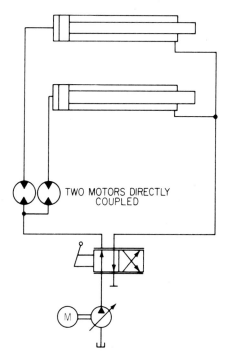

Fig. 15-33 Synchronization with flow-dividing motor assembly.

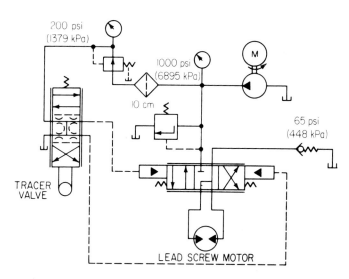

Fig. 15-34 Tracer circuit.

Industrial Hydraulic Circuits **407**

wise or counterclockwise motor motion. This pilot-valve circuit is used because 1000 psi is needed to power the motor and the tracer valve cannot function properly at pressures in excess of 300 psi.

VARIABLE-VOLUME PUMP CONTROL

There are several servomotors available to reverse and vary the delivery of a pump in response to circuit demands (see Chaps. 3 and 14). The circuit shown in Fig. 15-35 represents the basic operation of servomotor control. When fluid under pressure is delivered to the servo cylinder at a controlled pressure, as shown in Fig. 15-35, the servo control can amplify a signal to provide the needed force to move the pump displacement element. Movement of the pump mechanism can create pressurized fluid flow in either direction at infinitely variable rates and pressures.

A small fixed-displacement pump is used for supercharging, for leakage makeup, and to power the servo cylinder. The servo control cylinder, a portion of the lever mechanism, and some of the conducting lines may be integrated into the pump package. The servo control valve is shifted in response to the input. Fluid then moves the servo cylinder, and as the servo cylinder moves into the proper relationship with the controlling input, the servo closes.

EQUALIZING CYLINDER MOTION

Figure 15-36 shows another way to equalize cylinder motion in addition to those described earlier in this chapter. This simple servo circuit uses an equalizing valve of the three-way-spool type. In the neutral position of the spool, the lands cover all but a very small portion of the outlet ports. The pressure port remains uncovered at all times. If one cylinder starts to move more slowly than the other, this movement is transmitted through the linkage to the equalizing valve, causing the spool to shift in the direction which will uncover the port to the slower-moving cylinder.

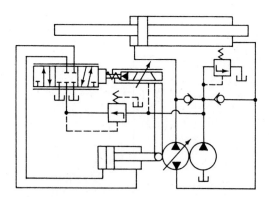

Fig. 15-35 Servo-type, variable-volume, reversible pump control.

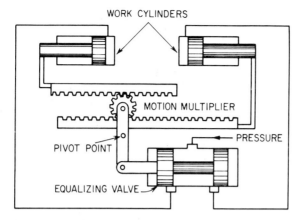

Fig. 15-36 Equalizing cylinder motion with simple servo control.

15-15 REPLENISHING CIRCUITS

Figure 15-37 shows the circuits that are needed when a pump and motor are connected in a closed loop. Makeup for leakage is supplied through the replenishing valves. Fluid is also drawn, by the motor, from the tank through suitable replenishing valves during braking if the supercharge or

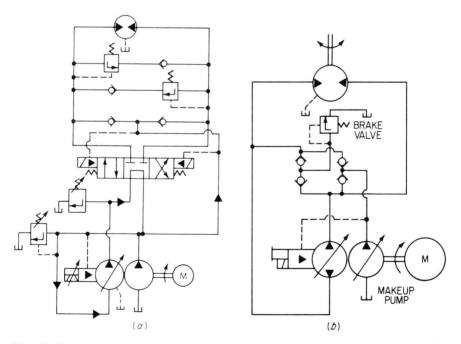

Fig. 15-37 Replenishing circuits.

Industrial Hydraulic Circuits **409**

makeup pump is of small capacity and unable to supply sufficient fluid. A small modification of the circuit to provide these additional makeup check valves will satisfy this need. Often, as indicated in the circuit drawing, the pump in the closed circuit is supercharged by a pump which also supplies replenishing fluid. A circuit without a supercharging pump draws replenishing fluid by suction from the tank.

The supercharging circuit illustrated in Fig. 15-37a uses a nonreversing variable-volume pump; a four-way valve is used to obtain motor reversal. The network of check and relief valves provides for replenishing and braking in either direction.

A reversing-type pump is used in the circuit of Fig. 15-37b. While the replenishing network is simpler in this circuit, brake pressure must be the same in both directions of rotation. Independent adjustment is not possible with the use of only one pressure-control valve at this functioning point. Mulitple remote-control pilot valves may be used with a piloted relief valve for this function with suitable isolation check valves.

15-16 DESIGNING CIRCUITS

The design of hydraulic circuits is predicated on certain basic needs and conditions, such as (1) how much force is needed, (2) how fast the circuit must function, (3) environmental conditions, (4) sophistication of control desired, (5) economic considerations—available drive force or input energy, (6) duty cycle, (7) life expectancy, and (8) reusability of components. There are, perhaps, other considerations that will affect the choice of circuit components, but they will usually fall into special categories.

The first concern is the *force needed*. An analysis of the operation to be performed will yield some indication of the force requirements. The amount of information available and the degree of accuracy that can be expected in analyzing this information will determine some of the other circuit requirements. For example, both minimum and maximum force requirements must be known before force can be translated into square inches of cylinder area, inch-pounds of torque on a motor, and pressure per square inch to supply these devices properly.

After the force requirements have been established, cylinder and motor size can be determined. Pressure level can be estimated on a preliminary basis. There is a definite relationship between the required pressure and the size of the motors and cylinders. As a basic rule—as pressure is increased, cylinders and motors may decrease their effective areas for equal force values. The envelope of the components will require increasingly stronger members as the pressure increases. Heavier wall sections and other design considerations may make the higher-pressure components lose some of their size advantage as pressure values are increased.

The determination of component size may hinge heavily on the second factor—*actuating speed*. The simple ratio of pressure to area for force determination must be related to the gallon-per-minute fluid flow that will determine speed of movement of the cylinder or speed in revolutions per minute of the motor. A determination of how fast a device must function will provide basic figures on the rate of fluid flow necessary to satisfy the quantities required by the chosen cylinder or motor.

Environmental conditions cover a multitude of basic factors, such as possible salt-spray contamination in marine service, nonmagnetic construction in certain military applications, resistance to shock in mobile usage, cast-iron particles in the dust of the average machine shop, and many other conditions peculiar to certain areas or applications. These environmental conditions may require special surface finish, special materials, or possibly protective devices of more than average capabilities.

Potential fire hazard is one important environmental condition. The fluid media can have varying degrees of resistance to combustion. Even pneumatic systems can support combustion if there is an oil mist in the system for lubrication. Although fluids are being developed to approximate closely the noncombustible characteristics of water without some of its undesirable features, systems using raw or mechanically filtered water are the only ones at present that can be considered absolutely fireproof. However, materials used to manufacture components for service with water are expensive, and service life is comparatively short. Less expensive materials for the components require rust protection and minimization of fluid-medium abrasive qualities. A careful analysis of the problems involved will determine the type of fluid most compatible with the chosen components and the expected environmental conditions.

The *sophistication of control* desired will determine the signal sources to create the cycle pattern of the power-transmission system. Manual actuation is the ultimate in simplicity. For example, the hydraulic jack, used to lift an automobile, has three basic components: a cylinder or ram, a pump, and a valve.

The signal to raise the car and the resulting cycle are all manual. The bypass valve is manually closed; the reciprocating-piston pump is actuated manually; the operator stops his pumping action when the desired height is attained. The device remains in a static state until the bypass valve is opened manually to release the contained oil within the cylinder or ram and permit the vehicle to drop. The rod end of the cylinder may be the reservoir.

The size of the cylinder and the energy the operator is willing to expend in actuating the pump will determine, within certain mechanical limitations, the load-lifting abilities. Speed of lift will be determined by the rate at which the operator actuates the pump handle. Speed of descent will be controlled by the degree of opening of the passage through the valve that returns fluid from the end of the ram or cylinder to the tank; it it also affected by the weight on the jack and the type of fluid used.

The simple laboratory press circuit in Fig. 15-38 uses a manually controlled plug-type needle valve for direction, speed, and pressure control. With the valve opened wide, the full volume of the pump will be directed to the tank. The cylinder or ram will descend by gravity plus the weighted-accumulator energy potential at the lip of the ram. As the manual valve is closed, the flow will be restricted until the energy available from the gravitational drop and the accumulator is balanced out, and the cylinder will stay in a static position. If the manual valve is further restricted, it will cause the ram to rise. When resistance is met on the platen, the orifice created by the manual valve will determine the pressure level. The safety valve will establish a maximum pressure in the system while the check valve will prevent fluid flow back through the pump if it is stopped during a holding cycle. The crank drive may be replaced by another type of power source, or the pumping system may be replaced by multiple-piston reciprocating pumps or rotary pumping mechanisms.

Circuit control can be by a multitude of more complex devices, as discussed earlier in Chaps. 10 to 12. The type of control chosen is directly affected by *economic considerations*. The farmer who uses his front-end loader on his tractor 10 times a year will not expect to pay for the sensitivity of control, heavy-duty features, and *life expectancy* of a bulldozer that works around the clock on construction jobs. Drive forces will vary from the manual force used to operate hand pumps to power takeoffs

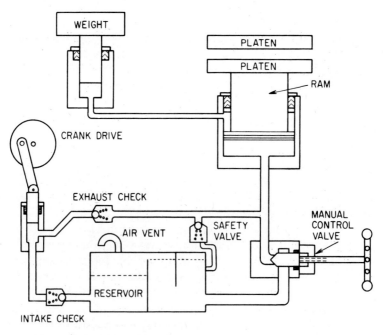

Fig. 15-38 Simple laboratory press.

from piston engines to various types of rotary motion from electric motors and turbines. Each type of drive force directly affects the type of pump that can be used and the circuit needed.

The *duty cycle* may require high performance for a limited time, as in certain aircraft components. Or it may require very heavy construction for long service life and minimum maintenance, as in some machine tool and construction-machine circuits.

Certain production machines are equipped with *reusable components*. At the end of a production cycle, the machine can be altered for new functional operations at much lower cost than would be possible with all new components. The use of gasket- or manifold-mounted equipment has made it possible to alter systems quickly by simply substituting components with different characteristics. Hydraulic plumbing with tubing as the conductor permits reuse of both valves and fittings in many installations.

15-17 CHOICE OF COMPONENTS

After these eight basic circuit considerations have been evaluated, the circuit can be designed. At this point the circuit design involves pressure and flow and controls for both, plus direction of flow and the end-use device (such as a rotary fluid motor or linear cylinder).

PUMPS
The choice of pump will be determined by flow needs (either constant displacement or variable displacement), operating-pressure levels, and the drive forces available. If steam or compressed air is the power medium available, a reciprocating-piston pump or turbine-driven pump might be chosen. With an electric-motor drive, there is a wide choice of pumps. With an engine drive, such as a power takeoff or diesel, the range of rotating speeds at which the device operates best and other characteristics peculiar to that drive must be considered.

PRESSURE CONTROLS
The control of pressure may be incorporated within the variable-delivery action of the pump. It may be a function of a relief valve specifically provided as a maximum-level control valve or as a safety valve.

FLOW CONTROL
This also may be a function of pump delivery, or it may be accomplished by a restrictive device or a diversion action or a combination of these.

DIRECTIONAL CONTROLS
The choice of directional-control valves is closely associated with the degree of control desired. The signal source and the method by which the

signal is transmitted into control vary widely with different systems. To make a competent decision, the designer should have a thorough knowledge of directional-control valves.

ACTUATED END-USE DEVICE

Choice of either a rotary motor or linear actuator will hinge on many basic requirements. The manner in which these requirements are to be met may have much to do with all the other components in the circuit.

SPACE REQUIREMENTS

The physical space within which a cylinder must be located may dictate high pressure to keep the size within specification.

SPEED OF OPERATION

The speed at which a rotary motor must operate may dictate elevated pressure to keep conductor sizes within certain boundaries.

PRESSURE CONSIDERATIONS

Low-pressure components may be inexpensive in small, widely used sizes. In larger sizes the low-pressure components may be more expensive than the equal-performance high-pressure components of a smaller size. Sensitivity of control may dictate low pressure in certain applications.

Efficiency will vary with different types of service and different circuit needs. Duty cycle can have much to do with relative efficiencies in circuits.

The design of fluid-power circuits can be summarized as being an art that translates force, speed, and control-function needs into components that will take a stated input energy and deliver it to the end-use device at the desired force and speed of actuation, with the desired degree of control. A knowledge of basic fluid mechanics and operational capabilities of available components makes possible the economical design of a circuit without reference to previous applications in similar machinery.

15-18 CHOICE OF SIGNAL SOURCE AND TRANSMITTING SYSTEMS

The choice of signal source and transmitting systems is based on the following factors:

1 Sensitivity required:
  a Acceleration and deceleration requirements.
  b Repetitive accuracy required.
  c Activity relative to speed of movement (velocity switching may be needed).
  d Signal strength (proximity switching may require amplification).

2 Available space and environment:
   a Temperature considerations (mechanical switches may not be usable in the location where the signal is to be generated, and pressure variations may have to be used for a signal source).
   b Fire hazards (electrical signals may have to be in explosion-proof housings or other types of signal used).
   c Localized contaminants may indicate the need for static switching.
   d Interference from moving parts may require the use of a signal like that from a light beam and photosensitive receiver, with the necessary amplification mechanisms.
3 Dependability and service life required:
   a Production level.
   b Ease of maintenance.
   c Is periodic cleaning or limited service needed?
   d Will *plug-in* units be best suited?
   e Are manifold-mount or cartridge units available?
4 Economic limitations:
   a Do short-run characteristics dictate "breakdown" maintenance?
   b Is the cheapest component the least expensive when installation costs are calculated?
   c Will the product be used in an area where trained service personnel are available?
   d What is the cost of downtime?
   e Is this a reusable or portable unit?
   f Is on-the-job maintenance feasible, or must easily replaced units be used?

A knowledge and evaluation of the available signal-transmitting devices will ensure a fluid-power system with the efficiency and dependability needed for the job requirements. Before a signal-transmission system is chosen, each new application should be tested with the four points mentioned.

From research into the nerve center of the highly versatile fluid-power train, better power-transmission systems have been developed for use in practically every industry.

SUMMARY

A hydraulic circuit is a connected group of components used to transmit power. The transmission of power is through a fluid medium. Power can be transmitted by hydraulic systems at varying speeds and in many different planes with a high degree of simplicity.

Circuit drawings present the arrangement of the components within a machine to provide the desired functions. These drawings make it possible to visualize malfunctions without disassembling any components or parts.

Circuit design must consider the mechanical movements desired and the availability of basic power sources that are to be transmitted. The type of basic power available will affect the choice of components and signals that may be used.

REVIEW QUESTIONS

15-1 What is the function of a replenishing circuit?
15-2 Can cylinder motion be synchronized with servo valves?
15-3 What is a flow divider?
15-4 What is meant by *pressure intensification*?
15-5 Is more than one feed rate used during a drilling or boring operation? Why? How does this affect the valving?
15-6 What is a differential circuit?
15-7 How does a bleedoff compensated flow-control valve control the speed of a cylinder or motor movement?
15-8 Why is oil sometimes combined with air in combination circuits?
15-9 What is the advantage of hydraulic flow-control valves over pneumatic flow-control valves?
15-10 Is a counterbalance valve used to relieve pressure?
15-11 How can a relief valve be remotely controlled?
15-12 Why do pressure-control circuits often provide for a venting function of the relief valve?
15-13 Explain the difference between metering oil into a cylinder and metering oil out of a cylinder. Why is each function required? What prevents the use of either system for all circuits?
15-14 What is meant by *force multiplication*?
15-15 What is meant by *synchronization of cylinder motion*?
15-16 What is an energy-absorbing circuit?
15-17 Can servomechanisms be used to control variable-displacement pumps?
15-18 What is an open-center valve?
15-19 Can a relief valve be used as a compensating medium for a flow-control mechanism?
15-20 What is the purpose of a hydraulic circuit drawing?
15-21 What is the purpose of the check valve in Fig. 15-1c?
15-22 Why is pilot valve C in Fig. 15-3 provided with an open-center condition?
15-23 What does the small equilateral triangle at the lower surface of the three-position directional-control valve of Fig. 15-4 signify?
15-24 Why is a spring-loaded check valve installed in parallel with the heat exchanger shown in Fig. 15-9?
15-25 What type of prime mover is employed in the circuit of Fig. 15-12?

15-26 Why is the rod stationary in the cylinder of Fig. 15-20? How does this affect force values?

15-27 Which pump supplies pressurized fluid to valve G of Fig. 15-21? Why? How does this affect interlock functions?

15-28 What is the purpose of the solenoid-operated, two-way valve of Fig. 15-24? What will occur if the solenoid fails to function?

15-29 What is the purpose of the check valve at the blind end of the large cylinder shown in Fig. 15-26? Will the circuit function without it?

15-30 Would the equalizing valve of Fig. 15-36 be *open-* or *closed-*center type? Why?

Appendix A-1

PREFERRED METRIC UNITS FOR FLUID POWER

Quantities Name	Symbol	Typical Applications	Customary U.S. Units	Metric Units
Acceleration	a	system damper, shock absorber	foot per second squared	metre per second squared
	g	gravity		
Angle, plane	α, β etc.	swing arc, actuator rotation, conductor direction	degree minute second	degree minute second
Area	A	orifice,	square inch	square millimetre
		heat exchanger	square inch	square centimetre
			square foot	square metre
Conductivity, Thermal	λ	component material heating and cooling calculations	British thermal unit per hour foot degree Fahrenheit	watt per metre kelvin
Cubic Expansion, Coefficient (Note 9)	Y	closed system pressure changes	per degree Fahrenheit	per degree Celsius
Current, Electric	I	component operation rating	ampere	ampere
Density	ρ	hydraulic fluids	pound per gallon	kilogram per liter
Displacement (Unit discharge) (Note 3)	q	pumps, motors, rotary actuators, intensifiers, air and hand pumps	cubic inch	cubic centimeter

418 *Appendix A*

U.S. Unit	Conversion (Note 1)			
	Multiply By	To Get Metric	Multiply By	To Get U.S.
ft/sec²	0.3048*	m/s²	3.281	ft/sec²
° ' "	1.0* 1.0* 1.0*	° ' "	1.0* 1.0* 1.0*	° ' "
in²	645.2	mm²	0.1550 0.001550	in²
in²	6.452	cm²	0.1550	in²
ft²	0.09290	m²	10.76	ft²
$\dfrac{Btu}{hr \cdot ft \cdot °F}$	1.731	$\dfrac{W}{m \cdot K}$	0.5778	$\dfrac{Btu}{hr \cdot ft \cdot °F}$
$\dfrac{1}{°F}$	1.8*	$\dfrac{1}{°C}$	$\dfrac{1}{1.8*}$	$\dfrac{1}{°F}$
A	1.0*	A	1.0*	A
lb/gal	0.1198	kg/L (Note 2)	8.345	lb/gal
in³	16.39	cm³	0.06102	in³

PREFERRED METRIC UNITS FOR FLUID POWER (Continued)

Quantities Name	Symbol	Typical Applications	Customary U.S. Units	Metric Units
Energy	E	heat energy	British thermal unit	kilojoule
Flow Rate, Heat	ϕ	heat exchangers	British thermal unit per minute	watt
Flow Rate, Mass	q_m	process pumps	pound per minute	kilogram per second
Flow Rate, Volume (Note 4)	q_v	pneumatic	cubic foot per minute	cubic decimeter per second
			cubic inch per minute	cubic centimeter per second
		hydraulic	gallon per minute	liter per minute
			cubic inch per minute	milliliter per minute
Force	F	actuator thrust, spring force	pound	newton
Force per Length	F/L	spring rate	pound per inch	newton per millimeter
Frequency, (Cycle)	f	acoustic vibration, electrical period, pressure pulsation	hertz (cycle per second)	hertz
		mechanical oscillation	cycle per minute	reciprocal minute

	Conversion (Note 1)			
U.S. Unit	Multiply By	To Get Metric	Multiply By	To Get U.S.
Btu	1.055	kJ	0.9478	Btu
Btu/min	17.58	W	0.05687	Btu/min
lb/min	0.00756	kg/s	132.3	lb/min
ft³/min (cfm)	0.4719	dm³/s	2.119	ft³/min (cfm)
in³/min (cim)	0.2731	cm³/s	3.661	in³/min (cim)
gal/min (gpm)	3.785	L/min (Note 2)	0.2642	gal/min (gpm)
in³/min (cim)	16.39	mL/min (Note 2)	0.06102	in³/min (cim)
lb	4.448	N	0.2248	lb
lb/in	0.1751	N/mm	5.710	lb/in
Hz (cps)	1.0*	Hz	1.0*	Hz (cps)
cpm	1.0*	l/min (Note 5)	1.0*	cpm

Appendix A **421**

PREFERRED METRIC UNITS FOR FLUID POWER (*Continued*)

Quantities		Typical Applications	Customary U.S. Units	Metric Units
Name	Symbol			
Frequency (Rotational)	n	rotational speed of pumps, motors, compressors and transmissions	revolution per minute	reciprocal minute
Heat	Q	quantity of heat	British thermal unit	kilojoule
Heat Capacity, Specific	c	hydraulic fluid temperature change	British thermal unit per pound degree Fahrenheit	kilojoule per kilogram kelvin
Heat Transfer, Coefficient of	h	heating and cooling, thermal conductance	British thermal unit per hour foot squared degree Fahrenheit	watt per meter squared kelvin
Inertia, Moment of	I	links, rocker arms, walking beams	pound foot squared	kilogram meter
			pound inch squared	squared kilogram meter squared
Length	L	bore and stroke, orifice diameter component location	inch	millimeter
			foot	meter
		filter rating, particle size,	micron	micrometer
		surface finish	microinch	micrometer

Conversion (Note 1)

U.S. Unit	Multiply By	To Get Metric	Multiply By	To Get U.S.
rpm	1.0*	l/min (Note 5)	1.0*	rpm
Btu	1.055	kJ	0.9478	Btu
$\dfrac{Btu}{lb \cdot °F}$	4.187	$\dfrac{kJ}{kg \cdot K}$	0.2388	$\dfrac{Btu}{lb \cdot °F}$
$\dfrac{Btu}{hr \cdot ft^2 \cdot °F}$	5.678	$\dfrac{W}{m^2 \cdot K}$	0.1761	$\dfrac{Btu}{hr \cdot ft^2 \cdot °F}$
$lb \cdot ft^2$	0.04214	$kg \cdot m^2$	23.73	$lb \cdot ft^2$
$lb \cdot in^2$	0.0002926	$kg \cdot m^2$	3417.6	$lb \cdot in^2$
in	25.4*	mm	0.03937	in
ft	0.3048*	m	3.281	ft
μ	1.0*	μm	1.0*	μ
μin	0.0254*	μm	39.37	μin

PREFERRED METRIC UNITS FOR FLUID POWER (*Continued*)

Quantities Name	Symbol	Typical Applications	Customary U.S. Units	Metric Units
Linear Expansion, Coefficient (Note 8)	α	shrink fit, shaft, gears	per degree Fahrenheit	per degree Celsius
Mass	m	component weight	pound	kilogram
Modulus, Bulk	k	system fluid compression	pound per square inch	megapascal
Momentum	p	shock absorbers	pound foot per second	kilogram meter per second
Potential, Electric	V	component	volt	volt
Power	P	input and output of hydraulic and pneumatic systems	horsepower	kilowatt
		heat exchangers	British thermal unit per minute	watt
Pressure	p	component and system rating, atmospheric pressure	pound per square inch	bar (Note 6)
				kilopascal (Note 6)
Stress, Normal Shear	σ τ	normal stress, shear stress, strength of materials	pound per square inch	megapascal

Conversion (Note 1)

U.S. Unit	Multiply By	To Get Metric	Multiply By	To Get U.S.
$\dfrac{1}{°F}$	1.8*	$\dfrac{1}{°C}$	$\dfrac{1}{1.8*}$	$\dfrac{1}{°F}$
lb	0.4536	kg	2.205	lb
lb/m² (psi)	0.006895	MPa	145.0	lb/in² (psi)
$\dfrac{\text{lb} \cdot \text{ft}}{\text{sec}}$	0.1383	kg·m/s	7.23	lb·ft/sec
V	1.0*	V	1.0*	V
hp	0.7457	kW	1.341	hp
Btu/min	17.58	W	0.05687	Btu/min
lb/in² (psi)	0.06895	bar	14.50	lb/in² (psi)
lb/in² (psi)	6.895	kPa	0.1450	lb/in² (psi)
lb/in² (psi)	0.006895	MPa	145.0	lb/in² (psi)

Appendix A

PREFERRED METRIC UNITS FOR FLUID POWER (Continued)

Quantities Name	Symbol	Typical Applications	Customary U.S. Units	Metric Units
Temperature, Customary	θ	thermal operating limits of components, dryer dew point	degree Fahrenheit	degree Celsius
Temperature, Thermodynamic (absolute)	T	pneumatic system calculations	degree Rankine	kelvin
Time	t	component response	second	second
		system cycle	minute	minute
		maintenance cycle	hour	hour
Torque	T	motor, rotary actuator output, bolt tightening	pound foot	newton meter
Velocity, linear	ν	fluid flow speed	foot per second	meter per second
		cylinder actuator speed	inch per second	millimeter per second
Viscosity, Dynamic	η	system fluid properties	centipoise	millipascal second
Viscosity, Kinematic	υ	system fluid properties	centistokes (Note 7)	square millimeter per second
Volume	V	lubricators, accumulators, reservoir capacity (pneumatics)	cubic foot	cubic decimeter
			cubic inch	cubic centimeter

426 *Appendix A*

		Conversion (Note 1)		
U.S. Unit	Multiply By	To Get Metric	Multiply By	To Get U.S.
°F	$t_C = \frac{(t_F - 32)}{1.8*}$	°C	$t_f = 1.8(t_C+) + 32*$	°F
°R	1/1.8*	K	1.8*	°R
sec	1.0*	s	1.0*	sec
min	1.0*	min	1.0*	min
hr	1.0*	h	1.0*	hr
lb·ft	1.356	N·m	0.7376	lb·ft
ft/sec	0.3048*	m/s	3.281	ft/sec
in/sec	25.4*	mm/s	0.03937	in/sec
cP	1.0*	mPa·s	1.0*	cP
cSt	1.0*	mm²/s	1.0*	cSt
ft³	28.32	dm³	0.0353	ft³
in³	16.39	cm³	0.06102	in³

PREFERRED METRIC UNITS FOR FLUID POWER (Continued)

| Quantities | | Typical Applications | Customary U.S. Units | Metric Units |
Name	Symbol			
		hydraulics	gallon	liter
			ounce	milliliter
Work	ω	cylinders, linear actuators	foot pound	joule

	Conversion (Note 1)			
U.S. Unit	Multiply By	To Get Metric	Multiply By	To Get U.S.
gal	3.785	L (Note 2)	0.2642	gal
oz	29.57	mL (Note 2)	0.03381	oz
ft·lb	1.356	J	0.7376	ft·lb

NOTES TO APPENDIX A

1. An asterisk (*) after the conversion factor indicates that it is exact and that all subsequent digits are zeros. All other conversion factors have been rounded to four significant digits. See Reference No. 3 if more exact conversion factors are required.

2. The international symbol for litre is the lowercase "l" which can easily be confused with the numeral "1". Accordingly, the symbol "L" is recommended for U.S. fluid power use.

3. Indicate displacement of a rotary device as "per revolution" and of a non-rotary device as "per cycle".

4. For gases, this quantity is frequently expressed as free gas at Standard Reference Atmosphere as defined in Reference No. 8 and specified in Reference No. 9. In such cases, the abbreviation "ANR" is to follow the expression of the quantity, "q_v(ANR)" or the unit, "m^3/min (ANR)". See Reference 4, for more detailed information on the attachment of letters to a unit symbol.

5. Mechanical oscillations are normally expressed in cycles per unit time and rotational frequency in revolutions per unit time. Since "cycle" and "revolution" are not units, they do not have internationally recognized symbols. Therefore, they are normally expressed by abbreviations which are different in various languages. In English the symbology for mechanical oscillations is c/min and for rotational frequency, r/min.

6. The bar and kilopascal are given equal status as pressure units. At this time the domestic fluid power industry does not agree on one preferred unit. The pascal is the SI unit for pressure and a major segment of U.S. industry has accepted the multiple, kilopascal (kPa), as the preferred unit. On the other hand, the majority of the international fluid power industry has accepted the bar as the metric pressure unit.

The bar is recognized by the EEC as an acceptable metric unit and is shown in ISO 1000 for use in specialized fields. Conversely the bar is considered by both the International Committee on Weights and Measures and the NBS as a unit to be used for a limited time only. Further, the bar has been deprecated by Canada, ANMC, and some U.S. standards organizations and is illegal in some countries.

7. Viscosity is frequently expressed in SUS (Saybolt Universal Seconds). SUS is the time in seconds for 60 mL of fluid to flow through a standard orifice at a specified temperature. Conversion between kinematic viscosity, mm^2/s (centistokes) and SUS can be made by reference to tables in Reference No. 8.

8. The linear expansion coefficient is a ratio, not a unit, and is expressed in customary U.S. units such as in/in and in metric units such as mm/mm per unit temperature change.

9. The cubic expansion coefficient is a ratio, not a unit, and is expressed in customary U.S. units such as in^3/in^3 and in metric units such as cm^3/cm^3 per unit temperature change.

---

6. IDENTIFICATION STATEMENT

   Use the following statement in catalogs and sales literature when electing to comply with this voluntary standard:

   "Metric units are selected and used in accordance with NFPA/T2.10.1-19xx."

7. KEY WORDS

   The following Key Words, useful in indexes and in information retrieval systems, are suggested for this proposed recommended standard:

   fluid power

   metric conversion

   metric units, fluid power

   quantities

   SI units

   units, metric

Appendix A-2

SELECTED "SI" UNITS FOR GENERAL PURPOSE FLUID POWER USAGE

Quantity	SI Unit for Fluid Power
length	millimeter (mm)
pressure[1]	bar (bar,g or bar,a)
pressure[2]	millimeter of mercury (mm Hg)
flow[3]	liters per minute (l/min)
flow[4]	cubic decimeters per second (dm^3/sec)
force	Newton (N)
mass	kilogram (kg)
time	second (s)
volume[3]	liter (l)
temperature	degrees Celcius (°C)
torque	Newton-meters (Nm)
power	kilowatt (kW)
shaft speed	revolutions per minute (RPM)
frequency	Hertz (Hz)
displacement[3]	milliliters per revolution (ml/rev)
kinematic viscosity	centistokes (cSt)
velocity	meter per second (m/s)

NOTES.
1) pressures above atmospheric
2) pressures below atmospheric
3) liquid
4) gas
5) @ 38°C; factor is 4.667 @ 99°C.

"Customary US" Unit For Fluid Power	Conversion
inch (in)	1 in = 25.4 mm
pounds per square inch (psig or psia)	1 bar = 14.5 psi
inches of mercury (in Hg)	1 in Hg = 25.4 mm H
gallons per minute (USGPM)	1 USGPM = 3.79 l/mi
cubic feet per minute (cfm)	1 dm^3/sec = 2.12 cf
pound(f) lb(f)	1 lb(f) = 4.44 N
pound(m) lb(m)	1 Kg = 2.20 lb(m)
second(s)	—
gallon (US gal)	1 US gal = 3.79 l
degrees Fahrenheit (°F)	°C = 5/9 (°F−32)
pounds(f)−inches lb(f)−in	1 Nm = 8.88 lb(f)−i
horsepower (HP)	1 kW = 1.34 HP
revolutions per minute (RPM)	—
cycles per second (cps)	1 Hz = 1 cps
cubic inches per revolution (cip)	1 ml/rev = .061 cip
Saybolt (SUS)	cSt = (4.635)(SUS)
feet per second (fps)	1 m/s = 3.28 fps

Index

Accumulator, pressure, 165–185
    circuits for, 174–177, 397–400
    cylindrical gas-charged, 173
    fluid-storage type, 178
    relaxation circuit, 178
    relaxation valves for, 179
    shock-wave absorption, 177
    spherical diaphragm-type, 171
    spring-loaded, 168–170
    types of, 165
    weighted, 165–168
Advantage, mechanical, 14
Air-exhaust line, 32
Angle check valves, 309–311
Automation circuit, 394
Axial-piston motor, 93–99
Axial-piston pump, 93, 95

Backwashing of filters, 211
Ball-type foot valve, 83
Barrel pump, 83
Bernoulli's theorem, 18
Bleed-off circuit, 386–394
Blocked port, symbol for, 43
Booster circuits, 400–402
Boosters, 133
Bourdon pressure gage, 22
Boyle's law, 17

Captive areas, 121
Cartridge-disk seals, 333
Cavitation, 108
Centrifugal pumps, 100–102
Characteristics, pump, 109–112
Charles' law, 19
Check valve:
    angle-type, 309–311
    four-way, 320–322
    symbol for, 35, 45–47
    (*See also* Directional-control valves)
Circuit(s):
    check points, 352
    diagrams, 349–352
    differential-area, 130

Circuits(s):
    elements, 31
    feedback, 371–374
    hydraulic accumulator, 165
    hydropneumatic, 322–324
    industrial hydraulic, 377–417
    integral bypass, 120
Cistern pump, 84
Closed system, 107
Components:
    for servos, 359–361
    symbol for enclosure, 58
    use in circuits, 413
Composite symbol, 60
Compound relief valves, 252–257
Compressibility, fluid, 17
Conductor, 31
Connector dot, 32
Control valves (*see* Directional-control valves; Flow-control valves)
Counterbalance valves, 266–279
    circuits for, 270
Crush fit, 152
Cylinders, 138
    double-acting, 119
    double-rod-end, 119
    pump, 83
    and rams (*see* Rams and cylinders)
    symbols for, 40

Dashpot, 120
Deceleration motion, 121
Deceleration valve symbol, 44
Degree of filtration, 210
Design considerations, 27, 113, 137, 162, 180, 206, 242, 336, 410–413
Diagram:
    circuit, 61, 349–352
    cutaway, 61
    pictorial, 61
Diaphragms, symbols for, 40
Differential area circuits, 130–133
Differential head, 105

435

Directional-control valves, 309–340
   check, 309–312
      angle, 309–311
      pilot, 312–313
      swing-gate, 311
   diversion, 319
   four-way, 320–322
   hydropneumatic circuit, 322–324
   pilot check circuits, 314
   poppet-type, 335
   seals for, 333
   for servo systems (*see* Servo systems, directional-control valves)
   shuttle-type, 317
   sliding-disk type, 333
   solenoid-controlled, 325–333
   symbols for, 48–51
   three-way, 318
   two-way, 316–317
   uses of, 309
Diversion valves, 319
Dual-pressure compound relief valve, 253
Dump valve, 328
Duplex pump, 99

Electrical devices for hydraulic circuits, 341–357
   choice of controls, 353
   circuit check points, 352
   circuit diagrams for, 349–352
   interlocks, 354
   maintenance of, 355
   protection of, 355
   safety considerations, 354
   solenoids, 341–348
   symbols for, 349
   torque motors, 348
Ethylene glycol, 23
Eye of impeller, 102

Feedback circuit, 371–374
Filters, 209–231
   cartridge, 210
   circuits for, 214–218
   degree of filtration in, 210
   flow ratings of, 213
   housings for, 218–221
   indicators for, 221

Filters:
   metal-element, 210–212
   nonmetal-element, 212
   operation and maintenance of, 226
   permanent magnets in, 213
   pressure drop in, 223–225
   starting new machinery, 227
   temperature ratings of, 214
   types of, 209
Fire-resistant fluids, 23
Flash point, 26
Flow, laminar, 213
Flow-control valves, 286–308
   basic two-way, 286
   check, 288
   circuits using, 294–300
   combined control and check, 288–291
   compensated, 291–294
      noncompensated, 287
   uses of, 286
Fluid(s):
   hydraulic, 11–28
   mechanics, 11–22
   pressure, 22–24
   properties, 26
   temperature control (*see* Temperature control of fluids)
   viscosity, 23–26
Fluid-power plumbing, 143–164
   changes in flow direction, 153
   cleanliness of, 153
   component mounting, 146
   conductor size, 155
   connecting lines, 154
   flow resistance, 154
   hydraulic hose for, 156
   maintenance of, 147
   manifolds for, 159
   pipe choice for, 157
   pipe threads for, 152
   purpose of, 143
   quick-disconnect coupling, 160
   requirements for, 144
   shock waves in, 147
   steel tubing for, 155
   strength requirements, 145
   terminal points of, 147
   welding of, 151

Fluid reservoirs, 186–207
  combination, 187
  construction of, 202
  overhead-type, 201
  piping for, 196
  power package, 191
  pressurized, 54, 190
  pump-valve-tank, 196
  pumps for, 198
  rectangular, 187
  scavenger tanks, 195
  settling tanks, 195
  sizing of, 201
  types of, 186–194
  vented, 190
Foot valve, ball-type, 83
Force, 2, 13
Friction loss, 102

Gate valve, 47, 286
Gay-Lussac's law, 19
Globe valve, 47, 286
Graphical diagram, hydraulic, 61

Head in pump system, 102–106
  differential, 106
  net static discharge, 105
  potential, 105
  static pressure, 105
  total static discharge, 105
Heat exchangers for hydraulic fluids (*see* Temperature control of fluids, heat exchangers for)
Hydraulic cylinders, 118–133
Hydraulic fluids, 23
Hydraulic horsepower, 20
Hydraulic motor, 94
Hydraulic press, 16
Hydraulic symbols (*see* Symbols, hydraulic)
Hydropneumatic circuit, 322–324

Impeller:
  blades of, 102
  eye of, 102
Industrial hydraulic circuits, 377–417
  accumulator, 397–400
  automation, 394
  bleed-off, 386–394

Industrial hydraulic circuits:
  booster and intensifier, 400–402
  component selection, 413
  counterbalance, 382
  design of, 410–413
  motion synchronizing, 402–406
    flow-control valves, 404
    flow-dividing motors, 405
    matching pumps, 404
    series pipings, 403
    tie cylinders, 402
  pressure-regulating, 377–379
  reduced pressure, 379–381
  replenishing circuits, 409
  sequence functions, 383
  servo circuits, 406–409
  signal source, 414
  speed-control, 384–393
    bleed-off, 389, 394
    closed meter-out, 389
    deceleration, 390
    hydraulic transmission, 392
    intermittent feed, 389
    meter-in, 384
    meter-out, 385
    variable volume, 390
    variable pressures, 381
  (*See also* Electrical devices for hydraulic circuits)
Inertia, law of, 100
Integral bypass circuit, 120
Integral jack rams, 124
Intensifier, 133
Intensifier circuits, 400–402
Interlocks, 354
Internally drained valve, symbol for, 51

Jet pump, 102

Laminar flow, 213
Law of inertia, 100
Limit switch, 167
Limited-rotation motor, 128–130
Linear motor, symbol for, 40
Lip-type seals, 122, 128

Major lines, 32
Manifolds, 159
Manometer, 21

Mechanical advantage, 14
Micron, 210
Mineral-base oil, 23
Motion-synchronizing circuits, 402–406
Motors, fluid, 89–99
    axial-piston, 93–99
    limited-rotation, 128–130
    radial-piston, 89–93
    symbols for, 39
Multiple-envelope symbol, 42

Needle valve, 120
Net static discharge head, 105

Oil, mineral-base, 23
Open system, 107

Packing, 127
Pascal's law, 13
Peripheral turbine pump, 102
Pictorial diagram, 61
Pilot check valve, 46, 312
Pilot check valve circuits, 314
Pilot line, 32
Pilot-operated check valve, symbol for, 16
Pilot piston, symbol for, 46
Pipe restriction, 36
Piston, mechanism of, 40
Piston movement, 14
Plugged terminal, 34
Plumbing, fluid-power (see Fluid-power plumbing)
Port, 42
Potential head, 105
Pour point, 25
Pressure, 1, 11–17, 20
Pressure-control valves, 247–285
    compound relief, 252–257
    counterbalance, sequence, and unloading, 266–279
    dual-pressure compound relief, 252
    electrically operated, 258–264
    in hydraulic circuits, 266–279
    hydraulic fuses, 279
    normally closed, 247
    normally open, 247
    piloted, 247
    pressure-reducing, 264

Pressure-control valves:
    pressure switches, 279
    solenoid-vented relief, 258–264
    spring-loaded relief, 251
    types of, 247
    uses of, 247
Pressure gage, 22
Pressure-reducing valves, 264
Pressure-regulating circuits, 377–379
Priority valves and circuits, 301–305
Propeller pump, 103
Properties, fluid, 26
Pump systems, 102–109
Pumps, fluid-power, 65–117
    axial-piston, 93
    barrel, 83
    centrifugal, 100–102
    characteristics of, 109–112
    cistern, 84
    cylinder, 84
    double-acting, 85
    duplex, 100
    generated-rotor, 73
    internal-gear, 72
    jet, 102
    peripheral turbine, 102
    propeller, 102
    radial-piston, 86–89
    reciprocating, 83–100
    reservoir, 197
    rotary, 66–83
    screw, 82
    single-acting, 85
    sliding-vane, 75–82
    spur-gear, 66–72
    suction stroke, 85
    symbols for, 37
    triplex, 99
    types of, 65

Quick-disconnect coupling, 37, 160

Radial-piston motors, 89–93
Radial-piston pumps, 86–89
Rams and cylinders, 118–142
    double-acting, 119
    dual, 124

Rams and cylinders:
  integral jack, 124
  main, 124
  single-acting, 118
Reduced-pressure circuits, 379–381
Relaxation circuits, 178
Relief valve, 42
  compound, 252–257
  spring-loaded, 44
Replenishing circuit, 409
Reservoirs (see Fluid reservoirs)
Rest position, 48
Rotating connection, symbol for, 55

Safety valve, 248
Sandwich-type valve, 197
Scavenger tanks, 195
Seals, 122, 127, 128
Sequence-function circuits, 383
Servo systems, 358–375
  circuits, 322–324
  components for, 359–361
  comparison value, 360
  pressure pickup, 359
  torque motor, 361
  directional-control valves, 363–365
    flapper valve, 363
    nozzle valve, 363
    pneumatic follower, 364
    two-stage circuit, 365–371
  feedback circuit, 371–374
  mechanical-linkage, 358
  oil-pilot, 365–371
  uses of, 359
Settling tanks, 195
Shuttle valves, 317
Signal source for circuits, 414
Single-acting pump, 85
Siphon or suction effect, 106
Sliding-disk valves, 333
Solenoid, 341–348
Solenoid-controlled valve, 48, 325–333
Solenoid-vented relief valve, 330
Speed-control circuits (see Industrial hydraulic circuits, speed-control)
Spool valve, 47
Spring-cavity drain, 179

Spring-loaded valves:
  check, 46
  relief, 252
  safety, 248
Static pressure, 105
Static suction life, 105
Suction pipe, 76
Suction stroke, 76
Switch, limit, 167
Symbols, hydraulic, 31–64
  blocked port, 43
  check valve, 35, 45–47
  circuit diagram, 61
  component enclosure, 58
  composite, 58, 60
  cutaway diagram, 61
  cylinders, 40
  deceleration valve, 44
  diaphragms, 40
  directional-control valves, 51–54
  electrical devices, 349
  gate and globe valves, 47
  graphical diagrams, 61
  hydraulic valves, 42–51
  linear motor, 40
  motor, 39
  multiple-envelope, 42
  pictorial diagram, 61
  pilot-operated check valve, 46
  pilot pistons, 46
  plugged terminal, 34
  pressure-reducing valve, 45
  pressurized tanks, 54
  pumps, 37
  relief valve, 42
  single-envelope, 42
  solenoid-controlled valve, 48
  test station, 35
  three-position four-connection valve, 48
  two-position three-connection valve, 48
  unloading valve, 44
  using, 61
  valve-port, 42
  variable flow-rate control valve, 47
  vented reservoir, 54
System International, 12

Tank, 33
Telescoping-type cylinders, 122
Temperature control of fluids, 232–246
  automatic, 239
  heat exchangers for, 233–239
Temperature effects, 19
Terminal, plugged, 34
Test station, 35
Torque motors, 348
Total static discharge head, 105
Triplex pump, 99
Tubing, steel, 155
Turbulent flow, 213

Unloading valve, 44, 266

Valve(s):
  angle check, 309–311
  compensated flow-control, 291–294
  compound relief, 252–257
  counterbalance, 266–279
  directional-control (*see* Directional-control valves)
  diversion, 319
  dual-pressure, 252
  dump, 328
  flow-control (*see* Flow-control valves)
  foot, 83
  four-way check, 320–322
  gate, 47, 286
  globe, 47, 286

Valve(s):
  intensifier-type, 133
  needle, 120
  noncompensated, 287
  pressure-control (*see* Pressure-control valves)
  pressure-reducing, 264
  priority, 301–305
  relief, 44, 252–257
  relaxation, 178
  safety, 248
  sequence, 383
  shuttle, 317
  sliding-disk, 333
  solenoid-vented, 330
  spool, 47
  symbols for, 42–51
  three-way, 318
  two-way, 286, 316–317
  unloading, 44, 266
Valve-port symbol, 42
Valving element, 42
Variable-displacement motor, 89–93
Variable flow-rate control valve, 42
Variable pressure circuits, 381
Vented reservoir, 54
Viscosity, fluid, 23
Viscosity index, 25
Viscosity measurement, 24
Volute, 102

Water-soluble oil, 23